AF208131

※∩ ※∩ ※∩※

COMMUNICATION:

DOES GOD HAVE A PROBLEM ?

(And Does It Matter ?)

※∩ ※∩ ※∩※

By Stuart John Cornell

※∩ ※∩ ※∩※

COPYRIGHT

TABLE OF CONTENTS

DEDICATION

※⌒ ※⌒ ※⌒※

※⌒ ※⌒ ※⌒※

TO MADELEINE, JAMES AND EDWARD

WHO HAVE BEEN PART OF THIS JOURNEY

(EVEN THOUGH THEY DID NOT ALWAYS KNOW IT)

※⌒ ※⌒ ※⌒※

※⌒ ※⌒ ※⌒※

ACKNOWLEDGEMENTS

※⌒ ※⌒ ※⌒※

※⌒ ※⌒ ※⌒※

I offer my grateful thanks to Margaret, Jim, Richard and Ray for their helpful and supportive comments, and their willingness to take the time to read earlier drafts of this manuscript and offer suggestions for improvement. I have taken full note of these but accept full responsibility for the finished product with whatever deficiencies and errors it still has and for any lack of improvement I might have made, but didn't.

※⌒ ※⌒ ※⌒※

Cover Design by James Cornell

※⌒ ※⌒ ※⌒※

FOREWARD

One wonders if the author was inspired by Bill Bryson's "Short History of Nearly Everything" writing this book, so wide-reaching is its coverage of Man's understanding of the physical world and its philosophical perception by mankind. It starts from the fact that the age of the world's larger religions (and their scriptures) are measured in Millennia rather than Centuries and, due to the constraints of the cultural environment from which they emerged, may benefit from being re-examined in the light of Man's continual pursuit of knowledge.

Inevitably, in matters spiritual, much of the subject matter is controversial, but also subjective and dependant on the reader's own religion. Herein is the only concession the Author has made to his wide-reaching scope, namely that he has approached it largely from his own, Christian perspective. Some people would argue that ever-increasing knowledge erodes and will eventually eradicate belief in a supreme universal entity, but the more is discovered, the more questions are raised; this book presents an alternative to this view.

If you want to learn about how Humanity's evolving understanding of our World can have a positive impact on belief, then this is the book for you.

James S. Croft

⁂∩ ⁂∩ ⁂∩⁂

⁂∩ ⁂∩ ⁂∩⁂

PREFACE

The idea for this book came out of discussions with fellow Christians, from different backgrounds and experiences. Many shared a common concern, namely, they were not sure of whether they were always believing the "right things" in terms of the Christian faith as taught by their respective Churches and indeed, they were not sure in fact, of what the "right things" were that they should be believing. This uncertainty transmitted itself both as doubt in their own personal faith[1] and in a lack of confidence in being able to explain their faith to others." What do I say ? How will I be able to answer other people's questions ?"

Such doubts were often associated with a sense of guilt or shame. Others expressed views that indicated they made a distinction between their faith and belief in God, and the 'man–made' rules and trappings of "the Church" [2] as organised religion. The awareness of these concerns then lead to further questions such as: why is it that there is such a variety of ideas about God and faith ? why do people interpret the scriptures in different ways ? why do some people have faith? what factors influence them to believe in God and others to reject the idea ? For those who do believe – how do they come to know what it is they should be believing ?

Although these questions arose within a Christian context, once I had started to think about such things, it became apparent that people of other faiths had similar concerns. So, although writing from a Christian perspective and sometimes perhaps appearing to have a Christian – centric focus, the arguments put forward in this book are, I think, applicable in many cases, to other faith groups and their members. Given this broader context, consideration of these questions seems to be important. Particularly given the consequences that follow from the way

1

human beings seem to respond to having such a variety of ideas about the nature of God. Namely, the frequent animosity, not to say violence, that has happened and continues to happen between proponents of different ideas or ways of understanding [3] Church, faith and scripture.

Such tensions between and within different religious groups are not, of course, solely related to the religious / doctrinal (stated beliefs) differences. Some being fuelled by the basic human drivers of power seeking, vested interests and personal gain and greed. Nevertheless, differences in doctrine and interpretation of scripture do themselves lead people to respond to disagreements in harmful and hurtful ways. If such differences are scaled up from the individual church or worshipping community, to the national body and then across the various Christian denominations [4] and then between the different faith groups – the scope of the variance is enormous. Thus, the opportunities for harm are also magnified.Through reflecting on these conversations, a number of questions arise, namely:

- how can some religious groups be so sure of their understanding of matters of faith and religion, that they claim they, and only they, know the truth and the will of God, and then openly condemn and exclude others who see things differently ?

- how can other groups justify going even further, by trying to impose their ideas on others, sometimes by force ?

- how is it that for some members of religious groups, their personal uncertainty relating to matters of faith leads to a feeling of guilt and or shame, which often goes un–acknowledged and indeed may be added to by the leading hierarchy of the organisation ?

2

- why is it that some people develop a faith and or belief in God and accept certain established religious practices whilst others don't, even though they may have had similar exposures to religious influences ?

Looking at these problems from a human point of view, and after all that is the only one we have, one might ask:

- why, if there is a God (and we will come to what I mean by God in the introduction), is there such a variation in belief and understanding ?
- why does such variation lead to such harm and distress ?

After all, it seems reasonable to suppose that God might wish to ensure that we, his creation, know that he exists and understand his nature, even if he gives us the freewill (see chapter 10) to accept or reject him. If God wants us to know him, and given the content of the various religions and their writings, it would seem that he does, then why would he not arrange things so that there was no doubt ? Presumably, making sure everyone is on the same page in knowing who God is, would not be beyond the capabilities of someone who could create a universe(s) ! Thus, the following further questions suggest themselves –

- does God have a problem communicating his nature with his creation ? or
- is it just that we have problems understanding his communications ?
- or both ?

3

- or have things been set up in such a way, that a variety of understandings of God's nature are either inevitable or deliberate ?

- In any event, whatever the causes for the differences in understanding God, faith and religion, it seems to me that the consequences do matter. It matters, that the way we respond to those differences can result in much distress, harm and misery. The personal day to day stress resulting from guilt and shame can be a life long burden, with the consequences of any violent conflict leading to irreparable damage for everyone affected, whether this be directly or indirectly. If it matters to me, perhaps it is reasonable to assume that such consequences will be of concern to God. If, through a better understanding of the reasons why these differences have come about and an acceptance that differences in understanding can be viewed positively, perhaps we can reduce the potential for harm. Thus, seeking such understanding seems to be a worthwhile enterprise. So as I considered these things, and discussed my thoughts with others, I was encouraged to think that some of my ideas on these matters were worth sharing more widely – hence this book.

The hope being, that anyone concerned, affected or afflicted by the problems set out above, will gain some encouragement and comfort from knowing that they are not alone in their uncertainty and that doubt and questioning can be ways in which faith is strengthened, rather than undermined. For those in positions of leadership of the Christian Church or of other faith groups, I hope the book will encourage them to welcome the diversity of theological understanding and belief and see it as a normal part of God's evolutionary mechanisms, whereby our understanding of faith and beliefs are strengthened and expanded. Thereby, they can both support their own members in their questioning of faith and encourage respect and dialogue within and between Christian denominations and between other faiths. Thus, perhaps it is possible to reduce the potential for harm and distress currently occurring in the name of religious belief.

By describing the evolutionary nature of our understanding of the world and demonstrating the patterns of evolution of knowledge in science, philosophy and cosmology, we can draw parallels with our understanding of God, faith and religion. This, I believe provides a sound and logical basis for legitimate questioning of prevailing religious dogmas. In so doing, I suggest that this is both a natural and healthy process which strengthens belief rather than undermining it. Thereby creating a sense in which it is felt OK to question the current tenets of belief of any religious framework, without guilt or shame.

To encourage the sense, that different understandings and ways of knowing, both within and between different religious systems, are healthy and can be managed, I argue that questioning can lead to new insights and understandings of the nature of God, which in turn, can strengthen personal faith rather than weaken it. It is possible to hold on to established and cherished beliefs whilst accepting, valuing and respecting other views and other insights. It is by questioning and refining our understandings in this way, that God seems to communicate his nature.

One assumption I do make in developing these arguments, is that God is unchanging. This is one of the characteristics that people of faith tend to ascribe to God, certainly in Christianity. We don't know this for certain of course. But the idea can be likened to some of the laws of mathematics and physics – gravity, the electro–magnetic force and the strong and weak forces (see chapter 4). We assume these have operated unchanged since the time of the "Big Bang" (see chapter 4) and are constants within the universe, but our understanding of them has emerged and changed, as our knowledge of science and the cosmos increases. Similarly, our understanding of religion, faith and God expands as our knowledge of archaeology, history and anthropology and our thinking in terms of studying in theology and philosophy, widens.

Notwithstanding the fact that this book is written from a Christian perspective, I do not wish to imply that I believe that God is necessarily "Christian", and that therefore other faiths are in error and or require "correction". On the contrary. given the development and evolution of the world and the resulting variety of life and cultures across the planet and the changing nature of our understanding of these processes, it is the claim that there is and can be "one absolutely correct" notion of faith, that seems to me to be the error. No one "knows" for certain. (I explore what constitutes knowing and knowledge in chapter 2.)

So, this book is written for all people of faith who share these concerns, and anyone else who may have been affected, either directly or indirectly, by the consequences of religious differences and understandings.

But in all this, how do we begin to understand the nature of God's communication of himself, without necessarily thinking that any particular religious group can have the monopoly on the truth or that such a position somehow gives them a right, or duty even, to impose those doctrines on others – through fear and guilt and even violence?

Thus, it seems to me that for all these reasons, the question of communication – between God and humans and between humans themselves on matters of faith and religion – is of fundamental importance. Trying to understand it or at least trying to set the discussion in some sort of framework of mutual respect, is key to mitigating the negative effects of such differences.

PLEASE NOTE:
I have included a lot of detail in some of the early chapters to support and demonstrate the underlying premise in the book regarding the evolutionary nature of our understanding and the development of the world. I also found it personally interesting and illuminating to do. However, if the detail is not particularly engaging for you, it is possible to go straight to the summary/conclusion sections in the chapter without losing the sense of the argument.

Also please note that the factual information in the book is based on research from a variety of sources. This is particularly true in relation to the major faiths where my personal experience is confined to Christianity. Therefore, any errors emanate either from my mistake or from those of the sources. But as mentioned in the book, even "facts" are not necessarily indisputable. I apologise for any such factual errors, with no offence intended, but I hope any other disagreement with information relating to the opinion or interpretation by the source, is respected as a different understanding and as such is in line with the spirit of what this book is ultimately intended to express.

7

INTRODUCTION

The spur to my writing this book, is concern for the harm and distress caused by the way various faith groups currently respond, and have responded in the past, to the diversity of understanding of faith, religion and God. This is both within and between such faith groups. As part of my exploration of these issues, the question arises as to whether or how much of these differences follow from the way God communicates. Is it possible that God has a communication problem ? By looking to better understand the reasons for this diversity and the consequences of it, my hope is that it will be possible to diminish future harm and distress by changing some attitudes and behaviours. But before proceeding further, I will briefly state both my personal faith position and what I mean by God.

Firstly, I am writing this book as a Christian and present the arguments and discussion from within that framework. However, readers from other faith traditions may notice parallels within their own faith frameworks and experience. I also acknowledge that it is not only people of faith who are harmed by the way faith groups respond and have responded to their diversity of understandings. Calling myself Christian, should not lead to any assumptions about what I may or may not believe, other than to say that I base my life on trying to be a follower, disciple of Jesus Christ. Within Christianity there are a variety of understandings of what or who God is. There are also a variety of opinions about how one gains such understanding in terms of how the Christian scriptures i.e., the Bible, should be read and understood. Consideration of the consequences of these variations form part of what this book is about.

Secondly, I will try to describe/define [5] what I mean by God. I say "try to" describe/define God, as from a human perspective this is an impossible task and itself represents an aspect of the communication problem referred to in the title. God's Communication is the main focus of chapter 10. But for now, what I mean by God [6] is a "something other than, both ourselves as human beings and everything we are aware of within the universe(s)." The fact that there is such a wide range of understandings of God and faith and their associated doctrines, affects everyone, not only people with some sort of religious faith.

For people of faith, we have mentioned the guilt and shame that may be a consequence of questioning aspects of doctrine and the statements of belief associated with a particular religious group. They can also be harmed and hurt by comments and pronouncements from certain faith leaders and members of some faith groups, that are critical of certain aspects of their character, behaviour and attitudes. For example, women wishing to enter the priesthood within some Christian circles have faced criticism, which at times has been quite vitriolic. Homosexuals are at the receiving end of much hostility.

This is not only in the form of verbal abuse but also violence and even death. Such responses of course, have not just been confined to religious groups. But discriminatory remarks, whether based on one particular interpretation of a religious text or a claim to be those that have the one true faith, can be harmful and hurtful to anyone – whether a person of faith or not.

As well as hurt caused at a personal individual level, populations at large can be caught up in major conflicts between warring religious groups, whether they have any connections with the combatants and their ideology or not and whether they are people of faith or not. So these issues have implications for everyone. It seems to me, therefore, that for all these reasons, the topics in this book are important to consider, as all of us can be affected by them, one way or another.

From a faith perspective, believing that God has created the universe (however that has been and is being accomplished), the question of how things have come to be the way they are, is intimately bound up with questions concerning God's communication with human kind and how much he involves himself in the ongoing evolution of the universe and whether or not he has an ultimate purpose for his creation. How we think about, understand and respond to these questions is also an important aspect of this. Plenty of room for more differences in understanding !

From a non–faith perspective, the fact that things are the way they are, is likely to be just accepted as "that's the way life is."

Either way, from whichever perspective one is coming, perhaps if people of faith can understand better the reasons for why they believe God has set things up the way they are, there might be a fighting chance of reducing the harmful consequences resulting from our responses to religious difference – for the benefit of everyone. Such a quest forms the purpose of this book. To fulfil this aim, it is necessary to explore the questions and consequences set out in the preface and the previous paragraph.

My exploration begins in chapter 1 by expanding and clarifying some of the arguments already made. A specific aspect of our quest relates to "how we know" whether something is true. How can we be certain that we know what we think we know. So chapter 2 considers what is meant by knowledge and whether we can ever be certain about knowing. Chapters 3, 4 and 5 deal with the evolution of science and philosophy, the cosmos and human beings – Homo sapiens, respectively. Chapter 5 also covers the changing geography of the earth, the development of language and the evolution of and the aggregation of lands into nation states. These three chapters set out the evolutionary development of their subject topics and are used as exemplars of how the universe, our world and its components and contents seem to evolve and change. That is, how God has set things up the way they are.

Chapter 6 looks at the origin of Religion and how it has evolved. Chapter 7 covers ethnic religions and four of the major faiths and their scriptures. Chapter 8 looks at the origins of ideas about faith and God. In chapter 9, I present the results and discussion of a small study, designed to look at what factors influence people to adopt or reject a Christian faith. It addresses the question – "Is there a difference between Christians and those that reject Christianity in terms of their exposure to Christian related influences ?" The associated questionnaire can be found as Appendix 2. The final chapter – chapter 10, gets to the crux of the whole enterprise – God's communication.

As a lay person, I claim no background in theological study or learning, beyond my own experiences as a Christian and fairly extensive reading. This has been rather ad hoc and directed by interest, opportunity and circumstances over many years, rather than as a systematic enquiry. Though, in researching the material for this book, I have looked into the literature regarding the origins of faith and humanity and of the "God Idea" and some of the particular aspects of theology. My personal religious experience is of trying to be a follower of Jesus Christ, since making a formal commitment to Church membership of the then

11

Congregational Church,[7] [8] at the age of eighteen or nineteen, almost fifty years ago.

So if you have got this far, and feel it worth reading on – good luck and "May your God go with you – Dave Allen."

CHAPTER 1

THE PROBLEM AND ITS CONSEQUENCES

The points raised in the preface and introduction are summarised as follows, as the core themes addressed in the book.

1. There are a wide variety of understandings of God, faith and religion arising from a combination of possible causes. From a faith perspective, these include: –

- inherent problems of communication between God and human beings, as an inevitable consequence of the way God has deliberately set things up

- the evolutionary nature of the processes whereby our knowledge is developed, necessarily involves changes in understanding and interpretation as our knowledge of the world increases, and so it is highly likely that similar false starts beset our journey towards a greater understandings of the nature of God and his purposes

- the nature of being human (also a consequence of how God has set things up) is that we are capable of manipulating and using what knowledge we do have, for our own purposes and so it is possible that in misusing that knowledge we have generated false understandings.

From a non– faith perspective: the differences in religious understanding will be assumed to have arisen just because we are human, and as a consequence of how the universe seems to be.

2. One of the consequences of the development of these differences in religious understanding is that human beings have responded to them in ways that have and do cause harm and distress. Such harm and distress may affect anyone – whether they have any personal religious belief or not, and so these matters potentially affect all of us.

3. The proposition of this book is, that if we can better understand the causes and processes by which these differences came about, we may be able to encourage responses to them that cause less harm and give us more awareness of God's communication and acceptance of different ways of understanding God. It seems to me that people of faith need to get to grips with this for the sake of everybody and the future of the world.

So the $64,000 question: So what ? How do we respond ?

SETTING THE SCENE

The uncertainty and the variety of understandings, that we have referred to, relating to matters of belief in God, faith and religion, present something of a conundrum for people of faith. Firstly, because of the variation itself :– why might God allow such variation to develop ? How did all this come about ? Secondly, because the response of human beings over the centuries to these differences, in causing harm to others, does not sit well with the notion of being representatives of a loving God. Thirdly, given where we are now, how might we make things better, in terms of reducing the harm caused ?

I do not intend to look in detail into the specific theological and cultural differences between all the faith groups, as for the purposes of this book, such detail is not the point. The more important question is, how do people of faith handle the fact that they understand their faith and the nature of God in different ways, without making such differences a focus of conflict between faith groups or a source of guilt and shame for some members within a particular faith?

THE VARIATION.
There are a number of what are called the "major religions" – Christianity, Islam, Judaism, Shintoism, Hinduism and Confucianism. There are others, maybe less well known such as, Sikhism, Zoroastrianism and though perhaps with fewer followers, have nevertheless well defined belief systems. And within each of these religions the followers are also likely to have a variety of understandings and beliefs. There are also hundreds more cults or sects and tribal religious practices. There are also other faiths that do not have a "God" or gods as an integral part of their belief system, such as Buddhism or Jainism.

Of the major religions, in which God or gods are an integral part, you may hear people say, certainly within Christianity in the "West", ".….but we all believe in the same God". In reality of course we cannot know whether this is true or not. At one level it will be the same God, there being presumably only one God of the universe, but at another level, none of us will have exactly the same picture or understanding of the nature of that God, and so for each of us, God will mean something different. But if we suppose that there is one God (and those religions which recognise multiple gods usually have one god that assumes a position of central overall importance), then how is it there is such a vast

15

array of ideas, beliefs and understandings across these religions ? How is it there is such antagonism between the people or institutions holding to these different understandings ?

Even within these religions, there are a range of understandings and differences of opinion about aspects of their teachings. So for example, within Christianity there are differences between denominations (reference 4) and also between members belonging to the same denomination. Similarly, there are also differences between the various groupings within Islam [9] and Judaism. [10]

THE HARMS

The harms relating to religious differences, result from the way humans have responded to the differences, not, I believe, because God has set things up deliberately to cause such suffering. Obviously I cannot know this for certain and there are a range of views about the relationship of God to the suffering of the world. In *God's Problem*, a book by Bart Ehrman,[11] he describes why the problem of suffering in the world caused him to loose his faith. He provides a number of insights into what is another difficult conundrum for people of faith i.e., if there is a loving God, why do people suffer ? Whilst I may not share his conclusions, the fact that he can share them and offer them for consideration, is part of what my book is trying to promote. Namely, the opportunity for people to offer their personal struggles in trying to understand faith, theology, God and religion in a climate of mutual respect. This applies both within any particular faith tradition and across the faith divide between different faith groups.

At their worst, religious differences can lead to untold human suffering and death, purely because of differences in belief. This can be particularly acute when one faith group claims it and it alone knows what God wants us to believe and how to practice the faith. We only need to think of the Crusades[12] and the Spanish Inquisition.[13] But such conflicts and suffering are not just a matter of history. They continue today. Witness the "ethnic cleansing"[14] for example, in Rwanda 1994 CE, Sri Lanka in 1983 CE, Bosnia & Herzegovina in the early nineties and of the Rohingya in Myanmar in 2017 CE. Although religious differences may not always be the initial cause of any particular conflict, once hostilities begin they can provide a focus and often an excuse for old rivalries and power struggles to re–emerge. We may think of "The Troubles" in Northern Ireland and the conflicts in Iraq and in Syria between different Muslim groups.

Less visible, but just as destructive for the people affected, the different ways of understanding religious teachings can also create a sense of shame, guilt and or fear, within the followers of any faith tradition. The particular emphasis differs, depending on the prevailing controlling cultural norms.[15] According to Benedict,[16][17] in the West, the predominant cultural form of discomfort is guilt, in the context of things being either right or wrong. Whereas in China, Japan and the Middle and Far Easts, such discomfort tends to induce shame, in the context of things being either shameful or honourable. However, these distinctions are not absolute and there is much common ground between guilt and shame for both individuals and organisations within these cultures.

There is general agreement that guilt tends to occur in societies that attach greater value to people as individuals and shame features more in those societies attaching more emphasis to the interconnectedness of individuals with that society. Thus, guilt tends to arise as a consequence of people feeling they have fallen short of their own internal value system in terms of some sort of perceived "wrong" behaviour, and have thus let themselves down. This often tends to be associated with a sense that others have also been affected by ones behaviour. The response to this can be beneficial, in that the individual can feel remorse and so attempt to try and "put things right". Perhaps by changing their behaviour or trying to remedy the effects of their actions or at the very least, offering an apology.[18] [19] However, such a positive response is by no means guaranteed.

Whereas, shame is a response to falling short of a cultural norm and there is thus, a sense that the group, community and society as a whole have been let down. The response within this cultural mindset is for the individual to focus on themselves rather than on the particular "wrong" behaviour. This may lead them to think they are a worthless or useless person. This can then lead on to feelings of low self esteem, depression, anxiety and resentment. It may then be associated with negative consequences for personal relationships and a withdrawal from and denial of personal responsibility. This may lead on to blaming others with expressions of anger and even violence towards them.[20] It may be that the individual commits suicide or is killed by members of the group in order to remove the shame, if a wise elder cannot intervene to arrange a compromise. [21] [22]

These characteristics under–pin some of the antagonism and mistrust between Christian and Muslim relations. With Christianity being seen to be more associated with the West and hence a guilt based society and Muslims operating within a shame based culture.[23] Guilt is seen by some as a more mature, superior or better and positive response than shame. This view then reinforces the divisions between cultures and the sense of being shamed.[24] However, even within certain organisations and groups

18

within western countries, shame rather than guilt can also be the predominant response. Guilt, as we probably know from personal experience, is not a pleasant feeling and by no means always results in a positive response. Both guilt and shame can arise out of a sense of personal failure, though may lead to different actions and outcomes. Beyond the individual, both emotions can also be used by institutions and societies deliberately, as forms of control. To ensure behavioural conformity of their members.[25] [26] [27] [28]

So guilt and shame can become a real experience for some believers, who, whilst still adhering to a particular religious ideology, may have doubts about whether they are believing or understanding the "right" things. Though still accepting the general tenet of the beliefs of their chosen Church, there may be aspects of the ideology with which they disagree or feel they cannot accept. One might think of contraception within the Roman Catholic tradition, with some members using artificial forms of birth control despite the teachings of the Church proscribing such practices. This therefore becomes a source of tension. Such tension can lead to feelings of disloyalty, anxiety, distress, shame and/or guilt.

Tensions can also arise from just thinking differently about certain aspects of faith and/or daring to question some of the basic precepts of that ideology. People may gain strength from their faith and so wish to still adhere to the general belief framework, but nevertheless, their lives can be adversely affected by acknowledging their doubts. Any such tensions may be aggravated perhaps, by a feeling they do not have a safe, non–judgemental forum in which to raise them. Trying to discuss these matters may risk ridicule, further shame and or condemnation. Religious hierarchies do not have a good track record of welcoming dissension and questions. [29] [30]

Within Christian cultures particularly, many people leave the Church. Having been brought up as children within its traditions, they may reach a point where it ceases to make sense to them. It may be that sometimes this is a result of being told "this is the way things are" and have been discouraged from questioning, or that their teachers were uncomfortable with anything that questioned the prevailing orthodoxy, perhaps because of their own doubts. It maybe that the national Church has failed to move on or convey any new insights and contributions from their theologians that could help contemporary society make sense of some difficult concepts. If things reach the point where leaving the Church seems to be the best and honest option, the act of leaving can itself also be a traumatic experience.[31]

Still other people are put off the Church by their assumptions of what they believe the Church is currently thinking. Their ideas may stem from distorted memories of what they think they were taught years ago or by thinking that the Church still adheres to old and outdated ways of understanding. They may not realise that sections of the Church have reviewed their way of understanding in the light of new knowledge, ideas and cultural needs, and in consequence feel more able to see the connections between their faith and contemporary culture. Yet, other people may be disillusioned by what they see as Churches squabbling within and between themselves over things – particularly sexually related issues, or the role of women in the Church – when they feel that society has largely (though not completely) moved on from such issues. Such disputes may seem trivial in the light of wars, poverty and inequalities which may be seen as more pressing.

Thus the Church is seen as being preoccupied with irrelevancies. Some people may be disillusioned by being aware of some of the atrocities committed by and in the name of the Church and religion. Perhaps not realising that many people of faith may feel just as uncomfortable about such things. In addition, in other areas of life over and above any problems relating to responses to religious diversity and irrespective of the causes, human beings do behave badly towards each other, when they perceive others as being different. Such differences, in a secular context may be – cultural, having different political views, supporting a different football team or all manner of things. This is an important social phenomenon in its own right, but not the focus of this book. However, considering secular approaches to behavioural change, when addressing a less confrontational approach to religious differences, needs to play a role in changing behaviour in the religious setting.

At its best of course, institutionalised religion can provide a liberating source of love, comfort, support and inspiration, and many readers will know of examples of loving, caring and supportive church members and leaders, through personal experiences in their own locality, nationally or internationally.

THE CAUSES – THE HOW
As for how the differences in understanding might arise, there seems to be two broad possible scenarios, depending on where one stands in relation to a religious belief. That is, either from a non–faith perspective or a faith one.

From a non–faith perspective, the differences would be considered to have arisen just because that is the nature of being human. Ideas arise and different people develop and use them as they see fit, either for the benefit of others or the benefit of self. So in this non–faith scenario, for the differences to occur as a result of the way human beings are, involvement of God, as previously defined, is not part of the equation. However, the notion of "god(s)," is an integral part of most religious beliefs and so becomes part of the different understandings of religion.

Here, the concept of "god" is regarded as a human creation – "a construct/concept/idea" and treated as a socio–cultural phenomenon. Such, "god(s)" will be incapable of independent action (see definition of God, reference six). Although we can offer many explanations for how things are, how they have happened, how things work and what makes the universe and the world as they are, there is no sense of why things began and so are now the way they are. Therefore, in this scenario, it is necessary to just accept that things are the way they are because they are the way they are, a fact of how the universe is as it is. In this scenario, opportunity for reducing harm, would lie solely in exploring the role of religion in the socio–cultural development of human beings. The role of "god" would be assumed to be confined to how "god," as a human construct, plays a part in human rituals, sociology and development. Any possibility of reducing the harm caused by how humans respond to the diversity of religious understanding, would be confined to behaviour modification through education and a better acceptance and understanding of difference. Interventions associated with this approach would also need to be an important part in harm reduction coming from a faith perspective.

From a faith perspective, God is deemed to actually exist as a separate "being" and the creator of the universe – however we may understand that. God is believed to interact with the world and its inhabitants. Again, however we may understand that. As such, God must have either deliberately engineered any differences in understanding or such differences have arisen as a natural consequence of the way he seems to have set things up. Including, that which is the nature of being human. If being a deliberate part of God's plan is the explanation, then that seems to cast God in the same mould as the mythical Greek gods of Mount Olympus – mischievous and delighting in tormenting human beings. Some may say that the god portrayed in some of the Old Testament books is not a million miles from such a picture! But from a Christian perspective, such a picture does not fit with the image of a loving God who is concerned about what happens to his people, as portrayed in Jesus. Thus, this explanation will not be considered further.

The way in which God seems to have set things up, appears to include endowing humans with "free will" to make their own decisions. This must still have been a deliberate act on God's part, but it allows for a variety of possible consequences for any given situation. Thus, in the context of our discussion, this includes allowing for different understandings of God. However, the concept of "free will", itself arising as one of a number of different understandings of God, whilst being a useful explanation of how things have got to where they are, also presents its own tension. This tension centres on the notion of our concept of God as being all powerful and all knowing on the one hand and the possibility of humans making decisions, over which God has no pre–determined control over the outcomes, on the other. Considering this paradox adds another complexity to the problem of God's communication, alluded to in the title of the book. We look at this in more depth in chapter 10.

But to even suggest that God may have a problem with communication, may seem a big presumption, and it clearly is. Even imagining that God, being God, could have any problems of any sort, seems to fly in the face of how God is generally understood – i.e., all powerful (omnipotent) and all knowing (omniscient). Particularly in the sense that we might recognise a problem, as being something one would have difficulties with and struggle to overcome. We will come back to this in chapter 10, but for now, from a human perspective "if it looks like a problem and behaves like a problem, then it is a problem" so we will approach the issue as if it is a problem.

So thinking about God may be having a communication problem, we can identify the following four dimensions.

Firstly, on the face of it, given all the different ways there seems to be of understanding God, it does not seem that he has given us the ability to fully understand his nature. The created does not fully know the creator. Inevitably, God will then have a problem conveying his complete nature and purpose to us and we, as human beings will have problems in understanding him.

Secondly, God clearly has not programmed us to unconditionally accept him without question, as if he had, the whole problem of any dissent from the party line would not occur and so harm would not be an outcome. So on both these counts a situation is created in which there is room, on our behalf, for doubt and misunderstanding of God's nature and purpose.

Thirdly, as we describe in chapter two, the evolutionary processes, whereby we acquire our knowledge, involves a series of what may eventually turn out to be misunderstandings, false starts and blind alleys. As what was previously thought to be known, is shown to be wrong or inadequate and is superseded. I suggest that similar processes occur in our understandings of God, faith and religion.

Fourthly, in the nature of human beings, greed and self serving are not unknown. These traits can lead to the manipulation of ideas for selfish purposes. One of the effects of this, for example, may be for one particular faith group to claim to have a monopoly on the "truth" about God.

Given this state of affairs, how can and does God communicate, in ways that we can grasp sufficiently to be aware and make some sense of ? We will come to this later, in chapter 10 but for now, we will deal with each of these dimensions in turn.

THE CAUSES – THE WHY

Although, from a faith perspective, the four dimensions just described, offer an explanation as to how these differences have come about, perhaps the rather more fundamental question is – why has God allowed any of these differences to occur in the first place ? Referring back to the second dimension, if the created cannot fully know the creator, then perhaps we cannot know the answer to this question.

We can speculate of course. Does God need "something" from us and is that "something," for us to believe in "something" that we cannot see or touch or prove i.e., "to have faith" ? It is possible of course but how can we know ? Is it a deliberate ploy, by God, in which to test our faith, if we have chosen to develop one. But again, how can we know ? Is it possible that God gets some satisfaction or pleasure from seeing us struggle and trying to work things out ? Is there some higher purpose that requires this type of preparation ? All these things perhaps are possible, but how can we know ?

But does it matter ? Isn't it just the way it is anyway and we will/can never know why, and so should just accept it ? Ultimately, I think we do have to accept that it is unlikely we will understand fully the reasons why there are a multiplicity of understandings of the nature of God. But that is no reason not to try, and by considering the four dimensions as just mentioned, and better understanding the causes of the ensuing diversity, there seems to me grounds for hope, that harm and distress can be minimised. By seeing the different perspectives of understanding God as a natural part of God's universe and the way he has set things up, we can embrace them positively.

Such a response, whilst requiring a change in attitude and behaviour, can be a catalyst to sharing and learning of insights within and between the different faith groups, in a climate of mutual trust and respect. Perhaps we might reach a point of mutual acceptance of the legitimacy of each others faith position ! We may then even live and work alongside each other, without being concerned about these differences or allowing them to get in the way of cooperation, joint worship and mutual respect. One never knows!

Firstly, let us consider that God does not seem to have given us the ability to fully comprehend his nature. Thus, allowing misunderstandings or misinterpretations of his nature and intentions to occur, as humans have "gone off" in different directions. From a human point of view, it seems reasonable to suppose that God might wish to ensure that we, his creation, know that he exists and understand his nature, even if he gives us the freewill to accept or reject him. Presumably, ensuring his creation has the capability and capacity to understand, would not be beyond him. (questions about why God would bother to create us and if he did, would want to bother communicating with us, are tangential to the substance of this book and are not addressed further, other than to say that, as we cannot know the "mind" of God such questions are impossible to answer – though some religious writings do offer comment on these questions).

However, it seems that although we do have the capacity to comprehend the idea of there being "something other" that is beyond ourselves, the precise nature of that "something other" is outside our ability to fully grasp. Thus, in our struggles to understand, there has developed a whole range of ideas and beliefs about what the nature of this "something other" is and how it relates to us and how we relate to it. Nevertheless, this does not preclude the possibility of continual and ongoing communications between God and human kind or that the previously stated harms are not of concern to him.

Secondly, God seems to have chosen not to have programmed us in a way that would leave us, like robots, with no choice but to believe in him. This seems to be something of a paradox. If God wants us to know him, and from the vast array of religious writings it would seem that he does, one would think he might arrange things to avoid the possibility of doubt. At the very least, he may have set things up so that human beings would be sure of his existence and understand something of his nature, even if they chose to reject it ? But perhaps there is an even greater paradox. If we could be sure of his existence and could fully understand his nature, then perhaps we could not reject him, even with free will.

Presumably, there could be no human construct that would be a better or more credible alternative to a "being" that was omniscient (all knowing), omnipotent (all powerful), omni–present (being present everywhere at the same time), omni–benevolent (showing perfect goodness) and being eternal (everlasting), i.e., perfection. If that indeed is his nature!

If we could fully comprehend this, then perhaps we too would be gods!

As to the third dimension, I suggest that differences in understanding of religious ideas evolve and develop as a natural consequence of how God has set things up. This statement stems from an analysis of the evolution and development of a number of exemplars, including – science, philosophy, cosmology, the geology and geography of the earth and of human development. These are described in some detail in chapters 3, 4 and 5. If we look at these different aspects of life we can see how they have evolved, changed and developed. If God has set up such evolutionary processes in these areas, it seems reasonable to assume, that in the same way, similar patterns might apply to the development and understanding of faith, religion, God and his communications.

All other aspects of our environment, including our own species, seem to change and continue to do so, and our capabilities to examine and understand have also evolved and increased and continue to do so. It would seem logical therefore, to assume that similar processes are applicable to the development of religious ideas and understanding. If this is the case, then the resulting differences should be viewed in the light of a God purposed natural evolution. Thus, variation can be seen as an inevitable part of evolution and development and hence God's intention, rather than an area of life that most people have got wrong and where only a few know the truth. This should then lead to a whole new perspective on intra– and inter– faith relationships.

If we accept this position, then differences in religious ideas should be seen positively and a source for good rather than as a source of distress and conflict. If we acknowledge this to be the case, then it seems inappropriate for any one faith position to claim absolutely that it knows the right and only way to believe in God. Consequently, it is no longer justifiable to impose any particular belief on anyone else, whether it be by pressure, guilt, shame or force. Of course, it is still important to share our individual and community insights, but in a climate of mutual respect and understanding, rather than one of attempted conversion. Only in this way, will we increase our empathy and care for each other and respond positively to opportunities for working together for the common good.

We are aware of the development of language, social organisation and cultural and religious characteristics and identities. We are also aware that although they differ across the world, they also have many similarities. Just as these attributes of the human race have evolved, so has our ability to find out about them and study and understand the features of this evolution. Our ability to get to grips with these ideas is a feature of what we call "intelligence" – shorthand for describing our human capacity for curiosity and our ability to observe, reason, think and identify and solve problems. These abilities being functions of our anatomical brains, which too, have evolved over time.

Not only have these abilities evolved but our capacity and capability to perform these functions has also increased with time. We have become able to do more and different 'intelligent" things. This seems to come as a consequence of identifiable physical changes in our brains and bodies, associated with an ability to learn. By appreciating that not only the nature of things change and evolve over time and our capacity to comprehend does too, but that our understanding of the nature of things also changes and evolves, then perhaps living with uncertainty will become more comfortable. Being "comfortable in one's skin" in this way may lead to increased tolerance and acceptance of other views and being less hung up on what is or is not the correct form of belief.

However, the evolution of ideas and understanding is not about a linear progression from ignorance to complex knowledge. Rather, it is that change occurs over time in different areas of life and at different times and rates. Change builds and develops on what has gone before. Learning to survive, and identifying and placing the first building blocks of knowledge, where there was previously an "empty space of unknowing" also required early hominids to use all the brain power at their disposal. It is likely that some were more able, insightful or creative than others. The ability of some present day human beings to understand quantum mechanics is only possible because of the discovery and thought and experimentation that has happened over the preceding millennia (although Richard Feynman is quoted as saying: *"It is my task to convince you not to turn away because you don't understand it [quantum mechanics]. You see, my physics students don't understand it either. That is because I don't understand it. Nobody does")*.[32] Our increased knowledge does not necessarily mean we are cleverer than past human beings or even our cousins – the Neanderthals. Though there is evidence that:

- our brains have evolved and grown compared to those of our lowest common ancestor (LCA)

- there is a link between brain size and ability

- our brains demonstrate an increased emphasis, being placed by evolution, on ensuring greater flexibility and efficiency in neural transmission, alongside any increase in size.[33][34]

Our increased capabilities stem also from the fact that our thoughts and ideas are starting from a point of more advanced knowledge. As Sir Isaac Newton said in a letter to Robert Hooke in 1676 – *" If I have seen further, it is by standing on the shoulders of Giants."* A similar idea had been expressed previously and attributed to Bernard of Chartres in the twelfth-century and also the Jewish scholar Isaiah di Trani in the thirteenth-century. Even today, I suspect that most of us, even if given a telescope, would not be able to trace the movements of the planets and calculate the distances to the sun and the moon. But there are scientists who have been able to use the ideas and thoughts of the ancients to take our understanding of the universe further and in new directions.

One also needs to be mindful of accepting the usage of epithets such as 'savage' or "primitive", as used by some in the pre-twentieth century literature. These descriptions are based on a presumption of our 'civilised intellectual superiority', arising from our greater body of knowledge. In line with the themes of this book, earlier belief systems need to be accepted as understandable within the context, knowledge and cultural environment in which they arose. But practices, that we would now see as cruel – human sacrifice or cannibalism for example – whilst they may be understood within the sociological or anthropological framework of their time, if they were known to be continuing now somewhere, we would, I believe, have a duty as developed nations, to influence their cessation. Much as there is an international movement to try and stop Female Genital Mutilation (FGM).

It can be a difficult balance to distinguish between practices and beliefs that should be accepted as part of a particular religious framework and culture and so be allowed to continue and activities that many would regard as cruel, harmful, inappropriate and unnecessary, and so justify interference to stop them. Some may put the guilt engendered by parts of western Christianity in this second category. Attitudes to women or homosexuality or abortion may also reasonably be included here. But addressing any of these issues of disagreement requires sensitivity and respect, but not necessarily acceptance. We will refer to this again in the final chapter.

As we shall see in chapter 2, whether new understandings of faith, God and religion can be classed as knowledge is another philosophical question. It is likely these understandings will be overtaken and replaced by fresh information and ideas in the future. But if such understandings allow us to move forward and help us practically, then we accept them as "knowledge" for as long as they remain useful and are not superseded by further discoveries.

I think it is amazing to think that many of our commonly held scientific beliefs are based on initial ideas originating more than 3000 years ago – atomisation, heliocentricity and gravity – to name only three. In those days, these concepts did not have the underpinning detail we are now able to provide through technology and experience, and so were not understood in the way we now understand them. Nevertheless, such ideas were stepping stones on the way to our current understanding. Other concepts – such as the four elements (earth, wind, fire and water) as the basis of all matter, have been discarded completely. We are not surprised by such changes in our understanding, as we assume new "knowledge" accumulates over time and with increasing investigation and thought. But, we should not be so arrogant as to assume it is because we are the cleverest human beings so far. We are just fortunate (perhaps, depending on your view point !) to have been born now.

So we have witnessed changes and developments in our general understanding of the world, its contents and how it works. We have uncovered ideas concerning the development of our human species in terms of language, communication and information. Why then would we think that such a process of development does not also apply to the religious aspects of life ? Is it sensible to think that understanding and insight stopped with the creation of the Bible or in the writing of the Qur'an or the Hindu scriptures etc. ? Why should we think scriptures suddenly become set in stone (though they were, literally, in some cases) whereas other areas of life experience move on ? Why do we think or want our relationship with God and our understanding not to develop and evolve too ?

Why should we not wish to learn from other cultures and groups on matters of faith, when we learn and share ideas on other aspects of life – trade, science, fashion, music etcetera ? Why should we think that religious writings (scripture), written centuries/millennia ago would be the last word on how to live and what to believe now ? Is it that the nature and quality of God's communication changes over time ? That at a specific point in history his communication changed and took on a special significance that it had to be written down (and called scripture) and treated with particular reverence, such that any and all subsequent communication would be of a lesser importance ? This would seem unlikely, given the way every other area of our experience of the world and the universe appears to develop within an evolutionary and changing framework.

Questioning is important. It is the very process whereby knowledge and understanding develops. As Paul Tillich, a twentieth-century. German theological expresses it, in his book – *Dynamics of Faith* [35] – doubt is a necessary and essential component of faith. I think that recognising and acknowledging doubt, paradoxically increases the confidence to talk about faith. One doesn't have to feel one has all the answers. This in turn helps towards becoming more open to hearing about other ways of understanding, which may otherwise be stumbling blocks to acknowledging other perspectives. Such openness then removes the drivers of the guilt and shame resulting from legitimate doubts. At the same time, if we see religious differences as an inevitable part of God's plan, then it is clear that a violent response to these differences is also not appropriate.

The Church has a role in supporting such questioning in a number of ways. By acknowledging that they don't have a monopoly on the truth, though they may have some uniques insights to share. By admitting they are still working it out, they will better support their members, who are also struggling to understand their own personal faith. Also, the church has an obligation to share new insights from their theologians/scholars as a way of contributing to conversations and debate as we all continue our journey of faith. This may help reveal the folly of not only a violent response to religious differences but also, increase awareness of the hurt caused to others when expressing specific interpretations of scripture, in absolute and forceful terms.

If God is God, then he does not need human beings to defend "his teachings" or promote them in violent and repressive ways.

In the fourth dimension, we have a situation in which some people, whether at a local level or as national and international leaders of particular faith groups, have used theological interpretation to justify all manner of activities. Some of which may be seen as contrary to the main message of that faith. To cite just a few examples – the way some elements of the Christian Church in South Africa supported apartheid, how some Muslim clerics advocate violence against both other Muslims as well as non–Muslims, the crusades of the eleventh to the thirteenth-centuries and the Spanish Inquisition of the late fifteenth-century. In these circumstances, concepts of God and theology and faith prevailing at the time, have either been used to justify activities that we might now see as inappropriate or the concepts were deliberately manipulated or new concepts created to justify such actions. In other situations, doctrines have been used to suggest that one particular set of beliefs or faith group has a monopoly of understanding God and so know the right or only way of doing things. Such distortions are a problem for society, as ordinary citizens can be caught up in religious conflict.

CONCLUSION

Thus, it seems to me that our knowledge and understanding of God, faith and religion evolves in the same way as in other aspects of life and the environment. They do so as a natural consequence of the way in which God has set things up. Recognising this, creates the possibility of being able to embrace diversity and so live side by side in tolerant respect and mutual support, even whilst still holding on to cherished and important beliefs and differences. Such a state of affairs would enhance and strengthen our understanding of God, rather than undermining and weakening our faith. We can embrace the concept of diversity which is of God, but question the detail and how it is managed, which are of humankind.

CHAPTER 2

ORIGINS OF KNOWLEDGE

Epistemology – the study and nature of knowledge – is the focus of this chapter, but don't be put off just yet by the terminology! I don't intend this to be an in depth analysis of a complex field of study. Rather, to use it as a means of providing a basis for discussing our approach to knowledge of God and the nature of faith. The fact that the nature of knowledge is a complex field of study may seem surprising, after all, we know what we know – don't we ? However, we will all have had experience of things that were thought to be true some years ago that are now no longer thought to be so. The original understanding and explanation have been superseded. For example, when Pluto was discovered in 1930 CE it was classified as a planet. However, over the years its classification has been changed and is now designated as a minor planet. Our understanding of some of its characteristics have also changed – including calculations as to its weight.[36] But the nature of Pluto itself has not changed substantially since it was discovered, it is our understanding and our ability to observe and measure it that has evolved and changed.

Another clear example is that of Ignaz Semmelweis.[37] He was a Hungarian physician in the mid- nineteenth-century. He found that hand washing by doctors and nurses could dramatically reduce deaths from puerperal fever, an infection that affected women who had just given birth. He was not able to explain why, and the origins and nature of infections were not known or understood at the time. The "knowledge" about the causation of puerperal fever at the time was that it was related to specific individual characteristics of the person affected and was due to an imbalance between the four humours – black bile, yellow bile, phlegm and blood. The treatment was bloodletting. The exhortation to "wash your hands" was taken as a personal insult by many of the "gentleman" doctors, who thought that as they were "gentlemen," their hands could not possibly be dirty. So Semmelweis's ideas were rejected by the medical establishment. It was only years later, after hundreds more women had died unnecessarily, that the theory and causation of infection became better understood. With this more specific explanation of Semmelweis's findings, hand washing was introduced, and became widespread, though even today, there is still the need to have campaign drives in hospitals and care homes on hand hygiene.

We may look back with a rather superior sneer and think we now know better because we are cleverer than our forebears, but any such arrogance is surely misplaced. It may still take any new evidence based techniques or interventions twenty years to be adopted.[38] This is not altogether based on reluctance to change, as it is clearly necessary to be as certain as possible that any new treatment is both effective and safe. There are instances where new medicines have been introduced and the outcomes have been disastrous such as Thalidomide which caused limb deformities in babies, when taken in pregnancy. But there is a balance to be struck. Nevertheless, implementation and widespread adoption of proven treatments can, for a variety of reasons, still take a long time to become part of routine care. There is a lot of literature documenting the barriers that impede the translation of evidence into practice.

This is well summarised under "knowledge transfer" in Wikipedia.[39] People and experts still resist new ideas when they clash with long held cherished beliefs/knowledge. We now know more because of the evolution of research and technology, driven by new insights and new and different questions about how the world is. But it is likely that these 'new" ideas may well be superseded in the future, as more information is uncovered. Of course, not all new ideas turn out to be helpful, though all knowledge – even that which doesn't work – is useful in the pursuit of progress.

So what is accepted as knowledge at one point in time, we find can change with the passage of time. It is important to understand that it will and that it should. Otherwise we risk not only replicating what happened to the clinical discovery of Semmelweis, but also in persecuting and ridiculing those championing the new ideas. This sort of behaviour, in itself, can be a barrier to the presentation of new ideas.

The main proposition of this book is to suggest that we should accept that the evolutionary and changing nature of our knowledge of the universe and our understanding of it, should also apply to our understanding of the nature of God and his communications. It seems reasonable to think that how we might "know" or "authenticate" the if, what, when and how of God's communication, should have parallels with the ways we might "know" or "authenticate" other areas of acquired knowledge. Therefore, it seems to me that an exploration of the nature of knowledge may have lessons for the ways in which it is appropriate to talk about how we "know" God and what he is communicating with us.

Looking at some examples of how the substance of our knowledge has changed, particularly in the disciplines we have come to call science and philosophy, might point us towards appropriate ways of how we might think about our knowledge of God and his communication. By so doing, we may come to accept, that it is only natural and to be expected, that this too has changed, is changing and will/should continue to change. However, we also need to recognise that such acceptance of changes in understanding in no way demeans the nature of faith or diminishes the ways we may understand the power of God.

I initially sourced the information for this chapter from Wikipedia and then followed the links and references as seemed most useful for my purposes. I suspect that my use of Wikipedia as a source at all, will attract some criticism. There have been many concerns expressed in connection with the accuracy, bias and editorial processes used within Wikipedia. These too, are covered in some detail by another Wikipedia article ![40] The debates surrounding the veracity of some Wikipedia articles in particular and the standards of factual and un–biased accounts in general, themselves illustrate the difficulties of "knowing," both what is true, and what is the relationship between knowledge and belief. Of course, the authors providing the criticism, which, although it may be legitimate criticism, also come with their own bias, experience and interpretation of the "facts" ! Other books, papers, articles and media coverage are not immune from the same criticisms.

In my own field of medicine, the British Medical Journal periodically has articles entitled "Head to Head". In these, two people with an interest in a chosen topic are asked to express their view on a particular aspect. The people are chosen not only because of their expertise and experience of the subject but also because they have different perspectives on the question posed. Quite often, having diametrically opposing view points. This is probably not an unusual situation in most professions. People will hold a variety of opinions, which is accepted as the nature of professional judgement.

Such judgement is acquired through personal exposure to life, training, study and professional experience. However, when professionals have access to all the same literature and scientific evidence and when they use some of the same studies, as well as selected different studies to support their own opposite conclusions, it may cause confusion. Not only for fellow professionals, whose expertise may be in other branches of medicine, but also for the general public, who don't know who to believe or how to behave in the face of contradictory advice.

Can or should we say that the evidence used to form the expressed opinion is knowledge ? Can or should we say it is true knowledge ? Can it be either knowledge or true, if it is used to support different conclusions ? Can or should we say that the opinions expressed are knowledge or true or just belief, and if belief, is it true or a justified belief, given that it is based on material that is being used to support a contrary belief ?

This is all by way of both introduction to the complexities of the subject, but also in justifying my use of Wikipedia – not as an infallible source of truth or fact but as a reasonable starting point for exploring the topic. As we shall see, the establishment of an absolute and eternal veracity is not the point. A critical and respectful appreciation of information, which is quite likely to change – is !

SO WHAT IS KNOWLEDGE ?
From the Oxford English Dictionary, we have the following definitions:

- Knowledge: Facts, information, and skills acquired through experience or education; the theoretical or practical understanding of a subject, awareness or familiarity gained by experience of a fact or situation. In Philosophy: knowledge is a true, justified belief; Certain understanding, as opposed to opinion

- Fact – a thing that is known or proved to be true
- Truth the quality or state of being true, that which is true or in accordance with fact or reality. A fact or belief that is accepted as true: the emergence of scientific truths.
- To know – Be aware of through observation, inquiry, or information:

In addition to the above, there is a sense that "true knowledge" should have a "certainty and enduring" quality, for it be counted as knowledge. But what constitutes knowing ? What does it mean "to know" something ? How does one know ? How does one know what one knows ? Do we know what we know ? If what we accept as knowledge can change, is it really knowledge ? What is truth ? Such questions have been analysed and dissected by philosophers and theologians for centuries and continue to be so.

My purpose is not to explore or try to answer these specific questions directly, but rather to outline briefly how, what in day to day interactions we accept as "knowledge," has evolved and changed over time. Thereby, suggesting parallels and similar processes in our approaches to knowledge and understanding of God and his communications with us.

We all have an idea of what we understand by knowledge and truth and belief and for the most part, this gets us through life both on a personal level and in the world of science, technology and business. The Judiciary and the police are perhaps the ones most confronted by the idea that "not everything is always as it seems" and so must be prepared to change what they "know to be true". Their "truth" is influenced both by people

41

deliberately trying to deceive them but also by the fact that the initial, most obvious explanation for a set of circumstances may not turn out to be the case, after further investigation by detectives.

Thus, what we regard as truth or knowledge on a day to day basis, may not conform to any absolute definition that we could be certain of in all possible present or future circumstances. But we accept that in daily life, we just have to get on with it and work with the accepted concepts of truth and knowledge, whilst ever they seem to work for us. So in the popular mind, "we do know what we know." But the upshot of all the thinking, writing, research and theorising is really to indicate that, even though we think we know something, which when put to the test seems to work in practice in real life, there is always a chance that at some point in the future, even hundreds of years in the future, it may be overthrown or need to be modified.

In science, this process is crystallised through the testing of hypotheses and theories. These may then be modified according to the results of experimentation or further thought. Of course, not all science proceeds in such a purist, objective and unbiased manner. We shall return to this in more detail later in the chapter three. Naturally, not all knowledge is derived through scientific discovery, exploration and experimentation. Much of what we know derives from either our own life experiences or by reading what others have written about theirs.

Thus, it seems to me, that what we know and what we regard as true, at any particular point in time, is what accords with our current experience and what seems to work pragmatically, at that time. Although the word "knowledge" may not be the most precise terminology, it is a word in common usage and its meaning is generally understood. On a day to day basis, we have little concern for or perhaps understanding of, the nuances and heartaches experienced by philosophers as they struggle to find an incontrovertible definition. So in continuing to use the terminology – know, known or knowledge in the rest of this chapter, I am referring to that which is/was generally believed to be true at a particular point in time. This may be because it seems to be in accord with current experience and seems to work, or it may be that it is something most people believe because it is promoted by those regarded as knowledgeable and thought to be trustworthy. Or it may be because people in authority, particularly the Church in past times, said it was so.

Such authorities often had their own purposes for declaring what was so, irrespective of any other information they may have been aware of to the contrary. For example, at what point did the "knowledge" that the earth was the centre of the universe and that the sun moved round the earth, change to become the knowledge that in fact the opposite was the case ? We will come back to this particular question in chapters three and four.

In the meantime, before going on to discuss our current, broad and complex categorisations of knowledge, I think it is useful to think that initially there were broadly two types of knowledge, what I may call – pragmatic or practical knowledge and specific or special/explanatory knowledge. The essential difference between these two types of knowledge is that, the first is more closely related to the direct practical needs of day to day survival. The special knowledge is generally focussed elsewhere, though in earlier times there would have been substantial overlap between them in a way that is perhaps less obviously so today. I use these concepts to try and describe how knowledge may have come about and evolved from accepted experience (pragmatic or practical knowledge) to a more conscious enquiry (specific or special/

explanatory knowledge) of experience and the environment. This distinction is clearly artificial as they would have had much in common both in content and the methods of acquisition. Initially, in reality, all what was known would have been the body of knowledge that was known.

As time went on, new additions to the pragmatic body of knowledge would become less as the immediate problems associated with survival were overcome. The body of special knowledge would continue to grow, though the demarcations between areas of enquiry remained blurred as each specialism spawned more detailed offspring. But as the quantity of knowledge increased, artificial categorisations were developed to try and make it more manageable. People would have found it more and more difficult to cope with the ever increasing volume of all that was known. We clearly still have issues with survival both in the developing world and globally, but these now tend to be addressed through scientific approaches and specialist areas, rather than just through personal experience and trial and error.

So to the first type of knowledge. The first type of knowledge (pragmatic/practical), that early humanoids acquired would have been through their senses and their experiences. In this way, they would have amassed useful information that enabled them to survive. They would have passed on this knowledge as they established oral traditions and their language developed. Then, as these ancient people began to explore their environment and began to think and to ask "why" and "how" questions about their experience of living, myths and legends involving the concept of god or gods evolved as a means of explaining how things were the way they were. (the question of how the concept of God/god might arise in the mind of someone who only had the evidence of their own environment and experience is explored in chapter 8).

Thus, questions and explanations would be the beginnings of the second type of knowledge – the specific or special/explanatory knowledge. Hence, reality and imagination would be intertwined and became part of an oral tradition. Whether these myths and legends were understood as imagination and allegory at the time or were initially believed as truth by the people who originated them, is unknown. Who initiated them, and what was their position in the society is also unknown. But the general population would gradually come to accept and believe in them, though it is likely, as has happened throughout history, not everyone would necessarily agree. This combination of pragmatic information and the content of the myths and legends and other special knowledge would become the body of knowledge handed down to the next generation. This knowledge would have been elaborated and added to as they widened their horizons and experiences, through travel and exploration beyond their familiar locations.

Our understanding of what our ancestors knew and how that knowledge was used, comes from various sources. The first body of evidence, indicating that they must have learnt enough to have survived and prospered sufficiently to produce subsequent generations – is our existence today! There is also tangible evidence in the form of increasingly sophisticated tool making from around 2.5 million years ago.[41] Evidence of rock art and cultural artefacts and fossils have been found across the world, dating back some 40,000 years.[42] [43] The precise lineage of the peoples who produced these finds is a source of much debate amongst anthropologists and this is touched on later in chapter 4.

But at this point, it is sufficient to note that whatever body of knowledge any particular hominid grouping acquired, is likely to have been passed on and learnt by subsequent generations and groupings. Some may even possibly have crossed over between "species". How such oral information was amassed, retained and passed on to the next generation is the subject of the book by Kelly – *Knowledge and Power in Prehistoric Societies*.[44] She suggests that prior to the development of formal written language, information and ideas would have been passed along through what became a well developed oral tradition of knowledge accumulation, memorising and transfer across generations and between tribes and cultures.

She points out that the acquisition, retention and transfer of knowledge in pre–literate cultures was much more sophisticated and organised than has been generally acknowledged.[45] The degree of detail associated with any particular piece of information or mnemonic (a pattern of letters, phrases, ideas, activities or other association used to help remember anything one is required to memorise) was understood at different levels. Such knowledge was built up, maintained and preserved in the use of song, rock art, dance and theatrical performances. It is also thought that the great monumental archeological sites (such as Stonehenge) may have been used not only to facilitate these processes but also as mnemonics to aid the learning and memorising of knowledge.

Although a piece of rock art, wooden carving, monument, story or dance would be accessible to everyone, its meaning would be understood in different degrees of detail. This depended on whether one was an ordinary member of the group or chosen to be initiated as a guardian of the knowledge.

As memory is an imprecise form of record keeper, intense and lengthy training was required for those entrusted with the greatest detail to ensure the knowledge was memorised accurately. The survival of the group depended on it. Thus, the details of crops, seasons, navigation, geography, meteorology, edible plants, habitats and habits of animals and so on was in the hands of a select and revered few. This information was memorised with the help of stories, myths, legends, song, dance and other forms of mnemonic. It is thought that much of this deeper level detail was only available to the initiated of the group. So now it is difficult for researchers to gain a real deep understanding of such knowledge. Nevertheless, there are examples of the classification of crops and plants into groups and sub–groups contained even today within the narrative poetry of the Aboriginal people.

Examples of this type of knowledge classification and the similar use of mnemonics can be found across continents, though with specific and particular characteristics emphasised within different communities. For example the "Dreamtime" referred to within Aboriginal culture contains their understandings of how things came to be. It is passed down the generations through stories. These stories incorporate essential knowledge for living. Such as, directions to places, what plants are edible and when and where to collect them. Also, how to behave and the rules of the group and so on. Such information is partitioned into "women's work" and "mens' work" together with information common to both. The information is passed on in stages. With increasingly deeper levels of meaning for each story, being conveyed as children get older.

Once writing was invented, the earliest evidence of which is found in Mesopotamia dating from around 3200 BCE, much more detailed information about the cultures and their activities becomes available and accessible. The myths and legends being an integral part of this historical record.

It is likely that most people contributed to the pragmatic body of knowledge as everyone would be using their wits and learning from their experiences in order to survive. So, as this knowledge accumulated and passed between people and on to their offspring, it becomes more generally available. Initially, it is likely that it would be fairly local, but as people moved about and engaged with other groups in other environments, knowledge was gradually disseminated. As groups developed and began the transition from nomadic hunters to more settled communities, so societal structures developed, as discussed by Wade.[46]

It is probable then, that some members were given time to devote to pursuits other than those solely focussed on survival. So, as well as perhaps being able to concentrate more on making tools, pottery, clothing and art and so on to support the group, there would perhaps be time to stop and observe nature and think about the world around them. Perhaps to start wondering how or why their world was as it was. How night followed day and the changing seasons. Thus, a few individuals may be attracted to such pursuits and were given permission and time to develop them. In this way, the myths and legends were born. Such thinking and wondering would occur in parallel with the development of language – enabling communication with ones self as well as with others. The development of language is covered in chapter 5.

Admittedly, it is unlikely to have been quite as pure a process as just described. But one can imagine people multi–tasking. Motivated individuals, snatching time out of a busy day or at night, after their daily tasks were finished, spending time thinking about other things. They would gradually come up with new ideas, insights and inventions, some of which might have been shared with the general population. But, unless particularly useful in day to day living, perhaps would be kept secret or only shared with a few other like minded individuals. Hence a new body of "special" or "explanatory" knowledge evolved, which these "specialists" would talk about amongst themselves. The earliest examples of "special knowledge" would probably be of a practical nature, such as more specialist tools, materials and artefacts. Examples of these things have been found dating back into pre–history (designated as the time before the development of writing allowed the contemporaneous recording of history). As they tried to make sense of "how things were and had come to be so", initial explanations are likely to have been in the form of what we now know as "myths and legends."

This aspect is explored more fully in chapter 6 – as The Origins of Religion. However, having created the myths and legends as initial explanations for how things were the way they were, members of future generations would begin to look and think deeper and seek perhaps other explanations. This sort of speculation then added to and expanded the body of "special" knowledge. Thus, there would have been an ever increasing amount of knowledge and information necessary and useful to living. It was essential this be passed on from one generation to another. But then, how and what to pass on and to whom, become problems in themselves.

Initially, passing on pragmatic knowledge and skills could be done face to face on a one to one basis. However, as the content broadened and the population grew, additional means were required. Initially, the learning, as we have seen, was through memorisation. With the development of increasingly sophisticated systems for facilitating this, specific individuals were chosen to become the guardians of this knowledge. As writing developed, the act of both learning and passing on information become less dependant on the memories of a selected few. A widening out of the people "in the know" could take place. Here, the concept of schools evolved. Places where this information could be passed on to groups, rather than just individuals. The skills of reading and writing were also passed on to facilitate acquisition of this knowledge. Thus, the concept of education – the passing on of knowledge, came about.

In his book, *A History of Education*,[47] P Ellwood Cubberley gives a comprehensive picture of how our western approach to education evolved from the time of the ancient Greeks. How education became a necessary feature of all civilisations, though initially, it was often the rich and well connected youngsters who were chosen to receive some sort of formal tuition.

As the amount of knowledge grew, the question arose as to what and how much to pass on. That is, it became necessary to consider "curriculum" planning. It would necessarily have to be selective, as not everyone could know everything and so not everything could be passed on. So initially, it was the important life skills such as the language of court, fighting and hunting which were taught to the sons of noblemen. However, this problem eased once there were books. These were initially transcribed and copied by hand. With the advent of the printing press, more books on more subjects could be produced and circulated more widely. They became less expensive and with probably a higher degree of accuracy or at least consistent error. So although the gathering of new knowledge – pragmatic learning – and the processes by which it was acquired, was taking place all the time, in all places and in all cultures, the specialised knowledge would generally be more focused on the wealthy and privileged.

What to me is remarkable, is how over the centuries, similar concepts and knowledge developed independently in widely separate cultures across the world. Also, how cross fertilisation of knowledge between cultures has led to the emergence of worldwide consensus about certain things and concepts. This has happened without, seemingly, any person or organisation taking control of it to make it happen. For example, the use of a common numbering system and units of weights and measures has become accepted across the world. With English being adopted as an international means of communication.

For all knowledge/information, there is the possibility for it to be proven wrong or changed/ improved upon, in the light of later understanding and observation. Such "upgrading" also applied to the pragmatic knowledge, as more efficient ways were devised for doing things or just through trial and error during day to day usage. For example, dragging things along would be superseded by moving things on rollers and once the concept of the wheel was discovered, movement with carts and so on become the new "sliced bread" ! A different type of example might be that of Castor oil. It has long been used to help digestion, but it is known that one bean can be fatal ! The reason it is now not harmful when ingested, is because in the manufacture of castor oil, the poisonous ingredient – ricin – is removed.

Clearly at some point in time, the " knowledge" would be that castor plants were poisonous but at some later date, this statement becomes qualified. Thus, that original 'knowledge' turns out not to be universally certain or enduring but contextual. So whilst we may think that pragmatic knowledge is more likely to be compatible with a definition of knowledge that is more certain, true and enduring it is always possible that it too will be changed by future revelations. This is even more true for the special knowledge.

As well as covering the theory and meaning of knowledge, Epistemology, also includes concepts such as truth, justification and belief. Plato's definition of knowledge was as "justified, true belief (JTB)." By this, he meant that holding a true opinion or belief was not sufficient of itself to count as knowledge. There had to be a justification for why that belief was held. For Plato, knowledge came from thoughts and ideas. Whereas, Aristotle believed that observation of the external world or the experience of ones' senses was a more robust source of knowledge. Plato and others thought that knowledge based on the experiences of one's senses was unreliable. Although he had some reservations about the completeness of his own definition, this requirement of JTB was generally accepted, even up until the Renaissance in the sixteenth century.

However, with the onset of the Renaissance, Aristotle's views were re-discovered and gradually his use of observation became accepted as the forerunner of what was to become the "scientific method" (see chapter 3). Thus, the concept of JTB continued, but the evidence for justification came to be accepted as coming from external observation and the experience of one's senses rather than from internally generated ideas and theory alone. Plato used an inductive method – going from the particular to the general – whilst Aristotle's approach is said to be deductive – going from the general to the particular. Despite the adoption of Aristotle's views, there was acknowledgement that the senses could deceive and so the emphasis moved from mere observation towards the development of experimentation and investigative enquiry, to reduce the possibility of error.

In 1581, in a book entitled, *That Nothing Is Known* the philosopher Francisco Sanches challenged both the Platonic and Aristotelian theories of knowledge and suggested that humans cannot ever attain genuine knowledge. However, he argued that it was possible, to use "scientific method" (a term he introduced) to carefully gather empirical information and to make cautious judgments about it. Sanches's life and philosophy are discussed by Roland Pérez.[48]

Later, in 1963, Gettier,[49] suggested that JTB may also not be sufficient to count something as being "knowledge". He argued that an additional fourth condition was needed in order to be certain that any particular assertion could be regarded as knowledge. There is no need to go into this further, but other philosophers have both disagreed with this or have agreed that a further condition is needed, but cannot agree what such a condition should be. These arguments are chronicled by Shope.[50] Similar debates abound within philosophy and theology in relation to the definitions of truth.[51 52 53]

In this quest to define and acquire knowledge, numerous other philosophical stand points have developed, including; Idealism, Rationalism, Foundationalism, Constructivism, Fallibilism, Skepticism, Empiricism and Coherentism to name a few ! These all put forward different perspectives when trying to answer questions about what could define something as knowledge, or as being true or as justification of a belief.[54] Whilst I am not competent to discuss the details of these schools of thought and such differences are not particularly pertinent to the theme of the book, it is useful to be aware of two particular schools –

Skepticism and Fallibilism. Skeptics argue that there is no "absolute" or 'infallible" knowledge. That is, there is no knowledge that can be counted as knowledge, in terms of it being an absolute, infallible, enduring and sustainable truth.[55]

Fallibilism, on the other hand, doesn't deny the possibility of knowledge, it is just that we cannot ever be certain that at some point, the knowledge we have won't turn out to be false.[56] But Fallibilism itself is used and understood differently by different philosophers.[57] Lewis[58] argues that it is possible to steer between Skeptiscim and Fallibilism, though whether this adds clarity to the discussion you may wish to decide for yourself. However, the fundamental tenet of both these schools is that it is not possible to know with certainty that anything is eternally true.

These arguments support one of the themes of this book, namely that for anyone or any organisation to think they know with absolute and eternal certainty what God is about and what he wants of and for humankind, is an error. It is so firstly, because of the characteristics of knowledge just discussed. Secondly, because the consequence of holding such an absolutist and exclusive position leads to some of the harms indicated in the opening chapter. That anyone who questions the certainty is deemed to be wrong and those designated as being wrong, may be seen by some to be deserving of some kind of retribution – even death. Such punishment would then be justified as being in accordance with, what they determined to be the will of God.

The only two statements that come close to being unchallenged as certain and enduring are Socrates's notion that the only thing he knew was that he knew nothing with certainty,[59] though apparently he was not sure that he even knew that – and Descartes' – "I think therefore I am" – that is: if I can think, I know I must exist.[60]

Over the centuries, various classifications have been devised for organising the knowledge we have. These include not only a whole range of different subject based categories but also a number of definition based categories.[61] [62] [63] [64] For example, de Jong et.al.[65] refer to generic (or general) and domain specific knowledge, concrete and abstract knowledge, formal and informal knowledge, declarative, conceptual and procedural knowledge, elaborated and compiled knowledge, unstructured and (highly) structured knowledge, tacit or inert knowledge, strategic knowledge, acquisition knowledge, situated knowledge, and meta–knowledge. They then state they find it more efficient to introduce two other dimensions describing knowledge: type and quality of knowledge.

They also reference a paper by Reif and Allen (1992),[66] who in turn describe at least eight different knowledge terms: main interpretation knowledge, general knowledge, definitional knowledge, ancillary knowledge, supplementary knowledge, case— specific knowledge, entailed knowledge, and concept knowledge. Subject based classifications have also been created to try and organise the increasing number of new areas of enquiry that come with increasing specialisation. Beyond a high–level subject classification, there are sub–divisions and sub–categories together with a plethora of other academic subject areas.

In days past, areas of study tended to be fairly well demarcated and, for example, within science we still recognise a basic distinction between biology, chemistry and physics for teaching purposes in schools. However, with increasing knowledge and information it is evident there is considerable common ground across many topics which once were regarded as distinct. But as can be seen in Figure 1 (Leydesdorrf and Rafols)[67] just the complexity of the inter–relationships between the categories is "mind blowing", let alone trying to keep up with the content in any of these categories. Such a detailed diagram, whilst bringing some clarity by demonstrating these inter–relationships, at the same time also shows how bewilderingly complex our accumulated knowledge has become.

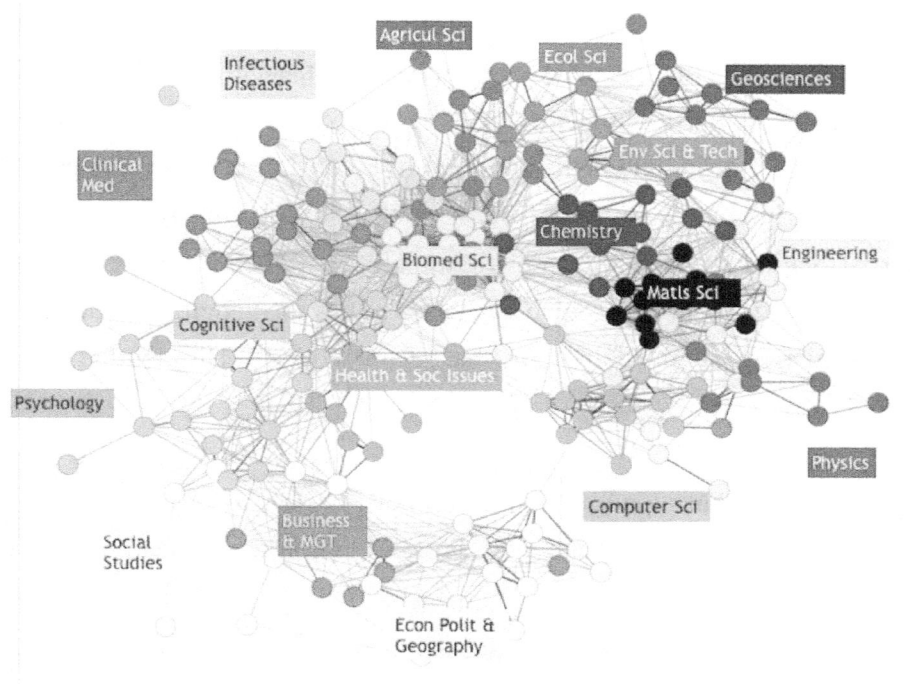

FIGURE 1: INTER–RELATIONS OF KNOWLEDGE

Recognition and acknowledgment of these complex linkages may suggest there should be opportunities for mutual cooperation across disciplines. Which in turn may reduce duplication and promote quicker solutions and more practical applications.[68] [69] But, this increasing volume of knowledge creates a conundrum. Whilst demonstrating the interconnectedness of many knowledge disciplines and hence the potential for increased cooperation, it makes it more difficult for individuals and teams of researchers to keep abreast of all developments even in their own fields, let alone those in other areas. This not only affects science but all aspects of life, including the study of theology and religion. Thus the management of knowledge is now a research field in its own right.

The idea of sharing knowledge has also generated a number of other concerns. Intellectual property – the question of "who owns" knowledge has become an important concept. There is competition for status and recognition in academia. This can lead to a conflict between the need to publish early for the kudos of being first, and delaying because of the fear of plagiarism or giving ones competitors a "heads up" about what you are doing. Benowitz [70] also discusses issues influencing the decisions of scientists to share their work, based on work by Blumenthal [71] and Campbell. [72]

They reported that in surveys of a number of scientists there were many examples where commercial and professional interests had interfered with the sharing of research data. Also, many academic and other professional journals charge for access to their published papers and so potentially deter wider dissemination beyond the circle of interested colleagues. Additional barriers to wider dissemination include the language, terminology and jargon often used in professional publications. This can make access more difficult for people outside these specialised fields. Who would repeatedly pay $40 for academic papers, just out of interest ?

Part of the definition of a 'profession' is that its members have particular/ special knowledge that distinguishes their field of interest and so potentially gives them power over others. The concept 'knowledge is power' ("Scientia potentia est") is popularly ascribed to Frances Bacon (1561 to 1626 CE), though Wikipedia[73] suggests the phrase was first attributed to Imam Ali (599 to 661 CE), as recorded in the tenth–century book *Nahj Al–Balagha*. The first known reference to the exact phrase appeared in the Latin edition of Hobbes' *Leviathan* (1668 CE). Seemingly, Hobbes was Bacon's secretary as a young man and had philosophical discussions with him. However, the relationship between knowledge and power had been understood long before this. In prehistory, before written records, Kelly[74] suggests that the oral tradition of knowledge retention and transmission gave power to those entrusted to memorise the cultural traditions.

With the advent of the internet as a vehicle for widely and freely disseminating information, there is increasing pressure from some organisations, institutions and researchers to have a free distribution of all knowledge.[75] [76] [77] There is still evidence of resistance to this.[78]

What is pertinent for our purposes, is that it is likely that some of these similar issues also apply to the publication of theological and philosophical research. Even though some of the concepts considered in theology may be generally familiar, the jargon used in these writings may be difficult for outsiders to grasp. All fields of enquiry will inevitably develop their own technical language to aid communication between colleagues. It is used as a type of shorthand to avoid perhaps long and complicated repetitive explanations of ideas. Although not peculiar to theology, I think it particularly necessary for theology, to use language accessible to everyone, as it is dealing with matters of faith, an area of life pertinent to many people.

In addition, this vast growth of knowledge also presents some purely practical difficulties in terms of dissemination. There are no clear processes for making decisions regarding what content and by whom and in what ways, new theological and philosophical insights could actually be made more generally available. Thus, making new ideas available to the faith community, let alone the community at large is a real practical problem, even if the Church hierarchies feel inclined to do so.

So, here we are, in the twenty-first century, in a scientific age, and there is still no universal agreement as to what constitutes knowledge. There is even doubt as to whether the possibility of attaining any certain knowledge at all, is possible.[79] Kvanvig discusses the idea of whether using the concept of "understanding" is a more useful approach than using the notion of "knowledge."[80]

This idea seems to me to be an extremely helpful concept when considering religious belief. Religious belief is an area of enquiry that is much reliant on interpretation and inference. Conclusions follow from considering information, that is itself open to varying interpretation, rather than on "facts," as normally understood in scientific investigation. By using this notion of "understanding" and taking a Fallibilism perspective i.e., that it is likely that any current understanding is likely to change in the light of future deliberation and information, ones' position can be seen as less dogmatic and less confrontational. Thereby allowing for more respectful dialogue over differing points of view.

Religious communities are likely to say that the ultimate or absolute truth resides in God or their deity. This then resolves (in their minds) the question of whether there is such a thing as an absolute or certain truth. However, many people of faith would wish to consider the intellectual aspects of trying to define truth. They are not content with just falling back on a statement that does not resolve the question of what is true in everyday life, nor actually defines what the concept "absolute truth" means. We are thus left with the situation that, whilst academics may question what counts as knowledge and whether in fact, we have any knowledge at all, in the purest sense of it being an absolute, infallible, enduring and sustainable truth, people are living and working with a pragmatic approach to truth and knowledge in the day to day business of their lives.

SUMMARY
Thus, it seems to me, notwithstanding all the philosophical debate mentioned earlier, "knowledge" is contextual. It is a product of its time and place and of the culture and of the men and women who describe it. This is not the same as Relativism [81] which argues there is no absolute or intrinsic value in any specific belief but that any value is only relative to or assessed against, another belief. That is to say the value of any particular knowledge or truth is in relation to some other truth or knowledge or within a specified reference framework. I would suggest that all knowledge and truth have value, if they accord with our current experience and work on a pragmatic day to day basis. However, this value only pertains until the knowledge or truth is superseded by some new information, thought or discovery. We might say "there is no knowledge, only understanding, which is transient." [82]

This suggests that I think there is no absolute truth. In terms of practical daily living, I do think this is true. But Christians, and I regard myself as a Christian, naturally argue that God is the absolute truth. I am not disputing this. But what I believe is open to question, is what this actually means in practice. It is the meaning of this statement that is open to different understandings. Such understandings will vary and change for different people and at different times, as new insights and new discoveries come to light.

For stuff that is not within our own immediate experience or awareness or ability we have to rely on others. This is why experts who lie, or commentators and journalists who provide biased reports deliberately to influence public opinion or politicians who lie and spin or historians who either deliberately put a particular spin on their work or who deliberately deceive, are so damaging. Not only for the particular content they lie about, but by the fact that they undermine the general trust we need to have in other experts and commentators, in order to function. We may not have any immediate means of assessing whether what they are saying is true or not, until someone else can demonstrate that, or at some point in the future it comes to light that, they have been lying. The concept of Fake news [83] is the latest example of these attempts to deliberately deceive. Lying and deliberate deceit are clearly different to presenting what is believed to be the truth at a particular point in time but which subsequently turns out to be in error because new information comes to light that changes it.

So, with the benefit of hindsight and the accumulated millennia of thought to help us, we can trace patterns of evolution of ideas and begin to appreciate the relationship of these early thoughts to the place we are in today. Using some of the written records of their ideas, let us journey, in the next chapter, with those people who were starting to look to the environment and the natural world itself for explanations of natural phenomena, rather than just following the accepted wisdom of their myths and legends. They may have just been curious, or perhaps they were looking beyond the explanation – "it was the gods that done it. !"

※∩ ※∩ ※∩※

CHAPTER 3

SCIENCE AND PHILOSOPHY

EXEMPLARS OF

THE EVOLUTION OF KNOWLEDGE

The purpose of this chapter is to show that even with subjects – science and philosophy – which we probably feel are well defined and well understood, their origins, in both terminology and content are uncertain at best. Not all the details of the progress of their evolution are clear either. What is clear and so silly even to state, though I will, is that there has not been a cosmic stenographer recording all the developments of humankind and recording everything that has been said and done and written or when. Therefore, anything we do know, stems from either the interpretation of the physical relics that have been found in many parts of the world or from oral transmission or written material.

Such material has often required much study to decipher ancient writing and symbols into known languages, which in turn have been translated into other languages. Often, in the case of the early, ancient thinkers, many of their writings have been lost or indeed they may not have written anything at all (e.g. Socrates). Information on their ideas has come from the records of other writers (e.g. Plato), which themselves may be incomplete. And of course, whatever has been written, has been done so for varied purposes by numerous people. But in general, not to specifically faithfully record the detail for consumption by later generations to help them understand what has gone off ! Even the diarists, who may write contemporaneously, have to select what to include and what to leave out.

Many factors therefore affect how history and the past is recorded, remembered and then subsequently interpreted and re–interpreted. Often, even the dating and timing of events is not certain. Thus, although what happened in the past does not change – recorded history does. The process of writing history – historiography – is also open to interpretation and is a field of scholarship in its own right. [84] [85] [86] [87] For example, the date range of certain designated "eras" of history – the enlightenment, the middle ages and so on are not precise or universally agreed. (My use of dates for eras of history are generally taken from Wikipedia and are an indication only of the time period covered). So the perspectives and subjectivity which affect what is written down at the time of the event – the primary sources – shape what we regard as the history we have today.

This history is also influenced by any subsequent interpretation of the how and why things happened, or any re–interpretation in the light of new knowledge and or the perspectives and understandings of new writers. These same factors and processes must surely also apply to religious writings, as much as any other forms of knowledge. But there are also particular additional perspectives when it comes to "religious writings" and I will briefly digress to consider these, though more will be said in chapter 7.

Religious commentators, theologians and leaders of the Church may counter the suggestion just made, that the factors affecting our understanding of history will apply to "religious writings". They may argue the belief that God dictated to or inspired the authors to write down his words. So scriptures are not subject to the same influences I have just outlined. A counter to this argument is, that God's communication could only be through the words and ideas people were capable of understanding at the time. Thus, what was understood, memorised and written down could only reflect people's understanding and comprehension at that moment. This in turn, would be influenced by the prevailing culture and experience of the time. Over time, with an increasing experience and capacity to understand, God's communications would most likely evolve accordingly. So we might reasonably expect therefore, that his communications are continuing to evolve, and so our understanding must evolve with it.

The Christian understanding of the nature of God, certainly since the protestant reformation, is that God speaks to and has a relationship with each and everyone of us directly. This is in contrast to what was the Roman Catholic position, in terms of the need for confession, through a priest. If it is the case that God communicates directly with us, it is reasonable to think that this has been the case all through human evolution. This realisation itself, forms part of the evolution of our understanding of God.

The Hebrew scriptures and the new testament of the Bible, record specific instances where God communicated his message to the Jews through the prophets and ultimately through Jesus. These recorded instances often relate to times in history when God perhaps felt the situation required his direct input. Such passages are written by the people of the time, in ways that reflect their understanding of God's relationship with them. Whether God actually intervened in the war like

64

way portrayed in parts of the Old Testament we cannot know, but the writings form part of the narrative that we seek to interpret as our understandings of God evolve over time. There are also examples of his communication with other significant people in biblical history – such as King David, Joseph, Samuel and so on. There seems to be no mention of God speaking to *"Mrs. Trellis of North Wales."* [88] Perhaps there are many more instances of God's communication that have been lost or just not recorded.

Other faiths have their own records of what they see as God's communication with them. But however and whenever he communicated he could only do so in ways that could be understood within the cultural and knowledge framework of those to whom he communicated. In pre–historic times, what form his communication would have taken is impossible to know. We might conjecture that with the development of language and greater knowledge, the religious rituals, myths and legends that we are aware of through paintings, other artefacts and eventually through primitive writing, indicate the types of communication. It is only with more extensive religious writings – scripture – that we have a greater insight into what form God's communication might have taken. So such writings could be seen as simply reflecting the best understandings of the writers of the time, rather than a "verbatim" total record of God's communication with humankind. As such, be seen to inform and guide rather than shackle all future thought.

So as part of the scene setting for the theme of the book, this chapter explores the origins of both science and philosophy. It aims to show how much uncertainty and serendipity have contributed to the evolution and understanding we have today, even for subjects we might imagine were more "objectively derived". The intention being to suggest that similar evolutionary processes are therefore also likely to apply to our understanding and knowledge of God, faith and religion.

I have chosen science and philosophy as exemplars of the evolution of knowledge because both disciplines are perhaps thought to be pretty well worked out and understood. The word philosophy was in use by the ancient Greeks and it has a long and intimate relationship with the ideas and processes that have evolved into what we now understand separately as science and also theology. Having chosen them as our examples, the obvious way to explore their evolution is to start with their beginnings and trace how they have changed over time. If we start with science, we can take the word and look at its etymology, origins and use, as far as we know it, and then consider the development of what we may call "scientific knowledge". The point being, to bring together both the evolution of both the word "science" and the particular knowledge content and topic classification as we know it today. Similarly with philosophy, starting with the word and following its evolution alongside the content knowledge that has come to be regarded as philosophy.

However, this is where we hit our first problem – terminology. This is perhaps not surprising. We only have to think of some of the words we are likely to be familiar with, which in our life time have changed their meanings. Gay for instance is really no longer used to mean happy and cheerful but relates to a person being homosexual. Other words take on additional meanings : "swipe" now relates to moving ones fingers over a computer screen or track–pad or even putting a credit card through a payment machine rather than just meaning to steal. A "mouse" no longer just refers to a rodent but to the gadget used to operate a computer, the "cloud" is a "place" to store information as well as moisture and a tablet is a computer as well as a form of medication.

We have also created new words for things that did not exist before – internet, aeroplane, telephone and so on. Such new words often have their roots in other similar words, or in other languages or may be totally created out of the blue – so called "nonce" words. Of course nonce also now has taken on a totally different meaning (if you ever knew the original – I didn't!) and is used as prison slang for a sex offender. For those interested, Bodle gives a good description of the various means by which new words come into the English language.[89] The point is, there is not a time in history, as far as we know, when someone thought to themselves," I'm at a bit of a loose end, I think I'll start to look systematically at how the world's come to be as it is and I'll call my enquiry – 'science or philosophy.' "!

The difficulty in identifying a beginning and a timeline for the use of these words and their meanings, is compounded by the way in which commentators, historians and authors apply modern terminology to the past. They use the terminology of to–day to describe things, actions and activities in time periods when these words and terminologies did not exist. This is referred to by Lloyd in the preface to his book – *Early Greek Science: Thales to Aristotle.*[90] Presumably this practice is done as a type of shorthand to indicate that such things, actions and activities of the past have parallels with things, actions and activities we recognise today. This may seem a convenient way of simplifying explanations, but it confuses attempts to understand the origins and meanings of words and ideas. Particularly if the meaning and understanding of the words themselves have also changed over time.

The story is further complicated by the fact that there is still not a clear consensus of how we understand and define science or philosophy even today.[91] [92] [93] So we cannot be at all certain that even when the same words are used, there is agreement that the meaning actually refers to the same thing.

And so back in pre–history, all that was known was what was known and there would have been no demarcation between areas of knowledge. The supernatural/myths and legends as explanations of the world and the pragmatic and increasing special knowledge would all have been accepted as an integrated body of knowledge that was known. However, as the content of what was known grew, some sort of classification became inevitable as a means of managing it. Even during the oral tradition, before writing evolved, Kelly[94] points to evidence of complex categorisation within the learned and remembered narratives. Then, as thinkers began to look at the natural world itself for explanations of why things were the way they were, that is they began to reason, the supernatural/myths and legends became separated, eventually giving rise to the branch of knowledge we now recognise as theology.

By the time of Aristotle (384 to 322 BCE), the range and span of knowledge was considerable, in areas we now know as: anatomy, astronomy, embryology, geography, geology, meteorology, physics, zoology, plants and animals. In his treatise "Meteorology," Aristotle discusses various weather related phenomena – hail, earthquakes, thunder and lightening. He also wrote on aesthetics, ethics, government, metaphysics, politics, economics, psychology, rhetoric and theology and studied education, foreign customs, literature and poetry. Aristotle is also credited with the development of formal logic as a means of enquiry, analysis and argument. The pre–socratics (philosophers and thinkers before Socrates – 470 to 399 BCE) had also shown evidence of the use of logical processes in developing proofs of certain ideas.[95] So, at that time,

Aristotle's works pretty much covered the whole range of Greek knowledge.[96]

But although we recognise the areas of study mentioned above, as subject areas with which we are familiar, they were not distinguished by headings/labels of their own until our current nomenclature came into being during the eighteenth and nineteenth centuries CE.[97] Whilst many of the actual labels just listed, were not in use in ancient Greece and as I have said, using them does complicate any discussion of their origins and time lines, I acknowledge they do help convey the wide range of topic areas stimulating the curiosity of the Greek thinkers, between about 600 and 200 BCE. It is a convenient way of avoiding giving long lists of "items" such as – rocks, animals, plants, natural disasters, planetary phenomena and so on. But at the same time, we need to remember that the ancients did not categorise "areas of knowledge" in this way.

Philosophy in ancient Greek referred to the "love of wisdom/ knowledge." So it covered the search for special knowledge, as opposed to the pragmatic knowledge referred to in the previous chapter. In addition, a more systematic exploration of nature and the environment began. This was known as 'Reason'. It was applied to questions about why the world was as it was, seeking for explanations within nature itself, rather than falling back on supernatural causes. The process and the topic areas became known as "natural philosophy," which eventually became our science. The range of topics covered by natural philosophy were brought together by Pliny the elder, who wrote in Latin, around 79 BCE.

His comprehensive *Natural History of the World* (another source of confusion as natural history here covers the field of natural philosophy, not the restricted fields covered by natural history as we understand it today – see below) included, in our modern terminology: astronomy, mathematics, geography, ethnography, human physiology, anthropology, zoology, botany, agriculture, horticulture, pharmacology, mining, mineralogy, sculpture, painting, and precious stones. Gradually the scope of natural history became more focused on botany and biology and palaeontology, and the other topics became separated into fields of enquiry in their own right.

The concept of Philosophy also changed and eventually came to relate more to the meaning of things, rather than just the love of knowledge. Mathematics and medicine also developed as separate areas of enquiry. Mathematics becoming a means of explanation and solution to problems identified in other fields of study. For Galileo, mathematics was the language of "The Book of Nature" as he called it. He said that without the symbols of mathematics – circles, triangles etcetera, the philosophy of nature could not be understood. He was one of the scientists, alongside Newton and Francis Bacon who felt the "Book of Nature" and the 'Book of Scripture' should be studied separately.[98]

The foregoing is just a broad brush approach to help conceptualise the evolution of the early main areas of knowledge as they emerged from the general "soup" of what was known. A more detailed discussion of their gradual evolution into the complex differentiated categories described in the previous chapter, occurs as the chapter unfolds. So, before we deal in any more detail with the content of what became science and philosophy, we will look a little more closely at the words themselves.

Notwithstanding the difficulties outlined above, of identifying the origins and timelines of our subjects, for the purposes of my argument and the proposition of this book, it is still necessary to make an attempt to uncover something of their origins and evolution, if we are to gain some perspective of the evolutionary nature of knowledge. So let us start with the word – science.

Etymology and Definitions:

Science

Here are some definitions of science from a number of different but respected sources:

The Science Council,[99] (and you would think they would be the arbiters of any modern definition of science) define science as:

- Science is the pursuit and application of knowledge and understanding of the natural and social world following a systematic methodology based on evidence.

The Collins English Dictionary defines science as:

- Science is the study of the nature and behaviour of natural things and the knowledge that we obtain about them.

The Oxford Living Dictionary defines science as:

- The intellectual and practical activity encompassing the systematic study of the structure and behaviour of the physical and natural world through observation and experiment.

The Cambridge English Dictionary defines science as:

- (knowledge from) the careful study of the structure and behaviour of the physical world, especially by watching, measuring, and doing experiments, and the development of theories to describe the results of these activities

Wikipedia defines science as:

- Science (from Latin scientia, meaning "knowledge") is a systematic enterprise that builds and organises knowledge in the form of testable explanations and predictions about the universe.

With most of these definitions there is a sense of a methodical/systematic approach to the gathering of knowledge, but they vary in the means they emphasise. There is the observational, experimental, measurement, testable explanations, predictions and the development of theories. Some talk of science being the study of nature or the natural world or the physical world or the universe or of its structure and or its behaviour. The inference from these definitions is that science is perhaps more of a methodology than a body of knowledge per se. But there does seem to be a conflation of the subject content and the processes by which that content is derived. Thus, it isn't clear that as we look back in history, to discover the origins of science, whether we should concentrate on looking for similarities in content or in methods or both or something else ? But, none of these definitions help us get back to the beginning.

There are varied accounts as to how the concept of 'science' emerged. Ross[100] suggests that the term "scientific" is thought to have been used by Boethius, in the sixth century CE. What he meant is not clear and nor is it clear at what point the translation occurred of the Greek "knowledge" into the Latin – "scientia". It may have been from around the twelfth to the thirteenth centuries CE, at the time of the great translations. As Grant indicates,[101] this is supported by translations from al–Fārābī entitled 'De

72

scientist' (On the sciences), and writings by Robert Kilwardby and Dominicus Gundissalinus entitled 'On the Order of the Sciences' and 'On the Division of Philosophy', respectively. These latter two authors both discussed the classification of the then designated sciences.

But even so, the meaning of "scientia" covered general areas of knowledge i.e., that which was known, rather than referring to some specific quality of that knowledge or method of enquiry. Some suggest that "scientia" possibly passed into the English language in the 1600s as the word science. Though the Merriam–Webster Dictionary suggests that the word science entered the English language in the fourteenth century. However, again, even at this point it still referred more to knowledge in general rather than in the more specific way we understand it today.

Again, we have Ross[102] who describes the etymology of the word science by beginning with the Greek word "epistome" – meaning knowledge. There were other words for knowledge but it is believed that Aristotle used this term to imply a more systematic, reasoned approach to knowing, rather than knowledge simply being ideas generated by intuition and thought. Ross describes how the Greek word for knowledge, in use at the time of Aristotle, was eventually translated as "scientia", but perhaps not until the early 1800s, around the time it was beginning to take on more of the meaning we understand today. Only in 1834, was the word 'scientist' coined. It was first suggested by William Whewell, as a collective noun for the people who did science. Even so, this suggestion created much controversy, which Ross describes in some detail.

This controversy related to the etymology of the word and to whether the suffix "ist" was grammatically correct. Similar controversy and arguments were applied to the word physicist – though naturalist, chemist and biologist had already been described and accepted by then. Eventually, with usage, scientist was accepted as the collective noun for men (it was mainly men at that time) who did science. In fact, Men of science was one of the preferred names by some and there were notable investigators who refused ever to call themselves scientist. By the time of the Renaissance (between the fourteenth and seventeenth centuries CE), the concept of the "scientific method" (see later under knowledge acquisition) was gaining traction. In this, experiments and hypotheses became the ways in which objective knowledge was acquired.

From the foregoing, it is clear that we cannot say when the word science was first used. Or precisely at what point in its evolution, its meaning moved from simply meaning "knowledge" in general, to the more specific but complex diversity of topics it covers today. Or, when within its current meaning, the implicit understanding that an underlying methodology (the scientific method) of acquiring the knowledge, became a necessary pre–requisite for that knowledge to be considered as science or scientific.

So another way to attempt to find out when science started, might be to look to individuals who have been thought of as being very influential in the development of science and philosophy. But again, there is no universal agreement over any single person who might be regarded as the founder of science. Numerous candidates have been suggested, covering many centuries, from – Thales of Miletus, Aristotle, Ibn–al–Haytham, Francis Bacon, Galileo Galilei to Sir Isaac Newton. So again, this does not take us back to a specific beginning point. Wikipedia, rather then indicating one person or giving a few options, gives a long list of people who they regard as the "fathers" for each of the many specific fields of science.[103]

Given the difficulties described, in pinning down when the word science came into use, using the term science, in relation to the activities of Aristotle or any of the other ancient thinkers, seems misleading. This is not to impugn the quality of their work or to suggest that it was somehow inferior, though this is the implication from some authors quoted in Grant – *A History of Natural Philosophy*.[104]

Aristotle did ask questions and used a systematic approach of observation, description, some experimentation and reason to draw conclusions and generalisations from this experience. So, in the light of our modern understanding, we may with hindsight, see this approach having parallels with our view of the scientific method. But it seems to me, it is better to avoid such retrospective labelling and just use descriptive terms to explain what he did and what objects he studied. Aristotle himself did promote the idea that knowledge derived from observation and experience and was of a different quality to that derived by pure internal thought through reasoning – as per Plato. This latter process came to be understood as Philosophy. But at the time of Aristotle, although the word philosophy was in use, its meaning referred to the love of wisdom/knowledge, rather than in the sense it is used now (see philosophy section).

So, did Aristotle or any of the Greek philosophers or any of the other great thinkers from either the ancient world or in the time before the renaissance in Europe – do science ? It is possible to find opinions both in support and against this notion. But such opinions are still too subjective and imprecise to definitively answer the question: "when did science start?" At a superficial level, we can easily say that they did not do what we would recognise as modern science. How could they, the word was not known at that time. What really matters is the reliability of the methods by which any knowledge is obtained, and that lends veracity to the information – at least temporarily until further insights displace it.

The question of whether they did science or not, thus becomes an irrelevance. The work they did and the ideas they had, were just what they were and of their time. What they did, how they did it and what they achieved stands alone and should be judged on its own merit as to its usefulness. From Aristotle, this usefulness was immense for 1000 years, until other understandings gradually superseded many of his ideas. Thus, for our purposes, this view allows us to by–pass the sterile academic discussions about what is and what is not science – the 'so called demarcation dispute" [105] [106] and concentrate on determining the value and utility of any given proposition or proffered solution or knowledge.

As Fara [107] says at the beginning of the chapter 'Interactions' in her book – *Science; A Four Thousand Year History* – *"There is no single form of science – what counts as science depends on where and when you're looking."* So the fact that it is difficult to say when science started, is not really of any consequence. She makes the point that we should not look back with hindsight and ridicule the past, (see later under "The Rise of Islam" for an example) as all of these thoughts and ideas have been stepping stones, without any of which, our current understanding of things might be quite different. In addition, Feyerabend's [108] view is that science should not be seen to have a distinctive approach that is separate from other sorts of knowledge, and just because something is classified as "science" it should not have an automatic right to be considered as true. Thus, each new idea should be evaluated on its own merits.

Similar problems of definition and terminology apply when trying to define many other things. We have discussed the problems of defining knowledge in chapter 1, but we could have equivalent debates about the definition of the universe, any of the science topic areas, beauty, a house, a home – probably almost anything one cares to mention. So trying to decide when a new concept or idea is actually "born" and when the name for it is actually introduced and accepted, is problematic. Hence, following its evolution, particularly if the meaning of the terminology also evolves, is correspondingly difficult. For a new technology or invention a new descriptor word is likely to be coined at the same time. But for new areas of study or methodologies, particularly if this is cultural and requires translation to communicate the concepts between countries, it may be some time – even centuries – for the activity and the descriptor word to become common usage.

What I think the preceding discussion demonstrates is that, not only is it difficult to answer the question of when science started but also to answer the question about whether the ancient thinkers "did science". It also shows how difficult it is to separate the etymology of the word and the processes, methodologies and content associated with it. This suggests that these are probably the wrong questions. What is more important, is assessing what contribution and influence the work of the ancients had on the development of later thought. This supports the idea of the evolutionary and changing nature of knowledge and understanding.

The importance of acknowledging the contributions of previous thinkers to current ideas is to recognise this evolutionary nature of thought and knowledge. That these things are not fixed at one point in time, but are subject to change, and that this notion applies as much to theological thought as to science and philosophy. All understanding is dependant on what has gone before and what may come in the form of new discoveries in the future. This is both necessary and inevitable, if we are going to accept the evolving intellect of humankind.

Thus, the definition of the word science and our modern understanding of it, has emerged from a whole evolutionary process of both, ways of thinking about the world and the content produced during this evolutionary process. Such evolution will continue and the methods and content will continue to evolve into the future.

Such a position, it seems to me, is perfectly compatible with a faith in God – whether it be from a Christian, Jewish, Islamic or any other religious view point.

Philosophy

And what of Philosophy? There is perhaps not quite the mystery with the word philosophy and its provenance, as with the word science. This is because the word was used by Aristotle and the pre–socratics, and has been adopted into English, rather than requiring re–interpretation. Its meaning has broadened however, from merely referring to the love of wisdom in Greek times to include a process of understanding the world through thought, logic and reason, applied to a wide variety of topics. The term "Philosophy" is thought to have been coined by Pythagorus (from around 570 to about 495 BCE), though apparently Aristotle regarded Thales of Miletus (from around 624 to about 546 BCE) to be the first philosopher.

※∩ ※∩ ※∩※

The Cambridge English Dictionary defines philosophy as:

- the use of reason in understanding such things as the nature of the real world and existence, the use and limits of knowledge, and the principles of moral judgment:

The Oxford English Dictionary defines philosophy as:

- The study of the fundamental nature of knowledge, reality, and existence, especially when considered as an academic discipline.

- The study of the theoretical basis of a particular branch of knowledge or experience. 'the philosophy of science.'

- A theory or attitude that acts as a guiding principle for behaviour.

- Origin Middle English: from Old French philosophie, via Latin from Greek philosophia 'love of wisdom'.

Wikipedia defines philosophy as:

- Philosophy (from Greek φιλοσοφία, philosophia, literally "love of wisdom" is the study of general and fundamental problems concerning matters such as existence, knowledge, values, reason, mind, and language. The term was probably coined by Pythagoras (from around 570 to 495 BCE).

These definitions tend to refer to the concept of modern Western philosophy, and the numerous different schools of thought that evolved from the pre–socratic and hellenistic (Greek) perspectives. But in accepting these ideas, we may easily forget there is a whole raft of other philosophies and approaches to life. These may conveniently be classified under the umbrella of Eastern Philosophy. However, the various branches encompassed by this are quite disparate and do not have the same basis and links to each other, in the same way that many of the

79

schools spawned by western philosophy have between them. Many of the Eastern philosophies are intimately bound up with the religions associated with them – including Hinduism, Buddhism and the Chinese and Korean philosophies.

The differences between western and eastern philosophies tend to be characterised as : Western society strives to find the truth by logical reasoning and places more emphasis on an individual's rights. In Eastern philosophy, there is more of an acceptance of the way things are. So it places more emphasis on finding balance in and awareness of the unity and mutual interrelation of all things. Seeing everything as integral parts of a cosmic whole.[109] [110] This is a very brief and superficial caricature of the different philosophies, but underlines the fact that western approaches are not the only game in town.

Although we do not know how the words science and philosophy were originally created as language evolved, for our purposes, this does not really matter. To some extent, it supports the point I am trying to make. Namely, that we can never be certain of what we "know" but only work with it, whilst ever it appears to be useful or until new discoveries indicate the need for change, and applies as much to our knowledge and experience of God and faith, as it does to all the other spheres of human experience.

The rather ad hoc unplanned evolution of knowledge suggests that no one should claim primacy over the meaning and understanding of such knowledge and that whatever we think we know today may be superseded tomorrow.

So as well as having difficulties defining the beginning of the trail of science and philosophy, defining the end or at least the point at which we are today, in terms of what does or does not count as science or philosophy, is also not universally defined and agreed. The fascinating paradox about this, being the "gap" between how the public may see things and the academic perspective. Most of the general public will probably feel they understand what it is that pertains to something being labelled as science and the notion that science is about precision. Even though they may not necessarily always fully understand the actual content. Whereas, within academia, there are still major professional and philosophical disagreements about what does count as science. About what it is that actually defines something as being science. A similar situation pertains to philosophy. Though perhaps there may be a less clear understanding within the general population of what might constitute philosophy.

So even the origins of the things we think we know, are not known with any certainty. Even what we think we know, is open to disagreement and argument between academics, about the precise meaning and significance.

Having explored the origins of the words science and philosophy, let us turn our attention to the development of their content or substance.

How did the knowledge of the pre–socratics become the science and the philosophy we understand today ?

Science – Natural Philosophy
The knowledge content, of what eventually becomes our modern science, has a long and complex gestation. As Fara points out, in her book –

Science: A Four Thousand Year History.[111] In the West, the usual narrative regarding the development of western thought goes something like this:

In the West, when considering the history of science and philosophy, the accepted wisdom is to start with the Greeks. This is usually with the pre–socratic philosophers in the fifth century BCE followed by Socrates, Plato and Aristotle. The narrative then demonstrates how their thinking formed the basis for modern western science and philosophy and upon which the Renaissance and the Enlightenment and our modern systems were built. In this way, European thought and intellect came to be the dominant force, not only in western civilisation but globally.

However, it is important to realise that, apart from the Greeks, we have evidence that other civilisations were also developing their own knowledge base – India, China, Babylon and ancient Egypt, both before and contemporaneously with the Greeks. It is highly probable that some of these ideas and knowledge will have crossed boundaries and so influenced the knowledge base and the evolution of ideas of other cultures. Though the "paper trail" and "time–line" of how all this knowledge meshed together is unclear. So for example, from other cultures we have evidence from:

- **Mesopotamia and Ancient Egyptian cultures, from around 3500 to 3000 BCE** – a more formalised inquiry into the nature of things, which produced the first known written record of an interest in astronomy, mathematics and other aspects of the physical world and so could later be classified as "natural philosophy"

- **Assyria, as early as 750 BCE** – optical lens making from polished crystal, often quartz.

- **Babylonia (1800 BCE), Egypt (1600 BCE) and India (1000 BCE)** – mathematical notation and usage as well as from China, Japan, Arab and Persian cultures.

- **India in the sixth century BCE, China in the fifth century BCE, and Greece between the fourth century BCE and the first century BCE** – evidence of sustained development of logic

- **Ancient China** :–

  ‣ Taoist alchemists and philosophers experimenting with elixirs to extend life and cure ailments.

  ‣ a focus on the yin and yang, or contrasting elements in nature; the yin was associated with femininity and coldness, while yang was associated with masculinity and warmth.

  ‣ an idea that the world comprised five phases – fire, earth, metal, wood and water – later also understood by Aristotle, but without the metal.

  ‣ philosophers and doctors exploring human anatomy and understanding there was a relationship between the pulse, the heart and the flow of blood, before it became accepted in the West

Advances in language, writing, medicine, agriculture, religion, science, weapons of war, art and culture occurred at different rates and with variations in degrees of sophistication, differences and similarities. Over time, with the advent of travel, adventure, trade and war these developments became more widely known and there was cross fertilisation between all these different systems. So the dominant influence in the West came from the ancient Greeks and the Arabic/ Muslim cultures. But western thought and knowledge were also influenced by Egyptian and Chinese culture and Indian mathematics and religions.

As we will see in later chapters, theological and religious ideas developed independently but often with remarkable similarities and also became part of such influences. So, although one Greek thinker could not fire off a quick email to his counterpart in India, China or Ancient Egypt or catch a jumbo jet to Thebes or Babylon, people did travel, and so knowledge flowed between different parts of the world. As travel increased and became easier, more people will have participated and hence been affected by the interactions and their own ideas would have influenced and been influenced by other cultures. Hypothetically, perhaps, it might go something like this:

A thought occurs to someone. Or a similar thought occurs to several people who may be unconnected and separated by geography, even on different continents. This thought is then developed by the person or taken up by others and is gradually disseminated, until there is a general realisation or acceptance that it is "out there" and is taken for granted. All this happens often without any control or plan or organisation. But also, sometimes due to clever marketing and self promotion by certain individuals or groups. Clearly there will be other thoughts that are discarded or forgotten, but these may occur to others in the future who may or may not be aware of their previous history. This time, because of other developments, the conditions may encourage the development of this idea.

Thus, ideas and activities spread between groups, tribes, cultures and nations and across the world, but all the time themselves being modified in the process. Thus, no one really owns them or has dominion over them, although some order and systemisation becomes imposed, as the development of the particular thought or activity becomes more complex and sophisticated and is taken off in new directions by other people. Later generations ask further questions and people challenge the underlying assumptions and conclusions. In this way, ideas develop and mature and are used to influence further ideas that affect peoples' lives.

This idea of the transmission of knowledge through travel and trade and exploration is referred to by Fara,[112] Huff [113] and Grant [114] and by Peter Frankopan in *The Silk Roads*.[115] There is a common expression "it is in the aether". This is not quite "the stuff, or substance" that was referred to as 'Aether' in Greek Mythology nor what Plato and Aristotle had in mind in suggesting a fifth element (after fire, earth, air and water). Nor is it quite what scientists have postulated over the centuries as "the material" by which light is transmitted, or that affects the movement of planets or is the 'nature' of the gravitational force. Nor, more recently is it the equivalent of "dark energy" as an explanation for an expanding universe.

However, I quite like to think of it as a concept by which thoughts and ideas are enabled to "float" around (in the "aether") for centuries, permeating the universe. Here, they wait for other people to catch hold of them and develop them, often as their own, and often without necessarily being conscious of their origins. Thus, they provide the building blocks for future developments. This idea is referred to again in chapter 7.

Clearly this is a fanciful caricaturisation. I am not suggesting some supernatural methodology that disperses concepts across the universe for the more perceptive to recognise and develop. Neither am I in any way belittling the effort, time and struggle in overcoming set–backs, ridicule, pain or even death, that has been the experience of many individuals or groups of people in creating new knowledge and new technological break throughs. But I think it conceptualises the means by which ideas are disseminated and "have their time."

Perhaps it doesn't really matter that we do not know or not know precisely when and how and by whom metal working, pottery, the wheel, writing or language etcetera began. It is just marvellous that they did. The man or woman who discovered that clay could be moulded and when dried hard, used as containers, has long been forgotten in the collective memory or were never known, but they were the foundation of thousands of years of human development. What does matter, I think, is to remember that all "new" ideas rarely emanate out of nothing and so all discoverers owe much to people in the past. For example, Newton drew on the work of Galileo, Descartes and Kepler in formulating his ideas on gravity.

Though even before this, Aristotle had had a notion of there being some sort of "cause" for the movement of objects towards the centre of the earth. Thus, it wasn't the falling of an apple that created a totally new concept in Newton's mind completely out of the blue. Whoever said "there is nothing new under sun" was probably not a hundred percent correct as at various points in time there must have been new ideas. Nevertheless, the point is that we should recognise the evolutionary nature of progress and knowledge and accept that where we are today is influenced by whatever has gone before.

I am not suggesting that there is anything wrong with identifying a particular person and giving them credit where credit is due but it is important to recognise that others have made contributions, whether known by name or not. If they are known by name then they deserve recognition, not airbrushed out of the historical narrative for the sake of self aggrandisement or the vested interests of others. Not by any means an unknown occurrence.

So bearing all this in mind, let us use the structure of the "usual narrative" to follow the development of our knowledge of science and philosophy to the present day. Whilst bearing in mind, that the journey of any particular idea that is ascribed to either a particular time or person, has been influenced by cultures other than the Greeks and that a linear time–line does not adequately convey the timing and complex interchange of influences that have affected the evolution of that idea.

From written records, it is possible to recognise that for more than 5000 thousand years, people have been concerned with finding out how the world is as it is. Probably initially by supernatural and mystical explanations (the myths and legends) but then by looking for causes from within nature and the world around them, as they experienced it. Some of the earliest detailed evidence of this relates to the pre – socratic Greek philosophers. They lived between about 650 and 300 BCE, all dates approximate and included:

Thales of Miletus (624 to 546 BCE)

Anaximander (610 to 546 BCE)

Anaximenes (585 to 525 BCE)

Pythagoras (570 to 490 BCE)

Heraclitus (535 to 475 BCE)

Parmenides of Elea (515 to 450 BCE)

Leucippus (first half fifth century BCE)

Anaxagoras (500 to 428 BCE)

Empedocles (490 to 430 BCE)

Zeno of Elea (490 to 430 BCE)

Protagoras (490 to 420 BCE)

Gorgias (487 to 376 BCE)

Democritus (460 to 370 BCE)

Euclid (around 300 BCE)

Such thinkers/philosophers were forming ideas/theories/hypotheses (though not by these names) about how things operated in their environment and looking for causes and explanations from within nature itself, rather than invoking supernatural explanations. By observing the natural world and using rational criticism and debate, the pre–Socratic philosophers sought to explain what they observed. Rather than simply relying on or just accepting the common view that the "gods" caused things to happen.[116] This was a fundamental change in the way people had previously thought. Asking questions about how and why their environment was the way it was, was the beginning of the use of "Reasoning" as a means of enquiry into the natural world.

Unfortunately not much of their writings remain, but what we have provides evidence of ideas like :

- Thales of Miletus explained earthquakes by theorising that the world floated on water and that water was the fundamental element in nature.

- Anaximander disputed the Thales' ideas and proposed that rather than water, a substance called apeiron was the building block of all matter.

- Anaximander also speculated there may be more than one world. This idea was later rejected by Plato, Aristotle, Anaxagoras and Archelaus, but supported by Anaximenes, Heraclitus, Empedocles and Diogenes. Thus, the "modern" idea of multi–verses (multiple universes) is not new! [117]

- Around 500 BCE, Heraclitus proposed that the only basic law governing the universe was the principle of change and that nothing remains in the same state indefinitely. This observation made him one of the first scholars in ancient physics to address the role of time in the universe, a key and sometimes contentious concept in modern and present–day physics.

- Hippocrates in medicine, (from around 460 to about 370 BCE), and his followers, were the first to describe many diseases and medical conditions and developed the Hippocratic Oath for physicians, still relevant and in use today.

- Herophilos (335 to 280 BCE) was the first to base his conclusions on dissection of the human body and to describe the nervous system.

- In the third century BCE, the Greek mathematician Archimedes of Syracuse (287 to 212 BCE) – generally considered to be the greatest mathematician of antiquity and one of the greatest of all time – laid the foundations of hydrostatics, statics and calculated the underlying mathematics of the lever. A leading scientist of classical antiquity, Archimedes also developed elaborate systems of pulleys to move large objects with a minimum of effort. The Archimedes' screw underpins modern hydro–engineering, and his machines of war helped to hold back the armies of Rome in the First Punic War. Archimedes even tore apart the arguments of Aristotle and his metaphysics, pointing out that it was impossible to separate mathematics and nature and proved it by converting mathematical theories into practical inventions. Furthermore, in his work On Floating Bodies, around 250 BCE, Archimedes developed the law of buoyancy, also known as Archimedes' Principle.

- Euclid laid down the foundations of mathematical rigor and introduced the concepts of definition, axiom, theorem and proof, still in use today, in his *Elements*. He also did work on Optics

- The early physicist, Leucippus (first half of the fifth century BCE) adamantly opposed the idea of direct divine intervention in the universe, proposing instead that natural phenomena had a natural cause.

- Leucippus and his student Democritus were the first to develop the theory of atomism. The idea that everything is composed entirely of various imperishable, indivisible elements called atoms.

- Pythagoras applied Greek innovations in mathematics to astronomy, and suggested that the earth was spherical.

- Anaximenes is believed to have stated that the underlying element of the universe was air, (not water or apeiron) and by manipulating air, someone could change its thickness to create fire, water, dirt, and stones.

- Empedocles identified the elements that make up the world which he termed the roots of all things as Fire, Air, Earth, and Water.

- Parmenides argued that all change is a logical impossibility. He gives the example that nothing can go from nonexistence to existence.

- Plato argues that the world is an imperfect replica of an idea that a divine craftsman once held. He also believed that the only way to truly know something was through reason and logic not the study of the object itself, but that changeable matter is a viable course of study.

- Aristotle believed that change was a natural occurrence.

- Galen (129 to about 200 BCE) performed many audacious operations—including brain and eye surgeries— that were not tried again for almost two millennia.

Thales is generally thought of as the first philosopher, mathematician, engineer and astronomer and the first known to engage in what might be regarded as "scientific thought" (though the term scientist was not coined until 1833 CE). He sought to find reasons for the existence of natural things and phenomena by using theories and hypothesis rather than just accepting the prevailing, what we now regard as mythological explanations. He thought that such "natural" explanations were by way of there being a unity of everything from a single ultimate substance. For Thales, this single material substance, from which everything else derived, was water. Quite what he meant by this is not clearly understood, though it may be the forerunner of the twentieth century idea of a "General Theory of Everything." [118]Nevertheless, it is an indication that Thales was beginning to see the world in a different way. One that was perhaps capable of being understood by the human mind. [119]

Anaximander also speculated about the beginnings and origin of animal life. He thought that from warmed up water and earth emerged either fish or entirely fishlike animals. Inside these animals, men took form and embryos were held prisoner until puberty. Only then, after these animals burst open, could men and women come out able to feed themselves. Anaximander put forward the idea that humans had to spend part of this transition inside the mouths of big fish to protect themselves from the Earth's climate until they could come out into the open air and lose their scales. He thought that, considering humans' extended infancy, we could not have survived in the primeval world in the same manner we do presently.[120] In more recent times, there are now theories that some of the first humanoids may have lived very close to the shore line and could swim and dive, rather than all originating from their ancestors in trees and open savannah plains.[121] Naturally, not all anthropologists accept this idea. [122] [123]

Anaximenese disagreed with both Thales and Anaximander and proposed that the unifying substance was not water or apeiron but air. He then used this idea to explain the nature of earth, moon and sun and their motion around the earth, earthquakes, rainbows and lightening.[124]

As a counter to Parmenides argument that all change is a logical impossibility as nothing can go from nonexistence to existence, Leucippus and later, Democritus proposed a theory of Atomism (that everything in the universe is either atoms – tiny, invisible, particles – or voids; empty spaces).[125]

There are two things here. One is that Parmenides view was seen by some to be flawed and the second is that an alternative proposal, namely the notion of "atoms – Atomism " was proposed. The point being that a proposal/theory/idea was put forward and then an alternative idea proposed to counter aspects of the first that were not accepted. This process became a pattern, though at this stage, nothing like an accepted methodology. It would take until the Renaissance of the fourteenth century CE for the concept of a scientific method to be adopted. But here we have someone prepared to articulate their own thoughts on "how things were and how they came to be" and put them out into the media for others to comment on and then put forward their own take on things.

Thus, these pre–socratic philosophers were looking for a broader picture in nature. They were seeking for a unifying cause of diversely occurring events, rather than either treating each one on a case–by–case basis, or attributing them to the gods or a personified nature.[126] They were thinking about the nature of existence itself. As well as developing a body of specialised knowledge in mathematics and astronomy, for example, alongside the pragmatic knowledge needed by everyone for daily living.

So, as this special knowledge was accumulating, as a mixture of facts in mathematics and some of the astronomical observations and of theories and speculation in terms of the nature of things, along comes Socrates (from around 469 to 399 BCE). He says that the only thing he knows is that he knows nothing. Or at least, this is what he is said to have implied in one of Plato's dialogues – Apology. And although none of his writings, if he made any, has been found, we learn a lot about what he thought from Plato's (from around 428/423 to 348/347 BCE) dialogues. In fact, there were many other philosophers around this period whose contributions have not been picked up and promoted in popular culture, in the same way as Socrates, Plato and Aristotle (384 to 322 BCE). Nevertheless, their writings and legacies have provided some of the foundations on which philosophy has been built.

Plato records Socrates's use of what is known as his "dialectic" or Socratic Method, in which he would ask a series of questions. The purpose being to gradually dissect a person's belief or statement, trying to get to the basic truth or contradiction of what they had said. Though, according to Aristotle, Heraclitus was the originator of dialectic. Socrates applied the questions to nature and then to human beings. Both Plato and Aristotle built on these foundations and wrote a great deal, much of which has been preserved. Plato argued that one obtained knowledge through intuition and thought, whereas Aristotle advocated that knowledge was only reliable if obtained through observation and experience interpreted by reason.

This approach was what distinguished Natural Philosophy, which delved into nature and the universe, from other areas of enquiry. Plato's approach was carried on in the form of what we now know as philosophy and Aristotle's approach eventually evolved into what we now know as the scientific method. So between them, Plato and Aristotle, through the learning based in their respective Academies, laid some of the foundations of modern western science and philosophy.

We indicated the wide range of Aristotle's interests earlier, and Grant, in his book – *A History of Natural Philosophy*[127] – states that Aristotle's own classification of knowledge separated the universe into terrestrial (earth) and celestial (heavenly – the cosmos) and was as follows :

1. Metaphysics or theology which considers things that are unchangeable and therefore distinct and separable from matter or body, such as God and spiritual substances

2. Mathematics, which also considers things that are unchangeable, but unlike metaphysics, the objects of mathematics have no separate existence because they are abstractions from physical bodies

3. Physics, or as it was also called natural philosophy, which was concerned only with things that are changeable, exist separately and also have within them a source of innate movement or rest. This category embraces the whole of the physical world and includes animate and inanimate bodies in both the terrestrial and celestial regions i.e the whole of the natural world

Other sources categorise the state of knowledge at around this time in slightly different ways. For example, from Wikipedia[128] we get the following classification of knowledge in Classical Greek times:

- Natural philosophy ("physics") was the study of the physical world (physis, lit: nature); (ultimately evolving into the various natural sciences, especially astronomy, physics, chemistry, biology and cosmology.)

- Moral philosophy ("ethics") was the study of goodness, right and wrong, beauty, justice and virtue (ethos, lit: habit, custom); (ultimately evolving into the social sciences, but still includes value theory, aesthetics, ethics, political philosophy, etc.).

- Metaphysical philosophy ("logos") was the study of existence, causation, God, logic, forms and other abstract objects ("meta–physika" lit: "what comes after physics"). i.e., theology (ultimately evolving into formal sciences such as logic, mathematics and philosophy of science, but still includes epistemology, theology and others)

or:

(1) Logic

(2) Theoretical Philosophy, including Metaphysics, Physics and Mathematics

(3) Practical Philosophy

(4) Poetical Philosophy.

Any topic not regarded as theological, philosophical or mathematical would come under the heading of natural philosophy. Though again, according to Grant, Aristotle, seems to have regarded medicine, mathematics and music as being separate from this main classification, with mathematics – geometry and arithmetic – along with optics and astronomy being seen as "exact or natural sciences". That is, they were about measurement and not about motion, movement or change. So from around this time, up until the nineteenth century CE, the term "natural philosophy" was the common term used to describe the practice of studying nature using reason (analytical thought), observation and experiment. Grant, again, in his book – *A History of Natural Philosophy* (Pgs. 156 to 157) – describes the main components of natural science according to Gundissalinus (from around 1115 to post 1190 CE), – as being the underpinning notion of the study of motion and things that moved, and included in this was the nature of bodies and their constituent parts, the heavens and plants and animals – a similar definition as understood by Aristotle's natural philosophy.

However, to understand the evolution of natural philosophy, it is necessary to be aware of some of the other things that were happening during the middle ages, which helped create a cultural climate in which natural philosophy could develop. The list is necessarily selective, but each element is chosen for its significant contribution to creating the cultural climate of what became known sequentially as the twelfth century Renaissance followed by the European Renaissance of the fourteenth to seventeenth centuries, developing as the Scientific Revolution of the sixteenth to seventeenth centuries and the Age of Enlightenment by the eighteenth century.

The list is not constructed in any particular order and no suggestion of any causal association between them is intended. Neither is there any implied sequential temporal relationship. As previously indicated, the timing of historical events is often imprecise at best and there will frequently be a number of events taking place simultaneously, even as they each impact and influence each other. This characterises the context in which further developments occur, as highlighted by Grant in the introduction to his book *God and Reason in the Middle Ages*.[129] The book discusses the factors contributing to natural philosophy first becoming integrated into the mediaeval university curriculum, and followed by its metamorphosis and emergence as our modern day science.

Christianity
In the early years after the crucifixion of Jesus, Christianity spread slowly within the Roman Empire and was competing with the traditional

pagan beliefs. Organisationally, there was a gradual establishment of five main centres or "sees" – seats for bishops in Antioch, Jerusalem, Alexandria, Constantinople, and Rome. Although Rome regarded itself as the prime centre, all were designated to be on an equal footing and independent. This set up provided plenty of scope for controversy over theological differences and also over power and status. Various summit meetings/councils to discuss such matters were convened throughout the middle ages. In 312 CE Emperor Constantine reportedly saw a vision prior to his Battle at Milvian Bridge and he commanded his troops to mark their shields with the Christian symbol he saw (now known as the Chi Rho – see chapter 5). After winning the battle, Constantine was able to claim the emperorship in the West.

But he then effectively split the Roman empire into east and west by setting up his main administrative centre in Byzantium (originally a Greek founded city) – re–naming it Constantinople. In 313 CE, he issued the Edict of Milan, officially legalising Christian worship. This then encouraged the further spread of Christianity across the area around the Mediterranean. Under Theodosius I (Emperor from 379 to 395 CE), Christianity became the Empire's official state religion and other religious practices were proscribed. This ensured that Christianity would have increasing influence in the culture and politics and evolution of the European nation states over the coming centuries. The influence of the Church in developing education was crucial, as was its role – both positively and negatively – in the relationships between questions of natural philosophy and theology.

The Fall of Rome

In 410 CE Rome was sacked by the Visigoths. The sacking of the city led to arguments as to why this had happened and people blamed Christianity for failing to appease the traditional gods. The religious and political attacks on Christianity spurred Augustine of Hippo to write a defence – *The City of God,* which became one of the foundations of Christian thought. A more severe sack of Rome by the Vandals followed in 455 CE, and the Western Roman Empire finally collapsed in 476 CE, though continuing as the Roman empire in the East for another 1000 years. This ensured that the scholarship, engineering and learning was preserved and continued, even though it was dominated by Greek rather than Latin. The fact that Christianity had been designated the official religion of the Roman Empire meant it had been widely adopted, and so was able to survive the fall of Rome. However, this did mean that the power and influence of the Bishop – now Pope – in Rome was reduced.

The Rise of Islam (All dates are given in the Gregorian calendar format rather than the Islamic calendar – AH (anno hegirae = "in the year of the Hijrah"). See chapter 7.)

Into this mix came the birth of Muhammad (570 to 632 CE) in Mecca, and his founding of Islam. He initially encountered opposition to his ideas amongst many of the polytheistic Arab tribes and felt the call to re–locate to Medina. However, he created a large enough following to overcome further opposition by force and succeeded in uniting the Arab tribes behind him. Thus, he established the Muslim faith and a means of succession. So following his death, his closest followers took over as leaders – caliphs – and continued the territorial expansion. Within 100 years, Muslim armies had conquered almost the entire Middle East including the Levant (Palestine, Israel and Syria), Egypt, Persia and as far east as Afghanistan, North Africa and much of Spain. Of the established five main Christian centres, Jerusalem, Alexandria, and

Antioch fell under Islamic rule and only Rome and Constantinople remained Christian.

Trading, as well as war, played an important role in the spread of Islam and diverse Muslim cultures grew up in the Indian sub–continent, Malaysia, Indonesia and China. The Islamic scholars also became aware of the extensive libraries in both Alexandria and Constantinople and so were acquainted with the teaching of Aristotle and the pre–Socratic scholars. So the Abbasids, established the House of Wisdom in Baghdad (in modern–day Iraq) in 825 CE, creating a centre for study, translation, copying and education and facilitating the learning of the scriptures. It was the sense of being 'the bearers of the final message' that gave the impetus to attract scholars from all over the world, with different cultural backgrounds, including not only Arabs but Africans, Europeans, Central Asians, South Asians and Persians as well.

This provided a climate in which Islamic scholarship could flourish and ushered in what is called, the "Golden Age of Islam."[130] Whether this is an appropriate term, and the dating of it, is debated, nevertheless it is a useful concept within which to describe developments in knowledge and science taking place around this time – from about the eighth to the fourteenth century. Bearing in mind that the use of modern terminology is just for convenience and that many of the words had not yet been coined or meant something different to what we understand today, Islamic science covered : economic development, natural philosophy, mathematics, astronomy, medicine, physics, alchemy and chemistry, optics and ophthalmology, geography and cartography.

Although such activity was conducted under Islamic rule, not all were Muslims. There were Buddhists, Zoroastrians and what we would now call Hindus. Jewish and Christian scholars, that had 'the ancient wisdom', were invited to supervise and translate all of the world's classical knowledge, including the works of Galen, Hippocrates, Plato, Aristotle, Ptolemy (100 to 170 CE) and Archimedes into Syriac and then into Arabic. In Spain, the ancient Greek wisdom that had been expanded in Arabic was translated largely by Christian monks into Latin. Together with some original work in Hebrew, these contributed to the development of medieval European philosophy.

Some of the Muslim philosophers – Al–Kindi (around 801 to 873 CE), al–Farabi, around 872 to 950 CE) Ibn Sina (Avicenna 980 to 1037 CE), Ibn al–Haytham (Alhazen – 965 to 1040 CE) and Ibn Rushd (Averroes – 1126 to 1198 CE), incorporated this material into their own work. Also, by writing encyclopaedias and summaries and writing commentaries, particularly on Aristotle, they played a major role in saving the works of the ancient Greeks. With the advent of universities (see below) this significantly influenced European thought in natural philosophy, optics, medicine, psychology, metaphysics, logic and ethics. Avicenna produced his "Canon of Medicine" which was used as a standard text at mediaeval universities until the seventeenth century.

The physician al–Razi (Rhazes – 854 to 925 CE) was an early proponent of experimental medicine and recommended using control subjects for clinical research. He also criticised Galen's concept of the four humours. Al–Zahrawi (936 to 1013 CE) was a surgeon and is sometimes referred to as the "Father of surgery". He used catgut sutures, cautery and described the use of numerous surgical instruments and procedures, not previously recorded. Al–Kindi warned against the futility of alchemists attempting the transmutation of simple, base metals into precious ones.

The following paragraph is quoted from part of the entry in Wikipedia[131] under the heading "Islamic Science – significance" :

"Historians of science differ in their views of the significance of the scientific accomplishments in the medieval Islamic world. The traditionalist view, exemplified by Bertrand Russell, holds that Islamic science, while admirable in many technical ways, lacked the intellectual energy required for innovation and was chiefly important for preserving ancient knowledge, and handing it on to medieval Europe. The revisionist view, exemplified by Abdus Salam, George Saliba and John M. Hobson holds that a Muslim scientific revolution occurred during the Middle Ages. Scholars such as Donald Routledge Hill and Ahmad Y Hassan argue that Islam was the driving force behind these scientific achievements. According to Ahmed Dallal, science in medieval Islam was "practiced on a scale unprecedented in earlier human history or even contemporary human history".

Toby E. Huff takes the view that, although science in the Islamic world did produce innovations, it did not lead to a scientific revolution, which in his view required an ethos that existed in Europe in the twelfth and thirteenth centuries, but not elsewhere in the world. Will Durant, Fielding H. Garrison, Hossein Nasr and Bernard Lewis held that Muslim scientists helped in laying the foundations for an experimental science with their contributions to the scientific method and their empirical, experimental and quantitative approach to scientific inquiry.

The negative opinions in respect of the contribution of Islamic science expressed by Russell and Huff, and quoted in the previous paragraph, seem like good examples of a biased, western–centric perspective. They seem to miss the point that knowledge and innovation are of their time and the contribution they make to the knowledge and innovations that come after them are a key ingredient of how and why such future knowledge and innovation comes about. It may not always be possible to be specific about or be able to demonstrate a direct and unerring link, but nevertheless without intermediate steps, it is likely that the future knowledge and innovation would look very different. It seems to me, that a failure to recognise and accept the proposition that todays knowledge and understandings have their grounding in past scholarship, is contributory to the intolerance of other ways of understanding, that this book is trying to address.

James E. McClellan III and Harold Dorn [132] try to correct the tendency, well illustrated in the above Wikipedia quote, of denigrating past contributions from a present western perspective. In their review of the place of Islamic science in world history, they comment on the positive achievement of Islamic science and suggest that its contribution flourished for centuries, in a wide range of institutions from observatories to libraries, madrasas to hospitals and courts, both at the height of the Islamic Golden Age and for some centuries afterwards. By stating that : "It plainly did not lead to a scientific revolution like that in Early Modern Europe," was, in their view, just an attempt to impose "chronologically and culturally alien standards" on a successful medieval culture.

The Establishment of the Holy Roman Empire
There was tension between the East and Western centres of Christianity. This, together with the declining influence of the West after the fall of Rome, the expansion of the Muslim empire and the waning of the germanic tribes, convinced the Pope in Rome that he needed secular

support. King Charles Martel of the Franks had stopped the Muslim invasion beyond part of Spain, at the battle of Tours in 733 CE. This paved the way for Pope Leo III to engage with the Carolingans and for Charles' grandson, Charlemagne (742 to 814 CE) to be crowned Charles the 1st, the first emperor of the Holy Roman Empire, in 888 CE. This established a relationship between the empire and the papacy that lasted until the empire was dissolved in 1807 CE. The relationship was one of varying tension as each side vied for power – both secular and papal. As variously, each side needed the other. By 1066 CE, the two largest cities of the empire were Constantinople and Baghdad. Each had its own centre for the study of theology, public speaking, maths, science and philosophy and so was able to preserve and perpetuate the ancient Greek philosophical traditions and manuscripts.

Developments in Education
Following the fall of Rome and the closure of the Neoplatonist Academy in Athens by Justinian I in 529 CE, western Europe lost contact with the knowledge of the past. Meanwhile, the Eastern Roman Empire (Byzantine Empire) still maintained its connection with Greek philosophy through its learning centres, such as Constantinople. However, in the West there had emerged the Monastic Movement. This built upon examples of hermitic ways of life that had arisen by the third century.

This movement was also driven by concerns that the Church, since Christianly had been adopted as the official religion of the Roman Empire, was adopting too much liking for material wealth and power. So a variety of different orders were established to try and get back to the roots of Christianity – a simple life of serving the poor. Some of the best known included: St Benedict and the Benedictines and St Francis and the

Franciscans. In 1098 CE, Robert of Mosleme left the Benedictines and with St. Bernard of Clairvaux (1090 to 1153 CE) founded the Cistercians. St Dominic founded the Dominican Friars together with a large number of orders for women. These orders each developed a number of houses, monasteries and Abbeys. Educating the members of the order and training men who wanted to be ordained clergy was the prime objective, but was gradually extended to the sons of the nobility and to other "lay" students. Also, by copying manuscripts, the monks helped preserve established knowledge for later generations.

Education was taken one step further when Emperor Charlemagne, around 789 CE, required the establishment of Christian monastic schools attached to every abbey in the empire. Spirituality and asceticism were initially the main focus of learning but this was expanded, and the basic curriculum of these schools came to comprise Latin, rhetoric and grammar. Medicine and healing was also a big part of a monastery's concerns as was agriculture. But, supported by the Church, schools were also established alongside cathedrals in order to train the clergy and provide literate administrators for the courts of the Renaissance kings. This was initially through educating, primarily, the children of noble families. Later, "laymen" seeking a latin education for careers in the law, medicine and civil administration would be the main applicants. In cathedral schools, educated teachers were hired in, rather than the teachers being the clergy, as the monks were in monastery schools.

The curriculum was extended to cover the trivium (logic – dialectic, grammar, rhetoric) and the quadrivium (arithmetic, geometry, astronomy and music). These were the seven subjects of the liberal arts but law and medicine may have also been included as additional areas of study. These studies were taught in latin. Some classical Greek texts and poetry were also available but there was by no means a "national curriculum." So what was on offer varied from place to place. The actual content of these subjects was fairly meagre, as there was not much access at that time, to the wealth of Greek and Islamic material. Students would travel long distances across Europe to study at specific schools, under specific well respected teachers. Peter Abelard (1079 to 1142 CE) particularly, developed a reputation as a teacher, scholastic philosopher and theologian.

By the eleventh and twelfth centuries a number of factors came together which took education a further step forward, by stimulating the concept of universities:

Expansion of educational institutions was required as demand was gradually outstripping supply. Cathedral schools were often financed by the Church and often run by one person and so could not cope with the new demand from people seeking education to further their careers. As students at cathedral schools were covered by canon law, they were not subject to normal civil laws and so their behaviour was often a source of tension between the schools and the town – drunkenness, gambling, prostitution and neglect of their studies caused trouble with local inhabitants. Thus, in some places they were forced to move away.

At this time the legal system was being overhauled. Both the civil law – by reviewing the original Roman Law, and the Church or canon law. Tensions between the two were highlighted as both sides struggled for dominance. One such tension was the fact that only the state could grant "cooperation" status to an organisation or group of people. Cannon law stated that this was not necessary and that no state approval was required. Being a cooperative granted a number of advantages to an organisation by allowing them – to be independent and govern their own affairs. This encouraged masters and students to organise themselves into separate autonomous groups – equivalent to a guild arrangement (Grant – *God and Reason in the Middle Ages)* and so came together as "universitas" – a co–operative in learning.

With the Great Translations (see below) and increasing trade, there was a major influx of and access to classical Greek manuscripts, particularly the work of Aristotle. But it also included : the Greek philosophers, the works of Ptolemy, Euclid, Hippocrates, Galen, Isidore of Seville and the Islamic philosophers including Alhazen, Avicenna, Averroes, and others. This provided a major increase in material on which to base a curriculum. This resulted in the "relegation" of the seven subjects of the liberal arts to a more introductory status, undertaken before entering into the main study of logic, natural philosophy and the exact sciences (mathematics, optics, mechanics and astronomy). [133] These themselves then became necessary pre–requisites for anyone wanting to go on to study medicine, the law or theology. Theology being regarded as the pinnacle of academic study. Metaphysical and moral philosophy tended to be taught at graduate level.

Thus, Aristotle's natural philosophy became a central part of the university curriculum. Everyone had to do this before getting a degree or going on to study law medicine or theology. Thus, all theologians and those who reached positions of influence in the Church all received a grounding in natural philosophy and its analytical methods. Not only were the original works an object of study but also were the many treatise and commentaries that other scholars wrote on their interpretation of Aristotle. Other major works on theology and the Church fathers were also studied, such as Peter Lombard's Sentences

(1095 to 1160 CE) which were published between 1155 CE and 1158 CE and became the text book of theology until the seventeenth century.

All this study was facilitated by a new form of analytical teaching known as scholasticism. It was founded by Johannes Scotus Eriugena, (815 to 877 CE), Peter Abelard, Archbishop Lanfranc of Canterbury and Archbishop Anselm of Canterbury. It was the platform whereby "Reason" could be applied to questions of natural philosophy and mathematics and thence to theology and the other academic subjects. It was a method of critical thought which dominated teaching by the academics ("scholastics," or "schoolmen") of medieval universities in Europe from about 1100 to 1700 CE.

All these structures, processes and learning were gradually consolidated and extended. However, there is then a bit of a lull in terms of the development of new knowledge, as the Great Famine, lasting from 1314 CE to 1317 CE and the Black Death (1347 to 1351 CE) exacerbate the struggle of everyday life. However, towards the end of the fourteenth century a period of optimism and growth returns to Europe. It is centred on Florence initially, but soon spreads to other parts of Europe over the next 300 years. Art, literature and music flourish. Funded by expanding trade, all begin to prosper in a period that later historians have designated the European Renaissance.

The Great Translations

From about the fourth to the eighth centuries CE, a group called the Latin Encyclopedists,[134] translated some works by Plato and Aristotle into Latin, and added their own commentaries. In addition, there was Boethius (480 to 525 CE), Cassiodorus (480 to 575 CE), Isidore of Seville (560 to 636 CE) and Venerable Bede (672 to 735 CE) who also helped keep the works of Plato alive. This ensured that some awareness of Greek thought was available in Europe after the fall of Rome. In addition, schisms within the Church in the fourth and fifth centuries CE also led to the translation, of some of the Greek texts into Syriac. By the eighth and ninth centuries, this had extended to include translations from Persian, Coptic, Hebrew, Hindu and Greek into Arabic. These together provided a ready source of material for European scholars, who by the tenth century, were looking to access a wider body of knowledge, than just the Latin they had been left with. Spain was a favourite destination. With its history of Muslim colonisation and its Islamic scholarship, there was access both to original Greek texts, Arabic translations and the commentaries by Islamic philosophers.

Aristotle's works on natural philosophy had been translated and studied across the Byzantine Empire even during the rise of Islam in the Middle East. The translation activity increased during the twelfth and thirteenth centuries not only in Spain but in Sicily and across Italy. During this period, not only were the original Greek manuscripts re–discovered and translated into Latin, but also the Arabic and other translations were themselves translated into Latin. Thus, the ancient Greek works of natural philosophy and philosophy were brought into Europe, eventually becoming the foundation of the rise of European scholarship.

The Fall of Constantinople

In 1453 CE, the Ottomans attacked Constantinople and over ran it, naming it Istanbul. Many of the Christian and Islamic scholars fled to Italy and Europe, taking their knowledge and many of the ancient documents with them. This added to the further exposure of European scholars to the works of the ancient Greek thinkers and gave them access to many more manuscripts.

Printing

Movable type printing had been invented in China around 1000 CE. Later developments include the movable type invented by Bi Sheng around 1040 CE and which was being used in Korea by 1234 CE. There is evidence of the creation of text by reusing individual characters in Europe by the twelfth century CE. But the various techniques employed (imprinting, punching and assembling individual letters) were not efficient for widespread use. But in 1456 CE, Gutenburg developed moulds for a new metal alloy for making letters and a new oil based ink that proved better for printing on paper. He used these in an adapted press and so was able to create and disseminate papers and books more quickly and effectively. This allowed natural philosophy to flourish and become more widely spread and accessible.

The Development of Science

All these developments gradually happened, simultaneously and or sequentially, but obviously in no particularly planned or orchestrated fashion. They produced an initial flourishing, with the "Great Translations" of renewed Western access to Greek philosophy, known as the twelfth century Renaissance. Then as the scholastic method took off, "Reason" was applied to this material and a culture of questioning and analysis and disputation was created. Thus the natural philosophy of Aristotle and the other great thinkers, including Galen and Ptolemy and the commentaries of the Islamic philosophers, were subject to the scrutiny of the scholastic method. Then, the natural philosophy, contained in these texts, was extended and expanded by notable scholastics such as Robert Grosseteste, (1175 to 1253 CE) Roger Bacon, (1219 to about 1292 CE) Albertus Magnus(from around 1200 to 1280 CE) and Duns Scotus.(1266 to 1308 CE) Thomas Aquinas (1225 to 1274 CE) and William of Ockham (1287 to 1347 CE).

110

We have mentioned that the early signs of a systematic approach to the discovery of the natural world had begun with the ancient Greeks. But these were developed more formally in the Muslim world, where significant progress in methodology was made. This began with the experiments of Ibn al–Haytham (Alhazen) on optics from around 1000 CE, in his *Book of Optics*. The use of experiments to distinguish between competing scientific theories was one of the most significant features of the scientific method. Scientists would then form hypotheses to explain the results of these experiments. The hypothesis was tested for veracity or falsification by comparing their predictions against the observed results of their new measurements and experiments.[135] The natural sciences continued to be called natural philosophy, but the adoption of the scientific method took science beyond the realm of philosophical conjecture, introducing a more structured way of examining nature.

Roger Bacon (1219 to about 1292 CE) in his Opus Majus, and influenced by earlier contributions from the Islamic world, is noted as being a great advocate for this new approach of more formal and systematic enquiry – "theories supplied by reason should be verified by sensory data, aided by instruments, and corroborated by trustworthy witnesses". He also argued, for the first time, that science should be directed at practical inventions for the improvement of all human life. Similarly, Grosseteste's emphasis on mathematics as a way of understanding nature, together with Frances Bacon (1561 to 1626 CE), gave their support to this scientific method. Frances Bacon regarded natural philosophy as the "great mother of the sciences[136] and so considered the exact sciences as sub–divisions of natural philosophy. This, according to Grant[137] allowed the broader concepts of natural philosophy to be applied to these exact sciences in a way that took them beyond a purely measuring function and paved the way for mathematics to be seen as an underpinning of natural philosophy.

111

This underpinning was exemplified, according to Grant, with the original publication of Newton's *"Mathematical Principles of Natural Philosophy"* in 1687 CE.

So, a cultural climate of questioning evolved which fostered the development of new ideas. This period became known as the "European Renaissance". It is said to have started in Florence in the fourteenth century. New ideas in art, architecture, literature and music and drama began to flourish. This climate of blossoming ideas also influenced the evolution of natural philosophy. With the adoption of the notion of a scientific method, there was a willingness to question the philosophical approach of the Scholastic method and the previously held truths of natural philosophy. With the adoption of mathematics as a means of describing and answering some of the questions of natural philosophy, and through experimentation and analysis, new knowledge was discovered. These activities led to the realisation that some of the teachings of Aristotle and Galen and Ptolemy and the other Greek scholars were not always found to match everyday observations.[138] In this way, the Aristotelian basis of philosophy and natural philosophy was undermined and eventually largely superseded.

Expansion of these methods of"scientific" enquiry came to be called the "Scientific Revolution." It began in earnest from the sixteenth and seventeenth centuries, paving the way for new topics to become areas of specialisation in their own right. Ultimately leading to the complex classification discussed in chapter one. Still, at this time, much of this work was being done by amateurs and people with private means and some who combined their work with teaching. It was really not until the nineteenth century that people began to make a living out of science, other than from teaching, and the discipline became professionalised.

Evolution of Some Specific "Science" Related Ideas.

Heliocentricity

The Polish astronomer Nicolaus Copernicus (1473 to 1543 CE), in 1543 CE, proposed a heliocentric model of the Solar system. The Greek astronomer Aristarchus of Samos (around 310 to about 230 BCE) had also suggested that the Earth revolved around the Sun, but Aristotle and later Ptolemy, in his great book on astronomy –*The Almagest* – had rejected this idea.

Galileo Galilei (1564 to 1642 CE), the Italian mathematician, astronomer, and physicist found that his astronomical discoveries, empirical experiments, his improvement of the telescope and his mathematical analysis of astronomical observations, all supported Copernicus' ideas about the relative motions of the Sun, Earth, Moon, and planets.

At this time, the Church believed the geocentric (the earth being at the centre of the universe) understanding of the Solar system. So, Galileo's support for helio–centrism provoked controversy and he was forced by the Church (Pope) to recant and spend the rest of his life under house arrest.

Planetary Motion

Tycho Brahe (1546 to 1601 CE), a Danish nobleman and astronomer made very accurate astronomical measurements over many years. This detailed data helped him develop a hybrid version of planetary relationships. He thought the moon and sun orbited the earth whilst at the same time the other planets orbited the sun. The German astronomer

Johannes Kepler (1571 to 1630 CE) working with him, disagreed and supported the Copernican theory.

Kepler also used Brahe's observations to formulate the idea of elliptical planetary orbits rather than spherical ones, which had been the assumption until then.

Motion

Galileo also performed mechanical experiments, insisting that motion had consistent characteristics that could be described mathematically.He also contradicted the Aristotelian idea that heavier objects fall faster than lighter objects, by finding that bodies do not fall with speeds proportional to their weights.

Light

Newton (1642 to 1726/7 CE) in 1687 CE published his *"The Mathematical Principles of Natural Philosophy"*, or *"Principia Mathematica"*, which, according to Grant[139] attempted to expose the mathematical basis of nature – the immutable rules it obeyed. In doing so, natural philosophy and mathematics were brought together for the first time. This both created a basis for the development of modern physics and for encouraging the principles learned in the exact sciences of optics, mechanics and astronomy to be applied to questions raised by natural philosophy. Newton explained light as being composed of tiny particles and developed a theory of colour, published in Opticks of 1704 CE. This was based on the observation that a glass prism separates white light into the many colours which form the visible spectrum.

Christiaan Huygens(1629 to 1695 CE) a Dutch astronomer and mathematician, developed a rival theory of light, in 1690 CE. He explained the behaviour of light in terms of waves. However, for various reasons coupled with Newton's reputation, this wave theory saw relatively little support, until the nineteenth century.

Then we have visible light being recognised as electromagnetic radiation, as part of the electromagnetic spectrum [James Clerk Maxwell (1831 to 5th November 1879 CE) – after Michael Faraday 1791 to 1867 CE)]. Max Plank (1858 to 1947 CE), a German theoretical physicist, proposed the concept of "packets" of energy (which he called "quanta") moving as a wave at the atomic and subatomic level — quantum mechanics. It is now accepted that light can behave as either particles or waves or even both, depending on conditions, and that light is not fully understood at all. Neither "does anyone understand quantum mechanics"![140]

Atoms

Although Leucippus and his pupil Democritus had postulated the idea that matter was made up of atoms, this was not really conceived in the way it is understood today. In Indian philosophy, Maharishi Kanada developed a theory of atomism somewhere between 600 and 200 BCE – his dates are uncertain. It was further elaborated by the Buddhist atomists Dharmakirti and Dignāga during the first millennium CE. Kanada suggested that everything can be subdivided, but this subdivision cannot go on forever, and there must be smallest entities (parmanu – "atoms") that cannot be divided. These then aggregate in different ways to yield complex substances and this is the basis for all material existence.[141]

Modern atomic theory of matter was postulated in 1811 CE by the chemist John Dalton (1766 to 1844 CE). Lorenzo Avogadro, (1776 to 1856 CE), an Italian scientist, later corrected some of the ratios in which certain atoms combined. He also coined the word molecule, for the entity created when atoms combine to form another compound. These ideas were followed by the discovery of sub–atomic particles – the electron by JJ Thompson (1856 to 1940 CE) in 1897 CE, the nucleus and proton by Ernest Rutherford (1871 to 1937 CE) in 1911 CE and 1919 CE respectively and the neutron by James Chadwick (1891 to 1974 CE) in 1932 CE.

Alchemy to Chemistry

Alchemy had existed from around 3500 BCE in Egypt but also in China and India. It was based on the belief that substances were made up of the four Aristotelian elements, fire, earth, air, and water in different proportions. It was supposed that gold was the noblest metal, and that other metals were ranked in order, down to the basest, such as lead. It was also believed, that a fifth element, the elixir of life or philosophers stone, could transform a base metal into gold.

Robert Boyle (1627 to 1691 CE) an Anglo–Irish natural philosopher and chemist, in his work *The Sceptical Chymist*, in 1661 CE, distinguished chemistry from alchemy. However, many of the methods and much of the equipment used by alchemists became the foundation for modern chemistry.

Antoine Lavoisier (1743 to 1794 CE), a French chemist, refuted the phlogiston theory, which posited that things burned by releasing "phlogiston" into the air. In 1780 Lavoisier discovered that combustion was the result of oxidation.

Age of Enlightenment (17th to 18th century)

Natural history had became a distinct branch of natural philosophy by the sixteenth century, referring more to the observation and description of

116

natural things. Carl Linnaeus, (1707 to 1778 CE), was a Swedish botanist, physician and zoologist whose 1735 CE taxonomy of the natural world is still in use. Linnaeus in the 1750s introduced scientific names for all his species.

This was a period of considerable expansion of knowledge through the developments in science, philosophy, mathematics and technology. There were also cultural changes in political thought, economics, theology and the arts – music, painting and literature, as the application of Reason and the scientific method became the means of challenge to accepted wisdom and convention. And so we begin to see a transformation of natural philosophy with the emergence of individual areas of enquiry – physics, chemistry, biology and so on, though still under its umbrella.

Historians have singled out a number of specific contributors to these developments, promoting the idea that new scientific knowledge comes about by strokes of genius from a particular person, leading to an heroic discovery. But as we have mentioned before, it is likely there were many more who, for various reasons have not been identified or remembered or recorded, yet have made their own necessary contributions along the way. Schaffer [142] discusses these issues within the context of the transition from Natural Philosophy, as an umbrella term for scientific endeavour, towards the modern scientific culture, in which individual areas of scientific study are classified as subjects in their own right. This coincided with the move to introduce appropriate training for undertaking a disciplined and a systematic approach towards uncovering new knowledge. This was to ensure experiments could be replicated and so authenticated as being genuine and reliable – at least as far as could be determined at the time of verification

By the middle of the nineteenth century, when William Whewell first suggested the use of the name "scientist," in 1834 CE, (though the word did not enter general use until nearly the end of the century) as a collective noun for people doing this new way of investigation, the study of science was moving from the realm of the interested amateur towards a professional status, within institutions equipped with the necessary apparatus. At the same time, the concept of natural philosophy gradually acquired the more modern name of natural science (gradually becoming the complex inter–relationship of categories and subdivisions shown in Figure 1 in chapter 2). In this way, science and philosophy came to be regarded as separate fields of enquiry and the epithet 'science' was only to be applied to those areas of study in which reason, combined with observation and experimentation could be applied. However, there are still universities that have departments and courses incorporating the title Natural Philosophy and there is an academic subject area entitled Philosophy of Science.

Natural science is broadly divided into life sciences (or biological science) and physical science. Physical science is subdivided into branches, including physics, space science, chemistry, and Earth science. But as well as new knowledge, new sciences are being constantly designated. So an internet search throws up lists of new emerging areas of study such as: Neuroparasitology, Quantum Biology, Exo–meteorology, Nutrigenomics, Synthetic Biology, Computational Social Science, Cognitive Economics, Organic Electronics and Quantitative Biology. More research reveals more questions and opens more avenues of enquiry and with this comes narrower and deeper fields of study. These in turn require more concentrated and dedicated exploration and so become designated branches and subdivisions and sciences in their own right.

For the purposes of this book, it is not necessary for me to discuss any of the specific areas of new knowledge and discovery that emerged during the last four hundred years, but for anyone interested, there are many sources on the internet and through Wikipedia, but the following I found particularly good reads : Bill Bryson's – *A Short History of Nearly Everything*,[143] *A History of Knowledge* by Charles Van Doren [144] and *Guns, Germs and Steel* by Jared Diamond.[145]

So we will leave science here and move on to the development of philosophy.

THE DEVELOPMENT OF PHILOSOPHY

As we have seen, philosophy originally referred to the love of wisdom, a search for knowledge in general. With natural philosophy, it was at the heart of the search for knowledge. Natural philosophy sought answers about why things were the way they were within the natural world, initially by observation, experience and analysis and then later through experimentation and the application of reason. Philosophy sought answers through thought and intuition and reason. In ancient Greece, both disciplines were often embodied in the same person. Gradually, people began to concentrate on one or the other. Mathematics, though classified separately as one of the exact sciences, became both a necessary tool for applying to the problems posed and also a concept worthy of study in its own right. Philosophers looked to understand the nature of mathematics, its relationship to logic and its role in helping to explain why the world was as it was.

We have indicated that the usual narrative is that western philosophy started with the ancient Greeks and in particular the pre–Socratic philosophers and with the philosophy of other cultures – short handedly labelled as eastern philosophy – also influential in its evolution. So for something like 2 – 3000 years there had been attempts to explain how things were the way they were, initially through myth and supernatural ideas. Then with the pre–socratic philosophers there is a change of view. They attempted to find explanations from within nature, from within the human experience of the world itself, without requiring to invoke an outside causation.

Into this culture, Christianity enters the frame and gradually establishes itself, finding itself in competition, not only with a number of other established religions but also with developing philosophies and their own associated religions. These philosophies, concerned with causation and the whys and wherefores of how things had come to be the way they were, were consequently concerned with the concepts of theology and religion. Different thinkers became fascinated by different aspects of these matters and so different "schools" of thought emerged.

Most people had a belief in a God, or gods or some sort of power, that either created or was influencing the world or both. Socrates, who in Plato's Apology [146] refers to an "inner voice" which indicated to him when not to do something. Plato referred to the "Demiurge" in the "Timaeus," [147] which was not a creator, but which ordered the cosmos out of already existing chaotic elemental matter. Aristotle, in his cosmological argument, proposed that at least one eternal "Unmoved Mover" must exist to create movement to support everyday change. Though Aristotle argued against a Prime Cause or creator.[148] Rather, he thought the universe was eternal without a beginning or an end. In addition to Platonism and Aristotelianism, other schools emerged, each with their own emphasis on different approaches to life, including – Sophism, Cynicism, Skepticism, Epicureanism, Hedonism and Stoicism.

Following the death of Christ, a new philosophy and religion develops – Neo–Platonism. This proceeds out of Plato's work, attributed to Plotinus (204 to 269 CE). It is largely a religious philosophy which became a strong influence on early Christianity (especially on St. Augustine), and taught the existence of an ineffable (beyond description) and transcendent (supernatural, beyond normal experience) God called "The One."149 Gnosticism was also a distinct movement at this time with links to Christianity but came to be regarded as a heresy, following criticism from Irenaeus (from around 125 to 202 CE). He was bishop of Lugdunum in Gaul, which is now Lyons, France. Irenaeus was born in Smyrna in Asia Minor, where he studied under bishop Polycarp (70 to 155 CE) who in turn had been a disciple of the Apostle John. Leaving Asia Minor for Rome he joined the school of Justin Martyr (from around 100 to about 165 CE) before being made bishop of Lyons in Southern Gaul in about 178 CE. Classical Neoplatonism ended with Emperor Justinian 1's closure of Plato's Academy in 529 CE but has continued in various forms since.

Neoplatonists did not believe in an independent existence of evil. They compared it to darkness, which does not exist in itself but only as the absence of light. So, too, evil is simply the absence of good. Things are good insofar as they exist; they are evil only insofar as they are imperfect, lacking some of the good which they should have.

Neoplatonists did believe human perfection and happiness were attainable in this world, without awaiting an afterlife. Perfection and happiness— seen as synonymous— could be achieved through philosophical contemplation. They believed that everyone returned to "The One", from which they emanated.These views later influenced St. Augustine and his theology.

There was controversy at the time as to whether "The One" and the "Demiurge" were the same or the "Demiurge" was subordinate. The controversy continues amongst academics, as to whether Neoplatonism is sufficiently distinct from Platonism for it to be regarded as a separate movement or not.

As Christianity began to gain ground, certain of the Church Fathers saw the benefits of incorporating natural philosophy and its methods into Theology, in so far as it might illuminate theological understanding. But it was thought that natural philosophy should not be studied for its own sake. Justin Martyr (about 100 to 165 CE) and Saint Basil (from 329/330 to 379 CE), who was the Greek bishop of Caesarea Mazaca in Cappadocia, Asia Minor (modern–day Turkey) both incorporated ideas from natural history into their own writings. Tertullian (150 to 225 CE), on the other hand, was against the use of natural philosophy by theologians. However, the idea was embraced by St. Augustine of Hippo (354 to 430 CE), but with two caveats – that the truth of scripture was inviolate and that where there were different ways of explaining a particular point, then the accepted explanation should not be held so rigidly, that if it is later shown to be false, people should not hold on to the belief in the face of this and so bring scripture and the way of faith into disrepute. Thus, by using philosophy and natural philosophy to illuminate theological questions, the concept that philosophy and natural philosophy were handmaidens of Theology,[150] was formed. This idea prevailed until eleventh to the twelfth centuries.

From around the fourth or fifth centuries CE, Europe entered the so–called Dark Ages, during which little philosophical development occurred. Within Christian theology, various clarifications or disagreements solidified, as the Church held a number of "councils' to resolve specific areas of controversy. For example:

- The First Council of Nicaea in 325 CE formulated the original Nicene Creed.

- The First Council of Constantinople in 381CE reinforced this and condemned Arianism – the idea that Jesus was the son of God but not God. It also expanded the Creed.

- The Council of Carthage in 397 CE approved the sixty six books of the Old and New Testaments plus the Apocrypha, as the set canon of the Bible. There were to be no additions from that time on.

- The Council of Ephesus in 431 CE proclaimed the Virgin Mary to be "Mother of God".

- The Council of Chalcedon in 451 CE defined the two natures (divine and human) of Jesus Christ. "We teach unanimously that the one son, our lord Jesus Christ to be fully God and fully human." It rejected Nestorianism (after Nestorius : 386 to 450 CE), who held that though Jesus was divine and human this was a loose arrangement, a duality, rather than being both at the same time. This disagreement led to the establishment of a separate Nestorian Church.

- The Second Council of Constantinople in 553 CE and the Third Council in 680 CE re–affirmed the pronouncements of the first Council of Constantinople.

- The Third Council of Toledo in 589 CE, is thought to be the point when the Filioque clause was accepted. This was the interpolation of the phrase "and the son" into the creed - "I believe in the Holy Ghost, the Lord, the giver of life, who proceeds from the Father {and the Son} Who with the Father and the Son is adored and glorified"– To consolidate the concept of The Trinity i.e., that God, The Son and The Holy Spirit are all one and the same, yet distinct, all at the same time.

- The Second Council of Nicaea In 730 CE outlawed pictorial presentations of Christ and the saints, creating the first iconoclasm. This was the last Ecumenical Council to be accepted by both Eastern and Western Churches.

- The Fourth Council of Constantinople (Catholic) in 800 CE. saw the coronation of Charlemagne as Emperor of the Holy Roman Empire by Pope Leo III and the handing over of the keys to the Tomb of Saint Peter. By this act, the papacy acquired a new protector in the West. This new relationship empowered the Pope, and so increased tension between the Western and Eastern parts of the Church. This led to the schism, between the two. In the Eastern part, the emperors and patriarchs of Constantinople regarded themselves as the true descendants of the Roman Empire dating back to the beginnings of the Church.

As we approach the end of the Dark Ages, by the eleventh century, we enter the era of the Great Translations and the development of the universities. At this time there was a renewed flowering of thought, both in Christian Europe and in Muslim and Jewish Middle East. Most of the philosophers and theologians were mainly concerned with reconciling faith, be it Christianity or Islam, with the classical philosophy of Ancient Greece (particularly Aristotelianism). The university curriculum became geared up to subordinate the seven liberal arts to the importance of natural philosophy and philosophy, and theology became the pinnacle of academic study. Not only did the translations take centre stage but so did the commentaries from current and earlier philosophers, particularly the Islamic ones – Avicenna (eleventh century, Persian) and Averröes (twelfth century, Spanish/Arabic). This curriculum was delivered through the Scholastic methodology.

In this climate of questioning and analysis, many theologians and philosophers were keen to apply the principles of natural philosophy to questions of theology and matters of faith. So we have Anselm of Canterbury (1033 to 1109 CE) producing an ontological argument for the existence of God (by reason alone) and Peter Abelard (1079 to 1142 CE) arguing that posing questions and applying reason and logic to theology and philosophy will lead to truth through enquiry. He introduced the doctrine of limbo for unbaptised babies (a concept whereby the souls of unbaptised babies who died, would rest in a state of "limbo" until the intervention of Jesus would allow them entry into Heaven). Also, Hugh St. Victor (around 1141 CE) and John of Salisbury (1115 to 1180 CE) argued for the use of logic and reason to be applied to philosophy and theology and matters of faith. This went against the prevailing doctrine – that faith was a matter of revelation not reason. Thus, it was a key turning point in how people thought.

Albertus Magnus (from around 1200 to 1280 CE – Albert the Great) also advocated the application of natural philosophy to theology. He was concerned more about what God could do through the natural world rather than the supernatural. His pupil, Thomas Aquinas (from around 1225 to 1274 CE) was also one of the theologians who advocated the application of natural philosophy to matters of theology and faith. He is known for his five rational proofs for the existence of God, and is generally considered as having the greatest influence on the theology of the Catholic Church. He argued that theology should be regarded as a science.[151] Though both were clear that the study of natural philosophy and theology should be separate.

We thus have a situation where the Church, over a period of 1000 to 1200 years, through its councils and responses to prevailing religious and philosophical challenges, had gradually reached some sort of agreed understanding of what is to be believed in doctrinal terms. That is, an agreed orthodoxy. Of course, this was not universal and there was still the schism between the East and West, as well as between theologians within the Western Church itself. The influx of Aristotelian natural history into the university curriculum, and the various commentaries on it – Christian and Islamic – was not uncontroversial. The incorporation of its methodology by theologians, especially in considering Lombard's Sentences,[152] lead to questioning of Church orthodoxy and to different understandings of scriptural interpretation. The sentences of Peter Lombard summarised the orthodox interpretations of specific aspects of scripture, and students were expected to discuss these as part of their studies. The flirtations of theology with natural philosophy created opposition. Bernard of Clairvaux opposed Abelard, and others argued that faith was being threatened by this emphasis on Aristotelian philosophy.

The principle argument was that Aristotle was adamant that a vacuum could not be created and that there was only one world and there could not be any more. This implied that God could not create a vacuum or create any other worlds, even if he wanted to. Thus, accepting such

Aristotelian arguments was tantamount to saying God was neither all powerful nor in total control. The Church could not have that.

St. Bonaventure (1221 to 1274 CE) said that the Church believed that God created the universe out of nothing, as a first cause. But Aristotle thought the world was eternal and had not had a beginning nor will it have an end. Thus, we have a tension between the Church and some theologians arguing that God is all powerful and can do anything and that the truth of God is revealed by revelation and other theologians arguing that it was necessary to apply reason and the methods of natural philosophy to reach a true understanding of the nature of God.

Also at this time, there was renewed interest in Hermeticism. This was a philosophy based on writings attributed to Hermes Trismegistus, who was some sort of composite figure based on the Greek God Hermes and the Egyptian God Thoth. It possibly originated around the first century CE, alongside Christianity and Gnosticism, but its roots are unclear. Part of its basis was the idea of the existence of a single, true theology that is present in all religions and that was given by God to man in antiquity.[153] Its other aspects included references to occultism, magic, alchemy and astrology. The alchemy aspects were what attracted the scientists during the Renaissance. It then fell foul of the Church and went underground.

It is understood that all three Abrahamic faiths (Judaism, Christianity and Islam) may all have been influenced to various degrees and influenced it in turn. The Corpus Hermeticum is the main body of writings, copies of which were found in Nag Hammadi, a town in Upper Egypt, in 1947. Hermeticism still continues and is linked with a number of other movements[154] including Rosicrucians, Free Masonry, Kabalah (mystical aspects of Judaism), Gnosticism, Alchemy, Wicca and Esoteric Christianity.[155]

As much of the new theological study was based in Paris, it was here that the tensions were most highlighted. There were a number of attempts by the bishops of Paris, with or without the acquiescence of the Pope, to restrict or ban the study of Aristotle. This first happened in 1210 CE and then again in 1215 CE but by 1255 CE there is evidence that all of Aristotle was again being taught. But his work had been re–instated to be used only in the Faculty of Arts and not the Faculty of Theology. Another attempt was made in 1270 CE to identify the most significant "errors" of Aristotle's work and to ban them, but without much effect. Then came the Condemnation of 1277 CE which identified 219 theses that were considered to be against the Church's orthodox teaching. These were taken directly from Aristotle and from the commentaries theologians and others had made on it. They also included statements derived through the application of natural philosophical methods to Lombard's Sentences and other writings.

As one of the key arguments was that God's actions could not be limited by Aristotle's natural philosophy or by human understanding and that God had the power to do what he liked, one of the effects of this Condemnation was to promote the questioning of Aristotle's philosophy in general. This resulted in freeing academics – natural philosophers and theologians – to think "outside the box".

This had the paradoxical result of effecting the separation of the three disciplines – natural philosophy, philosophy and theology. Natural philosophy could now be studied for its own sake without hindrance from the Church. This was achieved by the Arts masters swearing not to introduce any theological themes into their thinking, and if a question affected both theology and philosophy then it would be resolved in favour of faith. However, it was still possible for theologians, particularly those who were dually trained, to use the methods of natural philosophy to inform their studies. Thus, natural philosophy still influenced theology but was not really influenced itself by theology. Hence, the three disciplines remained linked, but were now free to develop independently. This set the direction for the gradual emergence of the sciences and the schools of philosophy as we know them today. And so the groundwork was laid for the expansion of ideas as the European Renaissance of the fourteenth century began.

We have discussed the effect this climate of questioning had on the evolution of natural philosophy, but it also extended into the religious realm, tapping into a stream of popular disquiet with the Catholic Church. This was both on theological grounds – the desire to read the Bible in one's own language and the reaction to transubstantiation (that the bread and wine at the eucharist/communion actually becomes the body and blood of Christ). But there was also concern about the corruption and greed witnessed in the behaviour of some of the senior clergy.

Similar concerns had previously been expressed by John Wycliffe (1320 to 84 CE) and the Lollards. Other concerns were being expressed within what became known as the Christian/Renaissance Humanist Movement. Theologians and philosophers like Desiderius Erasmus (1466 to 1536 CE), John Colet (1467 to 1519 CE) and Thomas More (1478 to 1535 CE) felt that over the years, errors had accumulated amongst many of the translations of the classic Greek texts. They wanted to return to the original and therefore "uncorrupted", Greek and Latin texts. By exposing the ignorance of the clergy and advocating for better education, with some of their writings criticising both the theology and the practice of the Catholic Church, they were thus in sympathy with some of the aspirations of what became known as the Protestant Reformation. This movement was led by people like John Calvin and Martin Luther.

The Renaissance Humanist Movement was also a reaction against the prescriptive approach associated with medieval scholasticism. Humanists sought to bring eloquence of speech and clarity of thought to the public square to help people live a good and virtuous life. This was to be accomplished through the study of the 'studia humanitatis', today known as the humanities. This was an expansion of the Trivium of grammar and rhetoric, excluding the logic and adding in history, poetry, Greek and moral philosophy. This reaction against Scholasticism was shared by the French philosopher and mathematician René Descartes (1596 to 1650 CE). Descartes, like Galileo, was convinced of the importance of mathematical explanations for the way things were. He developed the Cartesian coordinate system, and promoted the application of reason, and he and his followers were key figures in the development of mathematics and geometry in the seventeenth century. Descartes has been dubbed the 'Father of Modern Philosophy', and much subsequent Western philosophy is a response to his writings.

By The Age of Reason of the seventeenth century and the Age of Enlightenment of the eighteenth century, a climate of religious tolerance, with the co–existence of Roman Catholicism and Protestantism, and liberalism had emerged. This also provided part of the backcloth for the independent development of theology, philosophy and natural philosophy which itself evolved into our modern science. Within philosophy, two broad approaches sprung up – Rationalism (the belief that all knowledge arises from intellectual and deductive reason, rather than from the senses) and Empiricism (the belief that the origin of all knowledge is sense experience).

By this time, the Church's thinking had evolved by means of further councils. For example, the first Lateran Council in 1123 CE established the role of celibacy for the religious (includes both clergy and nuns), the third Lateran Council in 1170 CE prevailed on cathedrals to appoint teachers for the poor and the fourth Lateran Council in 1215 CE clarified the concept of transubstantiation. Others were held in Lyon in 1245 CE and 1274 CE and in Vienne (South East France – in 1312 CE).

The Council of Trent 1545 to 1563 CE was convened in part as a means of responding to the protestant reformation. The council issued condemnations of what it defined as Protestant heresies and defined Church teachings in the areas of Scripture and Tradition, Original Sin, Justification, Sacraments, the Eucharist in Holy Mass and the veneration of saints. It also specified what was the Catholic doctrine on salvation, the sacraments, and the Biblical canon. This prompted publication in 1566 CE of the Roman Catechism, in 1568 CE a revised Roman Breviary (Divine Office – prayers and readings etc.) and in 1570 CE of a revised Roman Missal (the book of the liturgy). This introduced the Tridentine Mass (from Trent's Latin name Tridentum – Latin Mass), and Pope Clement VIII issued in 1592 CE a revised edition of the Vulgate Bible.

So within and as a consequence of the new questioning climate beginning around the fourteenth century, all three disciplines – Natural Philosophy, Philosophy and Theology evolve their own thinking. From an initial position where all three were closely linked, the boundaries between them were blurred as all were concerned with the search for

131

knowledge, they moved to a situation where they become much more distinct and began to develop along different lines, though still having something to say to each other.

The Catholic Church still evolved its thinking through its councils. The First Vatican Council in 1869 CE issued definitions of the Catholic faith, the papacy and the infallibility of the Pope. The Second Vatican Council met from 1962 CE to 1965 CE and its decrees, which were mainly pastoral, advocated the necessity for ecumenical progress towards reconciliation with other Christian Churches.

As with the section on science, for the purposes of this book, it is not necessary for me to discuss any further specifics of the new philosophical thinking that emerged during the last four hundred years, but for anyone interested, I found the following particularly interesting and informative :– Bryan MacGee"s – *Confessions of a Philosopher* [156] and the website – The Basics of Philosophy.[157]

KNOWLEDGE ACQUISITION

What the discussion in this chapter so far demonstrates is the uncertainty surrounding both the origins of the words – Science and Philosophy – and the evolution and changing nature of their content.

If we think about it, perhaps this may not seem surprising. It is only natural, perhaps, that in the twenty-first century CE, we think we will know more than people did in the fifth century BCE. However, the process whereby our current knowledge base has come about, has not been a simple functional relationship between increasing knowledge with time. The development of knowledge is a consequence of the interaction of many factors. These include: the characteristics of the people doing the enquiring – their personal traits, relationships, whims, biases, assumptions, prejudices and the pressure of life and the need to make a living, the state of knowledge at the time they are working, the cultural climate – politics, and the environment – within which they are working, the responses to their findings of the people around them, the true veracity of the findings themselves and also how and what history records (or doesn't record) of any activities and how much is lost to future generations. Also, how future generations then view any new knowledge they come across and what they do with it in terms of preserving it, modifying it or discarding it altogether. These factors will apply, in varying degrees to all forms of knowledge !

Additionally, the means of acquiring the content has evolved. In ancient times, the quest for knowledge was seen as a finding out about things in general. By the time of Aristotle there was the beginnings of a minimal categorisation, as we have seen, though it was accepted that "investigators" would and could comment and contribute to whatever area interested them, with some being more eclectic than others.

The advent of the "scientific method"[158] in the seventeenth century was an important step, through into the "Enlightenment" of the eighteenth century, in the acquisition of what was then seen as "truer knowledge." This also paved the way for the gradual but deliberate separation of scientific, philosophical and theological enquiries. A consensus grew in support of the idea that what makes science different and valuable is that it is about dealing with areas and questions in which propositions are subjected to testing and that such testing is more systematic, measurable and detailed.

However, as the concept of "science" and the "scientific method" developed, there also developed in the minds of some, the idea that science was capable of providing all the answers to how things came to be the way they are. That religion and faith have been human constructs all along and now have little or no place in modern society. This notion of "scientism" holds that science is the only form of enquiry and hence the source of true knowledge.[159] This idea has been challenged, not only by religious commentators but others too.[160] [161] [162] [163] [164] [165]

Through the discipline of the philosophy of science, [166] the "scientific method" itself has come under scrutiny. The very notion of whether there ought to be something designated as "the scientific method" has now been questioned, as has the idea that conclusions resulting from its application should automatically be considered better or truer because of it.[167] [168] Even where the recognised steps in the scientific method would be most appropriate, they are not always applied in the standard, recognisable, sequential fashion. A useful overview on many of the questions over the scientific method is provided by Anderson.[169] The scientific method is now not regarded as a pure methodology applicable in all circumstances and to all questions. Discussions of whether there should or could be a unified approach to scientific enquiry or whether there are core principles that can then be applied differently to suit different circumstances, have largely become obsolete.[170]

There are also questions of whether the "knowledge" identified through scientific methodologies and whether knowledge gained through science is the only knowledge worth considering. *"The aim of philosophical inquiry is to gain insight into questions about knowledge, truth, reason, reality, meaning, mind, and value,"* to quote from Grayling.[171] Also, should, or can Science claim primacy over other ways of understanding.[172 173] Thus, the pure application of the scientific method is not seen as the only means through which we come to understand and know the world and the universe.

Whatever the scientific approach taken to the enquiry, in order for new ideas to be seen, discussed or adopted, it is clearly necessary to publicise them. This has always been the case. Even Socrates, who does not appear to have written or published anything, promoted his ideas through dialogue and the written evidence of others. By putting ones ideas into the public domain – whether this is just with colleagues or other researchers who may be seen as competitors in the first instance, means opening oneself up to criticism. Allowing ones ideas to be scrutinised is clearly an essential part of the scientific method. It is an essential part of the process for developing any ideas, not only scientific ones. But in science, it is the means whereby ones ideas can be replicated and verified, modified or rejected and the process by which knowledge can be advanced. However, this is not necessarily a neutral, value–free exercise. Being brave enough to put ones ideas into the public arena, opens one up, not only to genuine and appropriate scrutiny and criticism but also to ridicule.

In the past, it has also sometimes ended in physical pain, incarceration, exile and even death. Though this is less likely these days, people are still being tortured and imprisoned for their ideas. Socrates himself, was brought to trial and sentenced to death for sharing his ideas in the public sphere and for allegedly corrupting the minds of the young. [174] There are numerous other examples – we have already mentioned Ignaz Semmelweis. Copernicus only published his theory that the universe was heliocentric just before his death as he was afraid of the consequences of this suggestion. Both Keplar and Galileo were also disbelieved during their life time, when they supported Copernicus's idea. Kepler went further by suggesting that the orbits around the sun were elliptical rather than circular.

This too was largely ignored in his lifetime. But decades later, Kepler's work was the platform from which Isaac Newton discovered his law of universal gravitation. And Gregor Mendel discovered the rules of heredity which describe how physical traits are passed through generations of living things. The importance of Mendel's work was only properly appreciated in 1900 CE, sixteen years after his death, and thirty four years after he first published it. In more recent times we can mention Alfred Wegener. He proposed that Earth's continents move very slowly. Over millions of years they can move a long way. Between 1912 CE and 1929 CE he published a stream of fossil and rock evidence to support his theory. But his theory of continental drift was rejected by most other scientists during his lifetime. It was only in the 1960s CE, after he had died in 1930 CE, that continental drift finally became part of mainstream science.[175]

The idea that somehow, scientific investigation is a purely objective occupation that creates theories and explanations based solely on irrefutable observed and experimental fact, may be a nice comforting caricature, but it mis–represents the reality. Fara [176] describes how, even in the face of new contradictory knowledge, some ideas persist for years and even centuries and continue to influence the scientific community, until eventually a new perspective is adopted. The conventional wisdom is that, in medicine, it takes seventeen to twenty years for evidence to be incorporated into practice.[177 178 179]

Another aspect of this problem of getting new ideas recognised, involves barriers to acceptance of the work of scientists who are not part of the scientific establishment. For example, the great physicist Lord Rayleigh, when he discovered what turned out to be a "neglected" paper on the kinetic theory of gases by James Waterston, commented: *"a young author who believes himself capable of great things would usually do well to secure favourable recognition of the scientific world. before embarking upon higher flights."*[180] And William Harvey's experiences led him to an even more pessimistic view:

"But what remains to be said about the quantity and source of the blood which thus passes, is of so novel and unheard–of character that I not only fear injury to myself from the envy of a few, but I tremble lest I have mankind at large for my enemies, so much doth wont and custom, that become as another nature, and doctrine once sown and that hath struck deep root, and respect for antiquity, influence all men." [181]

In more general terms, both Robert K. Merton, [182] who remarks that *"the history of science abounds in instances of basic papers having been written by comparatively unknown scientists, only to be rejected or neglected for years"* and Moti Nissani,(1995 CE) highlight the same point. Another example is that of Cecilia Payne–Gaposchkin. in 1925 CE she concluded that stars were composed mostly of Hydrogen and Helium. She was persuaded by astronomer Henry Norris Russell not to publish this finding because of the widely held belief that stars had the same composition as the Earth. However, four years later, in 1929 CE, Russell vindicated this hypothesis and the discovery was eventually accepted.[183]

Neither is all criticism based on a genuine desire to progress and benefit humankind. There are many examples where either the ideas or the proposers or both have been subjected to ridicule, abuse and spurious "scientific" arguments, in order to discredit their findings. This may have been motivated by a competitor's personal interests to try and ensure their own status, income and reputation was not undermined by new ideas. Or the criticism may be levelled to protect an organisation's vested interests. For example, the Church has stepped in when they felt their beliefs and power were under threat. Such attacks have put back and delayed perfectly correct and good work for years or even centuries. We have mentioned how Galileo was affected and the condemnations of Aristotle's work in 1277 CE. But within scientific circles, we have the bitter disagreements between Newton and Leibniz about who had invented calculus. There are also examples where later scientists and philosophers have ridiculed some of the thinkers of the past for some of their ideas.

For instance, in *The Foundation of Modern Science in the Middle Ages*, Grant[184] cites Galileo's devastating critique of Aristotle's ideas and also John Locke's scathing remarks concerning the scholastic teachers. And in his book *A History of Natural Philosophy*, Grant cite's the views of Andrew Cunningham and Perry Williams and Flores Cohen who see Natural Philosophy as either having nothing to do with modern science or being a hindrance from which modern science was "freed" by the Scientific Revolution. For more examples, including well known named scientists, see http://amasci.com/weird/vindac.html#j34 – Ridiculed Discoverers – Vindicated Mavericks.

Thus, it is just as important to critically assess not only the veracity of the original work and ideas but also any criticism, to ensure it is based on equally valid science and thought, in a desire to advance understanding and not on personal prejudice and jealousy or for financial gain and prestige. Such careful scrutiny is important because of the demarcation problem – the differentiation between science and so–called pseudo–science. Calling something "science," particularly in the public square, may lead to its uncritical acceptance by the public, when in fact the methods and conclusions may be suspect. This is yet another means of manipulation which can happen for reasons of finance, prestige and politics.[185]

And this is not the half of it !

Not all scientific enquiry is as it seems. There are plenty of examples of research being falsified or plagiarised for personal monetary gain or professional status. For example, there is Andrew Wakefield falsely claiming that his research indicated a connection between autism and the measles–mumps–rubella vaccine. In 2006 CE, Korean researcher Hwang Woo–suk was found to have fabricated a series of experiments in stem cell research, and Bengü Sezen over a period of ten years manipulated and falsified his research data.[186] [187]

Is it possible that all these various behaviours could be found in relation to our religious writings, commentaries and interpretations of scripture ?

We do like to identify moments in history that changed the world or specific individuals who have totally changed our understanding of the world. The reality is that it is rarely a single action or decision that is the moment in history. It is the accumulated actions and decisions that happen prior to any one given moment, that are the cause. These actions and decisions may occur over months or years and may include people and places thousands of miles apart. These may then eventually coalesce into what we then seize on as the defining moment.

Any of the recent political actions across the world can be traced, not to single random acts of violence, rebellion or rejection of the status quo, but as an accumulated response to a whole host of decisions and actions taken by others, often over many years, leading to a sense of injustice, exclusion, exploitation or betrayal. Superimposed on this state of affairs is often the naked ambition, greed and cruelty of a group of individuals following a charismatic leader. The long standing historical, political, social and economic factors that provided the context for war in Afghanistan and Iraq for example, are covered by Bailey and Immerman and by Shahrani in their respective books.[188] [189]

In a similar way, the increases and changes in our knowledge and understanding of the world, occur in response to a whole variety of ideas and contributions from people and circumstances and contexts. Even if the ideas and concepts turn out to be erroneous, they will nevertheless be part of the fabric that others later weave into new concepts of understanding. History does recognise individuals that have had insights that perhaps have lead to a "paradigm" shift in the way we view the world,[190] but more often than not their ideas have arisen out of exposure to the contribution and thinking of others.

This is summarised by a quote from the proceedings of a scientific conference in 2008 CE held by the International Astronomical Union (IAU). It was convened to try and define the term planet. It restricted the definition to the eight largest bodies orbiting the Sun, thereby deleting Pluto from the list. The demotion of Pluto sparked considerable public controversy. Numerous planetary scientists and astronomers protested that the IAU's definition was not useful, while numerous other planetary scientists and astronomers supported the outcome. Recognising the need for further scientific debate on planet definition, more than 100 scientists and educators representing a wide range of viewpoints on the issue, converged for three days on the Applied Physics Laboratory of Johns Hopkins University (APL) for "The Great Planet Debate: Science as Process" conference. The conference was sponsored by NASA, APL, the Planetary Science Institute, The Planetary Society, and the American Astronautical Society.

"We all have a conceptual image of a planet. Therefore, we need a term that encompasses all objects that orbit the Sun or other stars," said Larry Lebofsky, Senior Education Specialist at the Planetary Science Institute in Tucson, Arizona. "The debate is a great teaching moment. Whether dwarf planets are grouped together with the classical planets is not as important as the process by which scientists arrived at their conclusions. Scientists look at the same information in different ways; there may be more than one 'answer'. Facts change. What we know now may not be what we know in two or three years. Learning to think critically and understanding how scientists organise facts to develop theories are lessons that will serve students for a lifetime."

This is not to romanticise the process and imply that there is a steady, linear like progress in knowledge and understanding from ancient to modern and that each new generation of thinkers simply extend what has gone before. Far from it. There have been eras where previous knowledge has been forgotten, cul–de–sac explorations, re–discovery of ideas years later and so on. Paradigm shifts when individuals have taken an idea that was not found to be of value in one subject area and applied

141

it in another where it has opened other doors of exploration. And this has happened in all areas of human endeavour.

Our history books like to paint this process as one of continual progress with "hero" like figures emerging that change the course of events by their own abilities and clever insights, searching for new knowledge to benefit mankind. However, this is rarely the case. Even some of the most momentous discoveries happened either by accident, or their significance was not immediately recognised or even the implications were denied by the originator and only realised by subsequent investigators. And so, as the previous paragraph implies, the process, in a climate of mutual respect and acknowledgment of all contributions – past and present, by which new knowledge is derived, is important, because it is likely that what one knows today will change by tomorrow.

SUMMARY

From the foregoing discussion, it can be seen that the evolution of science and philosophy is a patchwork of interconnections, ups and downs, paradigm shifts, dead ends and enormous expansions in knowledge and understanding. We have noted how specific fields of enquiry gradually emerged as science, philosophy and theology and how these have themselves spawned their own specialised branches of endeavour. We have also noted the changing relationship of the Church and its doctrines, to this process. As we will discuss in more detail in later chapters, it seems only natural therefore to believe that the nature of our understandings of faith, God and religion is also subject to similar evolutionary processes. Thus, any sense in which our ideas on such matters could reach a point where everything was known or fixed beyond question or that one group has "all the answers" and everyone not sharing such certainty is wrong, is surely the error. We will come back to this point in the final chapter

CHAPTER 4

COSMOLOGY

ORIGINS OF THE UNIVERSE

Cosmology is concerned with the origins and nature of the universe. It is therefore of interest in our explorations of the origin and evolution of religion and our consideration of God's communication.

Cosmology, simply put, is a branch of astronomy that studies the universe – its evolution and origins i.e., its past and present, but also speculates as to its future. Astronomy is the study of the "night sky" – the stars, planets and all celestial objects and their nature and movement. It is also concerned with the nature of space and galaxies and what goes on in them. Records of humankind's interest in these phenomena go back at least as far as the sixteenth century BCE, in Mesopotamia. Initially, any objective recording of celestial activity and the interpretation of their significance, was interlinked with religion, myth, astrology and magic. These are the ways which humankind have used to make sense of the questions of why we are here, how it all came about and why things are the way they are. Gradually, these linkages have been severed, though both cosmology and religion are still interested in the same fundamental questions. Many people still take astrology seriously enough to provide a living for those who describe connections between cosmological activity and our everyday lives.

We find the earliest descriptions of how our ancestors saw and explained the world and the natural phenomena they experienced through their senses, in the recorded myths and legends of ancient cultures. These are recorded in both pictorial and written forms and can be followed through history as they change and spread across different cultures into the present day. A topic dealt with in more detail in chapters six and seven on the evolution of religion. Traditionally, it is posited that science can provide a lot of detail of the mechanics of how things have come to be as they are, whilst philosophy, theology and the myths and legends provide insights into why things might be the way they are. They also comment on our relationship with the environment and the purposes and possibilities of the earth and the universe(s) and our role within them. Cosmology too, as a scientific discipline, is also concerned with these same fundamental questions about the universe(s) and the origins of life.

Firstly, we will begin with some of the common cosmological ideas of antiquity. These provide a variety of explanations, from different civilisations, for the origins of "God" and the universe. These are followed with a brief time–line of our scientific descriptions of how things began. The relevance of this to our theme, is summarised in the conclusion.

THE MYTHS

There is evidence from many civilisations and ethnic groups across the world about their stories of how the world/universe came to be. Wikipedia[191] provides a list of these, with examples from all continents and Stephanie Dalley, in her translation of *Myths from Mesopotamia – Creation, The Flood, Gilgamesh and Others*[192] gives further examples of creation myths. The History World website also provides a neat summary of some of the common content of these myths.[193] A quick overview of the concepts used by humankind to explain how things came to be the way they are, is provided by one of the characters, "Professor Langdon" in the novel "*Origin,*" by Dan Brown.[194] It also touches on aspects of the "science versus religion" debate.

The following paragraphs provide some specific detailed examples, and each represents a compilation taken from these sources.

There is much commonality between these stories, even though they are often widely separated by geography and time. Many, begin with a concept of empty nothingness, darkness or void or a situation of chaos. Other stories commence with an ocean or sea or water whilst others imagine some sort of cosmic egg as the starting point. In yet others, there is a "creature" that dives into and retrieves some primeval mud, out of which the universe is created. There may be inclusion of some sort of failure or deformity of the offspring who may be eaten or locked away or otherwise disposed of. Frequently there are giants of some sort, there is often a reference to floods and generally two figures emerge that come to represent the first humans. Usually they are male and female but not always.

How the rest of the cosmos and the plants and animals of the earth are derived is frequently not particularly specified, at least not in the same detail as in the Hebrew story. Some of the stories have a sense of a cyclical creation with spirits being reformed until reaching some sort of final form. A sense of an end time does not feature very frequently. Other cultures, which understand the world to be eternal, with no beginning or end, describe the origins of the world without invoking the idea of a creator.

So for the Hebrews, for whom God already existed, God begins creation directly out of the empty void. For many other cultures, the first step is for a God to emerge out of the void, chaos or water or cosmic egg (China, India) and then in various ways sets about making or separating the heaven and earth as well as creating other Gods. In Hinduism, there are a number of versions that incorporate some or all of these descriptions. In those cultures with multiple Gods, there is usually a complex narrative involving the interactions of the Gods amongst themselves and also with humankind. These interactions reflect both supernatural occurrences and anthropomorphic (human like) based activity using human characteristics such as – jealousy, anger, love, war, sex and so on. These Gods may have their own special skills or areas of responsibility or influence, and often, there is some sort of struggle to assume dominance over all the other Gods.

Hebrews

In the book of Genesis, in the Hebrew Scriptures, written between the seventh to the sixth centuries BCE, the Hebrews imagine a first moment when all is void, with darkness on the face of the deep. But God already exists and out of this darkness begins creation. This is a chronicled sequence over six days when the universe, the earth and its inhabitants are created. However, Genesis has two similar but distinct versions of this story. The second, possibly being the solution as to why the world as we experience it is not as perfect as the first description seems to indicate it ought to be. As creation stories go, this is a straightforward one. The Hebrews simply declare that God did it. As Christianity was born out of the Hebrew tradition, with Jesus Christ being a Jew, this monotheistic version was accepted and the Hebrew scriptures formed the basis of the Old Testament part of the Christian Bible.

Mesopotamia

In this story, we start with two watery beings – Apsu and Tiamat. They create a variety of sea monsters and gods. But as Tiamat tries to take control, her descendants rise up against her and choose – Marduk, the god of Babylon – to lead them. Marduk meets Tiamat and her evil companion Kingu in battle and kills them both. He splits the body of Tiamat into two parts, thus creating the heaven and the earth. In heaven he builds a dwelling for his colleagues, the gods. Realising they cannot do all the work, he creates a race of servants, from the blood of Kingu and creates the first man. The creation of rivers, plants and animals then follows.

Greek

The Greeks acquired many different gods as they gradually migrated as a group of Indo–European tribes, into what is now modern Greece. This multiplicity of Gods provides for a complex creation narrative, written around 800 BCE. From the Chaos, emerges Gaea, the earth. Gaea gives birth to a son, Uranus, who is the sky. So now the world and Gaea and Uranus go about populating it with their children – the Titans. And like in the Mesopotamian Creation story, though for different reasons, her offspring are not a blessing. Uranus, unhappy about the situation decides to shut them away. This upsets Gaea who persuades her youngest, Cronus, to castrate his father. This is followed by infanticide and incest as Cronus also eats his own offspring who he has had with his sister Rhea. Rhea saves her youngest child, called Zeus. Zeus then overwhelms his father, defeats all the other Titans in a great war, and settles upon Mount Olympus to rule over the world. How humankind come into the story is not clarified, but there are several versions of how it might have happened. One states that Zeus sends a flood to drown mankind but two humans escape in an ark. When the flood has subsided, the oracle at

Delphi tells them to throw a stone over their shoulders, from which a human being is created.

Islam

In Mediaeval Islam, we also have the concept of the creator God, who created the world and all its inhabitants at one point in time and so is associated with the concept of a finite world. But in addition, we have the idea of a duality; that is, there are two worlds – one Unseen Universe and one Observed Universe. There is also a notion that there may also be multiple worlds.[195]

India

In India, the complexity of the multitude of Gods, particularly in Hinduism, allows several creation stories to co–exist. There is one particular story about the god Brahma. First he creates, by thought alone, the waters in which he deposits his seed. This grows into a golden egg from which he is born. He splits the egg in two and the two halves become heaven and earth. There are also stories of dismembered giants as well as all the themes already mentioned. But in addition, in the Rigveda (a collection of Vedic Sanskrit hymns and verses and one of the four canonical sacred texts of Hinduism known as the Vedas), there is an acknowledgement of doubt:

Who really knows, and who can swear,

How creation came, when or where!

Even gods came after creation's day,

Who really knows, who can truly say

When and how did creation start?

Did He do it? Or did He not?

Only He, up there, knows, maybe;

Or perhaps, not even He.[196]

In contrast, Buddhism and Jainism have no need of a creator God, as for them the universe is eternal with no beginning and no end. Life is a continuous cycle of re–birth and re–death. Jinasena (eighth century CE), head of a monastic order, was the author of the texts of the Ādi purāṇa and Mahapurana, in which he is writes:

"Some foolish men declare that a creator made the world. The doctrine that the world was created is ill advised and should be rejected. If God created the world, where was he before the creation? If you say he was transcendent then and needed no support, where is he now? How could God have made this world without any raw material? If you say that he made this first, and then the world, you are faced with an endless regression."

China

Similarly, amongst some of the Chinese mythology, there is the story of P'an Ku, who is hatched from a cosmic egg with half the shell above him as the sky and the other half below him as the earth. He pushes them apart but the effort is too much so he falls apart with his limbs becoming the mountains, his blood the rivers, his breath the wind and his voice the thunder. His two eyes are the sun and the moon. The parasites on his body are mankind. This story is thought to have been compiled between the fourth and third centuries BCE.

However, this is only one story and the general view is that the Chinese do not have an overall creation story in the same way as the Hebrews. Rather, they have a number of narratives telling about the origins of the world, but not necessarily invoking the idea of one creator.[197] There is the notion of "duality' – the Yin and the Yang – opposites that are complementary and in balance and the "Tao" or the WAY – a notion of the right way to live and be in empathy with all life.

Japan

Again, there are a number of versions that vary in both detail and substance. For example, the Ainu people of the North Island of Japan, have several stories. In one, the creator deity sends down a water wagtail to create habitable land in the watery world below. The little bird fluttered over the waters, splashing water aside and then he compacted patches of the earth by stomping them with his feet. Thus forming the islands where the Ainu were later to live. In another, Kamuy (a spiritual or divine being) sends a heavenly couple to earth who have a son. The first Ainu, and he is believed to have given the people the necessary skills to survive. Yet another version has it that before God created the world, there was only a vast swamp in which lived a large trout, and the creator placed the world upon the trout.

From traditional Japanese culture we have a further version which starts with a floating amorphous mass, similar to the slithery substance of an egg but moving more like a jelly–fish. From this, emerges a reed–like object, which produces eight generations of brother–and–sister gods. Their first outpourings are flawed (a child which cannot stand at the age of three, an island composed of foam). This seems to be because the woman spoke first in their sexual encounter. By correcting this behaviour, they create many gods – including those of the eight islands of Japan. The gods proliferate and have numerous adventures and set the basic rhythms of day and night and the seasons. Eventually the Sun goddess sends her grandson, Ninigi, to rule the Central Land of Reed Plains. This is now Japan.

A variant of this, from about the eighth century, has a complicated procession of Gods that are created but then hide and do not take creation any further, until the eighth pair of gods. These are Izanagi (The Male Who Invites) and Izanami (The Female Who Invites). Standing on the Floating Bridge of Heaven, they lean down to stir the brine of the sea with a lance. The liquid begins to curdle and forms an island. The two gods come down on to it, and build a Central Pillar, from behind which they create more islands and gods.

Norse

Again we start with nothingness. But gradually this space fills with water, freezes and then partially melts. From the drops of melting water a giant in human form emerges – Ymir. From his armpit two giants appear – a man and a woman – who produce more giants. In the meantime, a cow licks the melting ice to reveal yet another giant and from whom the god Odin (or Wotan) comes down.

Then Odin and his brothers kill Ymir and make the earth from his flesh, the heavens from his skull, the sea from his blood and the mountains and trees form his bones and hair respectively. Odin builds a place for himself and the other gods to dwell in, linked to earth by the bridge of the rainbow. Then he and his colleagues breathe life into two tree trunks, turning them into Ask and Embla, the first man and woman.

A different version, as written in an epic poem – the Völuspá dates to the tenth century. Here we have the earth being lifted out of the sea by the sons of Burr. The god then establishes order in the cosmos by finding places for the sun, the moon and the stars, thereby starting the cycle of day and night. A golden age ensues until it is ended by three giant maidens. The poem, as recounted by a female enchantress to Odin,

151

includes references to dwarfs, folk wars and the destruction of the Gods by fire and flood. It prophesies a new world rising out of the ashes and the emergence of a dragon before the enchantress awakes from her trance. Some scholars think that some aspects of Christianity are likely to have had an influence, as it appears around the time of transition between paganism and the bringing of Christianity to Scandinavia.

SUMMARY SO FAR

These are just a selection of the myriad stories that different cultures have devised during their early evolution, to explain how things for them, were the way they were. However, as their cultures developed, people began to look for other explanations and as discussed in chapter 2 around 5000 years ago we start to see examples of this. And so as we meet the pre–socratic philosophers around 600 BCE we find evidence of a new way of thinking and the exploration of nature itself to look for answers – the beginnings of science and philosophy.

THE SCIENCE

In this section, we will concentrate on the changing nature of scientific explanations of the cosmos. Don't worry ! It is not intended to be an in depth consideration of the theoretical physics and mathematics that underpin our current scientific understanding of the cosmos. Nor indeed is it a whistle stop tour of all aspects of astronomy [198] and cosmological [199] theory. It is more to provide a brief glimpse of both the astonishing enormity of the universe and the mind–blowing conceptual ideas and thought that has provided us with this knowledge. We will also look at how both the universe itself has and is changing and how our understanding of it is also changing.

Not only is the world and the universe something wonderful and marvellous to behold, but the brains of those that have studied it and those continuing to study it, are equally wonderful and marvellous to contemplate. I find ideas such as the "Big Bang", string theory, the multi–universe concept, the complex maths of cosmology, the concept of space – time, breathtaking. Then there is the idea of gravitational waves which were only discovered in 2015/6 CE but theoretically predicted by Einstein 100 years ago. It is all totally amazing and for the most part beyond my comprehension.

The designation of the point at which the "Big Bang" happened as a singularity, (see later) talk of a Nano–second of time after the "Big Bang" and an expanding universe, leave my mind numb with amazement as I cannot even begin to picture what any of this means or picture what it might look like. However, I am happy to accept there are scientists who understand that such mind–blowing concepts offer the best explanations we have, at the moment, for our observations relating to space and the universe. I am equally happy to think that in time, other explanations may very well emerge.

The current estimate of the age of the universe is about 13.7 billion years, with the earth being about 4.5 billion years old. A time–line picture showing the events occurring during this span of time can be found in Figure 2, at the end of the chapter.

So how is the age of the universe and the earth estimated ? How do we "know" the information included in these time–lines ? How do the ideas and theories originate ? How do scientists do it ? I find it awe inspiring to even begin to think that people have been able to generate ideas and theories about the universe and its origins, let alone develop technology to observe and examine it to test their ideas. Not to mention the courage exhibited by some to persist with their ideas in the face of fierce opposition – not just by religious institutions, but by their peers too – when their ideas questioned the prevailing "norm" – for example Galileo, and Semmelweis.[200] [201]

But our modern ideas and theories have not suddenly developed. They have emerged from the knowledge and ideas of our ancestors over several thousand years. Then coupled with new information gained from developing new methods of study and new technologies for observation, theorising and experimentation. We have already mentioned some of the astronomical and cosmological views and ideas from the pre–socratic times and up to the Renaissance. Many of these have parallels in the ideas of modern science. They have then been elaborated upon, proven to be wrong or refined and described with different degrees of understanding. In addition, the evolution of these old notions or the process of their overthrow, has generated new ideas. So as we look at some of these ancient ideas, we will broadly follow a chronological framework, though there will be some moving back and forth as we mention different concepts.

The first concept is the shape or form of the earth.

In Mesopotamian cosmology, around the sixteenth century BCE, the idea was that the earth was flat. This was still the prevailing idea around the sixth century BCE. However, by Classical Greek times (fifth to the fourth centuries BCE), the idea that the Earth was spherical ("round") was common and by around 240 BCE, Eratosthenes (276 to 194 BCE) had estimated its circumference. Very accurately as it turned out, based on Aristarchus of Samos's published work on how to determine the sizes and distances of the Sun and the Moon. Even so, modern Flat Earth Societies were started during the twentieth century based on theories generated during the mid–nineteenth century.[202]

The second concept relates to the relationship between the earth, sun and other planets.

The Sumerians,[203] one of the first known civilisations along with the Egyptians and the people of the Indus Valley, settled in Southern Mesopotamia (now southern Iraq) from as early as 5000 BCE. There are records of their ideas of the cosmos and astronomy, in cuneiform notation on clay tablets, from as early as 3000 BCE.[204] Their ideas included: a concept of multiple heavens and earths (multi–universe), the concept of heaven and earth as a whole rather than as geocentric, and an ability to describe the positions of Venus through the year. They also used mathematics to predict periodic phenomena such as the variation in the length of the day, records of eclipses and the use of "omens" as a form of astrology.[205] Based on this knowledge, Thales was also able to predict the solar eclipse of May 28th, 585 BCE.[206]

In the fourth century BCE, Aristotle established the idea of the universe being geocentric. That is, the fixed, spherical Earth was at the centre of the universe and was surrounded by concentric circles of the planets and stars. Despite contrary ideas, which are referred to below, this idea of geo–centricity remained the dominant view of the cosmos until the sixteenth century CE.

Aristarchus of Samos (from around 310 to about 230 BCE) disagreed with Aristotle and developed an argument for the universe being heliocentric i.e., the Sun, not the Earth, being at its centre. Seleucus of Seleucia, who lived about a century after Aristarchus, also supported this theory and used the tides to explain heliocentricity and the influence of the Moon. He also suggested that the Earth rotated around its own axis, which, in turn, revolved around the Sun. In the fifth century CE the Indian astronomer and mathematician Aryabhata also believed the earth rotated around its own axis, though he still accepted a geocentric universe.[207]

Hipparchus (190 to 120 BCE), focusing on astronomy and mathematics and used sophisticated geometrical techniques to map the motion of the stars and planets. He even predicted the times when Solar eclipses would happen. In addition, he added calculations of the distance of the Sun and Moon from the Earth, based upon his own modifications to the observational instruments used at that time. The level of achievement in Hellenistic astronomy and engineering is impressively shown by the Antikythera mechanism (150 to100 BCE). This in essence is an analog computer for calculating the position of planets ![208]

In the second century CE, along came Ptolemy, a Roman–Egyptian mathematician and astronomer. He supported Aristotle's concepts and so reinforced the notion of a geocentric universe. He enshrined these ideas in a number of treatises but most notably in the *Almagest – The Great Treatise*. This was to be the prime source of astronomical knowledge for the next thousand years. In the Almagest, Ptolemy points out that any model for describing the motions of the planets is merely a mathematical device, and since there is no actual way to know which is true, the simplest model that gets the right numbers should be used. This principle enabled him to accurately build an armillary sphere, for modelling the movements of the stars and the planets. This concept, it seems, was created independently in both Greece, possibly by Eratosthenes in the second century BCE and in China in about the fourth century BCE.[209]

There are two basic versions, one in which the sun is the centre – now known as the Copernican and the other with the earth at its centre – known as the Ptolemiac version. So, although Ptolemy had the wrong

156

version, the actual measurements were correct. I find the concept of it mind boggling even now. The minds that imagined it and then did the measurements to construct it accurately, without the aid of our sophisticated technology, are something to wonder at.

Not surprisingly perhaps, we then get attempts to combine the geocentric and heliocentric systems. In the fifteenth and early sixteenth centuries, Somayaji Nilakantha of the Kerala school of astronomy and mathematics in southern India, developed a computational system for a partially heliocentric planetary model, in which Mercury, Venus, Mars, Jupiter and Saturn orbited the Sun, which in turn orbited the Earth. This was very similar to the Tychonic system proposed by the Danish nobleman Tycho Brahe later in the sixteenth century. A kind of hybrid of the Ptolemaic and Copernican models.[210] Whilst in many areas of life, "the middle way" is often the best in terms of finding a way through opposing views, in this instance it didn't quite work out like that, though it was accepted for a time.

In 1543 CE, the Polish astronomer Nicolaus Copernicus published his ideas for a heliocentric universe in his *De revolutionibus orbium coelestium*. He demonstrated that the motions of celestial objects can be explained without putting the Earth at rest, in the centre of the universe. Galileo Galilei is known for supporting this Copernican theory of heliocentrism, against Church opposition. This resulted in him being put on trial and sentenced to house arrest. The Church was opposed to this new concept of the sun rather than the earth being at the centre of the universe, as much because they had integrated the Aristotelian position into their world view, and change meant having to admit they might be wrong. So they used "religious" justifications for why the earth should be seen as the centre of things. However, as we have seen in the previous

157

chapter, they changed their minds about Aristotle, once they realised other religious implications of his views of the universe seemed to imply limitations on the power of God. Nothing is for ever – may be !

Within a couple of years of this revolutionary concept being generally accepted, in 1546 CE, the English astronomer Thomas Digges (1546 to 1595 CE) extended the Copernican system by replacing the concept of a limited universe with an outer edge, with one of a star–filled unbounded space – a multitude of stars extending to infinity. The Italian philosopher Giordano Bruno (1548 to 1600 CE) goes further in 1584 CE by proposing that in fact the Copernican solar system itself is not even the centre of the universe, but rather, a relatively insignificant star system, amongst an infinite multitude of others.

Then in 1610 CE, Johannes Kepler used the accurate stellar observations of Tycho Brahe, to propose that the assumption of circular orbits was out and that elliptical orbits were in. They were in, because they could explain the apparently strange movements of the planets much more accurately. In 1687 CE, Sir Isaac Newton came along and described the large–scale motion of planets and stars throughout the universe and introduced his theory of gravity as a force of attraction between all particles in the universe. Thus, these ideas and principles became the accepted bedrock on which future physics and astronomical considerations were based.

The third concept is the nature of the universe.
In the fifth century BCE, the Greek philosopher Anaxagoras believed that the original state of the cosmos was a primordial mixture of all its ingredients, which existed in infinitesimally small fragments of themselves. This mixture was not entirely uniform, and some ingredients

were present in higher concentrations than others, as well as varying from place to place. At some point in time, this mixture was set in motion by the action of "nous" (mind), and the whirling motion shifted and separated out the ingredients, ultimately producing the cosmos of separate material objects, all with different properties, much as we see today.[211]

We have mentioned the notion of a basic unifying substance of the universe. For Thales it was water, for Anaximander it was apeiron, for Anaximenes it was air. Aristotle then proposed the idea of "aether"to describe the void that fills the universe above the earth. In more modern times there has been speculation on the need for a substance of some sort to facilitate the transmission of light waves across the universe. This idea too has now been shown to be erroneous. However, we now have the notion of Dark Matter (see later) which may yet be found to have some sort of unifying function.

Around the later part of the fifth century BCE, was the notion that the universe was composed of atoms. (see chapter 2). Aristotle also believed the universe to be of a finite size, and that it was eternal, that is, it had existed unchanged and static throughout eternity and would continue to do so.

By the sixth century CE, the idea of the universe being infinite was being questioned. A number of medieval Christian, Muslim and Jewish scholars proposed that the universe must be finite in time in order to fit with the notion that God created the world. If God created the world, it must have been done at a point in time and so their must have been a beginning point. The Christian philosopher, John Philoponus of Alexandria (490 to 570 CE), and early Muslim theologians such as Al–Kindi (ninth century) and Al–Ghazali (eleventh century) also offered logical arguments supporting a finite universe. As did the tenth century Jewish philosopher Saadia Gaon and Immanuel Kant (1724 to 1804 CE) also offered arguments against an infinite past.

The opposition to Aristotle continued. In the twelfth century, the Islamic scholar – Fakhr al–Din al–Razi – uses verses from the Qur'an indicating that the universe has more than "*a thousand thousand worlds beyond this world, such that each one of those worlds be bigger and more massive than this world, as well as having the like of what this world ha*s". He argued that there exists an infinite outer space beyond the known world, and that there could be an infinite number of universes. This is later taken up by the Catholic Church and forms part of their opposition to the teaching of Aristotle in the new Universities – as mentioned in the previous chapter. This idea of multiple and infinite universe is the precursor to some of our modern ideas, which we will come too later, though based on different reasoning. These concepts of the infiniteness versus the finiteness of the universe still continue. Though now in more nuanced form with the idea of time itself being a linear progression, being regarded as meaningless. Again, we will come back to this shortly.

The fourth concept is the origin of the universe.
As mentioned in the previous section, the myths were the first attempts to think about the origins of the universe. But just to reflect. The records from the fifteenth to the twelfth centuries BCE in the Hindu writings, which speculate about a spherical world beginning from a Cosmic Egg containing the whole universe (including the Sun, Moon, planets and all of space) and then expanding out of a single concentrated point called a Bindu, is not a million miles away from our "Big Bang" and a singularity! (see later). They also considered the universe to comprise

repeated oscillating cycles between expansion and total collapse. Again, this idea is not dissimilar to the idea supported by Einstein and the idea of the "Big Bang" being followed by the Big Crunch or a Bounce. (see later).

So beyond the myths, there are the competing views about whether the universe is eternal and infinite in the sense that it has always been there and will continue to be so, or that it is finite in the sense that it started at some point in time, generally created by a God or similar. As natural philosophy evolved into science and the scientific method became the concept whereby knowledge was advanced, more objectivity and reasoning was applied to the concepts of the origin of the universe. Cosmology became a science in its own right. The role of "God" as the initiator or prime cause of the universe then came under scrutiny and ideas about the universe – "something" emerging from "nothing" attracted philosophical speculation.

Early in the mid–seventeenth century, the French philosopher René Descartes outlined a model of the universe, not dissimilar in character to Newton's later proposals. But according to Descartes, space was not an empty vacuum but filled with matter that swirled around. In 1687 CE, Sir Isaac Newton published his "*Principia*", which described the universe as being static, in a steady state and infinite. In this universe, matter on the large scale was uniformly distributed, and was gravitationally balanced but essentially unstable. Under the influence of Gravity, the universe had the potential to collapse in on itself. This became the new accepted position, until Einstein and then Hubble came along (see below).

In 1734 CE the Swedish scientist and philosopher Emanuel Swedenborg proposed and developed a Hierarchical Universe Hypothesis.[212] This is a concept whereby stars and galaxies cluster in to ever bigger groupings that can be identified as entities in themselves. Although still based on the Newtonian static universe, the matter in a hierarchical universe is clustered in ever larger groupings, and is continuously being recycled (not dissimilar to the Hindu notion of collapsing and expanding cyclical concept !). Such a hierarchical concept is now supported by the findings of clusters and superclusters of galaxies that congregate together to form substantial cosmic structures millions or billions of light years across.[213] These structures are described as walls or filaments or superclusters or large quasar clusters depending on their size.

The concept of a static universe continued into the twentieth century and was accepted in Einstein's model of the universe in his theory of gravity. That is, that the universe was both static and dynamically stable. That is; the universe was neither expanding or contracting. However, in order to demonstrate this mathematically, Einstein had to add in a "cosmological constant" to his theory of general relativity. This was in order to avoid his theory indicating that the effects of gravity would cause the universe to collapse in on itself. But in 1929 CE, Edwin Hubble (1889 to 1953 CE) definitively showed that the universe was expanding away from us in all directions, and therefore not static. This leant credence to the suggestion made in 1927 CE, by a Belgian priest called Georges Lemaître, that the universe began from a single primordial atom. This was not the first time that "a big bang" notion – a single event – had been suggested as the possible beginning of the universe.

A similar theory has been proposed in 1922 CE by the Russian Alexander Friedmann although it had not been taken any further. It was, however, not yet called a "Big Bang". That name was not coined until 1950 CE when Sir Fred Hoyle (1915 to 2001 CE) did so.[214] The concept of such a

"Big Bang" is that there was a point – a singularity – at which or from which, the bang or explosion occurred. At this singularity, all the energy and matter that is in the world today was concentrated into an infinitely dense point of radiation energy. Due to quantum fluctuations, (temporary changes in energy at a point in space) this then exploded and spewed out its contents, a bit like a volcano, and this spreading out – expansion – has continued ever since and still continues. With the expansion, the temperatures cooled and the particles (quarks), produced by this high energy situation, began to slow down and so coalesce.[215] By this coalescence of quarks, the building blocks of the atomic nuclei as we now understand them – protons, neutrons and electrons – were formed. Atoms then coalesced to form matter. This allowed matter to be formed into stars, planets and galaxies.[216] Life formed on earth as favourable conditions gradually came about.[217]

Sir Fred Hoyle himself did not support the idea of a "Big Bang" and always favoured the Steady State theory. But in an expanding universe, this theory required matter to be continually added in order to maintain a constant density. This constant creation of matter was thought to be no more improbable than the idea of the universe being started from nothing i.e., the "Big Bang", in the first place. Although Hoyle never accepted the "Big Bang" theory, further evidence for it was provided by Arno Penzias and Robert Wilson in 1965 CE when they discovered evidence of cosmic microwave background (CMB) radiation.[218] This is thought to be left over from the time of the "Big Bang". The radiation is like that used to transmit TV signals via antennas and is evidenced in some of the "snow" one sees on a television set that is not quite tuned to a specific channel.

Einstein later abandoned what he described as his greatest mistake – the cosmological constant – once it was realised that the universe was still expanding, and supported the idea of an Oscillating Universe. The Oscillating Universe followed from Alexander Friedmann's model of an expanding universe based on the general relativity equations for a universe with positive curvature (spherical space). This results in a universe which expands for a time and then contracts due to the pull of its gravity, thus forming a perpetual cycle of "Big Bang" followed by "Big Crunch". Although, some scientists think it would not go completely back to the singularity at the beginning and so we could get a "Bounce." effect.[219] More recently, the clarity of whether a singularity would re–occur in an oscillating universe has been questioned.[220] In these scenarios, the concept of Time becomes endless and beginning–less. Thus, the "beginning–of–time" problem i.e., the problem of what was happening before the "Big Bang" is also meaningless, and so is no longer a problem to be considered.[221] Again, this is not a million miles away from the Hindu concepts of a cyclical universe.

Further refinements of the "Big Bang" theory have since been proposed. One is that immediately after the initial event, there was a very short period – 10^{-35}– to 10^{-32} seconds – of Inflation, that is, a short period when the expansion of the universe was very rapid. This was proposed in 1980 CE by the American physicist Alan Guth. This model was put forward to try and explain the horizon and flatness problems of the standard "Big Bang" model.[222] This inflation has been postulated to explain a number of other questions: why the universe is so vast, why the universe appears to be the same in all directions, why the cosmic microwave background radiation is distributed evenly, why the universe is thought to be flat, and why no magnetic monopoles have been observed.[223] I won't even begin to explain what these problems are, let alone discuss how the inflation proposal helps solve them. A variation of the inflationary universe is the cyclic model developed by Paul Steinhardt and Neil Turok in 2002 CE.

Based on using state–of–the–art M–theory, superstring theory and brane cosmology, which involves an inflationary universe expanding and contracting in cycles.[224]

So although at present, the "Big Bang" together with the initial period of inflation is the favoured model, it still leaves many unanswered questions. One is about how a singularity could occur. There is a lot of literature that seeks to explore the "something out of nothing" idea or "why something rather than nothing?" A lot of these discussions are philosophical. But there is now a mathematical model that claims to demonstrate how this in fact is possible and that the universe could have arisen out of nothing. This is because of something known as quantum fluctuations within a vacuum bubble.[225] [226]

But this description of the singularity followed by the "Big Bang" with a period of inflation and then a continuing steady expansion has its detractors. Other scientists question the logic of the regression to a singularity and suggest that it would not happen out of nothing but follow a period of inflation and then lead on to a "hot" "Big Bang" rather than from something infinitely dense.[227] [228]This is indicative of the way our knowledge evolves. Suggested explanations are put forward, in the knowledge that they don't explain every last detail and then others, in addressing those details, offer additional or alternative explanations. In this way our understandings move forward and new insights emerge.

This seems to be an accepted and healthy way to progress (not withstanding the natural human resistance to such change and challenge as egos and vested interests may be bruised and knocked along the way). Hence, similar processes must surely apply to our understanding of religious faith, if we are to reveal new insights. But equally, there will be resistance from vested interests, wishing to maintain the status quo.

Another question, is what happened before the "Big Bang" – the infinite regression question. Some scientists have addressed this by dismissing it as meaningless. This is because time is now thought not to be quite the simple sense of moving from past to present that we assumed it to be. Even though it appears to work in that way on a day to day basis. Minkowski, Einstein's teacher, first described the idea of "space–time" in 1907 CE as he developed the mathematics in support of Einstein's theory of Special Relativity, published in 1905 CE. He called this space–time continuum – Minkowski Space. Time depends on where you are in the universe and in relation to what you are considering it – as May et al say in their book – *Bang*.[229]

We also now have concepts of multiple universes.[230] There is the idea of Infinite Universes in which it is thought that if the universe stretches out infinitely and space–time goes on forever, then it must start repeating at some point, because there are a finite number of ways particles can be arranged in space and time. Or there are Bubble Universes in which some pockets of space stop inflating, while other regions continue to inflate, thus giving rise to many isolated "bubble universes." There is also the concept of Parallel Universes which harbour the notion of "braneworlds" — parallel universes that hover just out of reach of our own. The idea comes from the possibility of many more dimensions to our world than the three of space and one of time that we know. In addition to our own three–dimensional "brane" of space, other three–dimensional branes may float in a higher–dimensional space. There are other concepts such as Daughter Universes and Mathematical Universes and various other classifications.

※∩ ※∩ ※∩※

Other Concepts

Another major and fundamental development that we have not yet mentioned is quantum theory and quantum mechanics. These are yet more examples of the mind blowing complexity of concepts that some scientists are capable of creating and thinking about. It also provides the frame of reference for considering how things operate at the micro level, in contrast to classical physics which provides the frame of reference for the macro – universe level.

Christiaan Huygens (1629 to 95 CE) theorised that light consists of waves. This contradicted Newton's idea of light consisting of particles. Initially, this was a source of controversy as each side tried to assert their explanation as the correct one. Later, it was realised that both concepts explained some of the characteristics of light but neither explained all. So in some circumstances the wave theory was employed and in others the particle theory was required. This provided scientists with a dilemma. As Einstein wrote:

"It seems as though we must use sometimes the one theory and sometimes the other, while at times we may use either. We are faced with a new kind of difficulty. We have two contradictory pictures of reality; separately neither of them fully explains the phenomena of light, but together they do."[231]

This so called duality (wave and particle), became an important aspect of the discussions considering how things operated at the sub–atomic level, particularly, as discoveries were made identifying numerous sub–atomic particles. A complex categorisation evolved as more such particles were identified and were either discovered in reality or postulated in theory, to account for certain observations. Thus, we identified and classified particles such as – electrons, neutrons and protons initially and then later fermions – quarks and leptons, bosons – photons and gluons and many others have been proposed theoretically, such as gravitons and axions but not yet observed. [232] [233] What becomes clear is that at this sub–atomic level the normal laws of motion don't work. The accepted laws of motion don't properly explain the way things operate at this level. A new system was required and gradually the idea, which became known as quantum mechanics, developed.

The term "quanta" is used to refer to packets of energy. "Quantum mechanics / theory" refers to the model describing how energy exists and moves both as discreet "packages," whilst also complying with wave theory. This became what is still the currently accepted explanation of the sub–atomic state. Some of the key thinkers behind this evolution include: Max Planck (1858 to 1947 CE) founded quantum mechanics in 1900 and then Albert Einstein (1879 to 1955 CE), Niels Bohr (1885 to 1962 CE), Werner Heisenberg (1901 to 1976 CE), Erwin Schrödinger (1887 to 1961 CE) and Louis de Broglie (1892 to 1987 CE) who developed these ideas and expanded their fields of influence. In keeping with the idea that this is a complex and contradictory area, Richard Feynman (1918 to 1988 CE) is famously attributed to have said something like *"if someone thinks they understand quantum mechanics – they don't."*

It was also discovered that both weak and strong forces operate at the sub–atomic level in order to keep the atomic structures stable. This means there are four natural forces of the universe – gravity and electromagnetic forces at the macro level and the strong and weak forces operating at the subatomic level.[234]

There is a hope that a unifying theory can be devised that link all these together, but at present only the electromagnetic and weak and strong forces have been brought together by the Standard Model. The inclusion of gravity, so far, has eluded these attempts. Gravity is the force associated with mass and matter and it is this that keeps the galaxies and planets in relation to each other. But it is understood now that there is insufficient known/visible matter within the universe to actually do this. So the concept of "dark matter" has been devised to explain the universe's stability. Such calculations demonstrate that in fact eighty percent of the mass–energy of matter in the observable universe, and about a quarter of the observable universe's total mass–energy in the universe must comprise such "dark matter" for the universe to be as it is.

"Dark matter" is so–called because it is not able to be seen as it does not react with any of the waves along the electromagnetic wave spectrum. To date it has not been identified. In a similar way "dark energy" is postulated to account for the fact that the universe is still expanding, and so some force, other than the four mentioned above, is required to overcome the tendency for matter – planets, galaxies etcetera to be attracted to each other by gravity. The Standard Model does not explain these requirements. A number of unifying theories, such as string theory, superstring theory and M – theory, are proposed but none have succeeded as yet.[235]

CONCLUSIONS SO FAR

In chapters two, three and four we have highlighted the uncertainty of what we know. We have considered the uncertainty of the origins of the words and definitions of science and philosophy which provide the context and framework for most discussions about the veracity of knowledge and what might be known. We have also looked at some of the factors which influence the production and discovery of knowledge, such as the controversies and sometimes the flawed methodologies and flawed personnel that can affect the claims that may be made. We have noted some of the ideas from the ancient and more recent past and seen how similar some of them are to our present day concepts, although they were not originally conceived in quite the way we now understand them.

Whilst it may not be possible to make direct connections between these "ancient thoughts" and our "modern concept/invention/discoveries" it seems more than likely that such ideas from the past, have influenced the present, whether consciously or indirectly. So that our current understanding of the problems and explanations and our ability to investigate them is now greater, as far as the details are concerned, This suggests that no idea should be summarily dismissed because it comes from another time or different culture. We never know when in fact, like fashion, they may come back into currency and throw fresh light on our current struggles. Even the act of demonstrating their current inadequacy, can be a positive step towards a new reality. Thus, progress happens, in spite of any backward steps made along the way or breaks and pauses in continuity that may occur.

So it seems reasonable to me, to think that those same factors, influencing our scientific, philosophical and every day knowledge, also apply to our understanding of theology and faith. There are parallels between the narrative as it relates to the development of scientific knowledge and as it relates to the development of theology and religious belief. With all faiths, there is uncertainty surrounding the authorship and timing of their written scriptural texts. There are a variety of interpretations of quite what these scriptures mean, in terms of application to a person's life and often there are definitional difficulties and disputes about translations. Such differences, both between and within different faith systems can become very acrimonious if not downright violent. In the light of all this, the idea that one faith or one group within one faith system know THE truth, seems at least to be an arrogant position to take, if not down right dangerous, if such a belief leads to actions that cause harm to others.

The paradox is, that all this uncertainty in science still allows the development of technologies, whose soundness and utility can be used with certainty on a day to day basis. Both for good or ill. Similarly with faith. The uncertainty of theology and the religious understanding of the scriptures of a particular faith, allows their followers to operate on a day to day basis with some degree of certainty or reassurance, again for good or ill. However, as with the good sceptical scientists, many people of faith have doubts, if only in private, and of course such doubts form part of the impetus or rationale for this book.

It will not have escaped your notice that within all this science and philosophy, the idea of whether a God is involved, and if so, in what ways, has not really featured. We will go into the origin of religious belief and the idea of God in detail in chapter eight. In the meantime, just to say that many scientists do believe in God and so incorporate God into their own personal understanding of how the universe came into being and what makes it tick. Many of the early scientists and philosophers we have already mentioned maintained their faith alongside their scientific work and looked for ways in which their science would remain compatible with their religion. This suggests that in those days, science and theology were not automatically at odds.

The modern scientists, who profess a faith, also do not see that science and religion are mutually exclusive and are able to accept and work within a scientific framework. I cannot say how each of them integrate God and science for themselves. However, because of other information – from the Bible, historical writings and the experiences related by thousands of other people – it seems to me to be more "scientific" to accept and try and understand and explain this information, rather than to just dismiss it out of hand. Clearly, many people take the opposite view. But for me, the crux of the matter is being willing to acknowledge that as no one truly knows whether God exists, taking a dogmatic atheistic position is inappropriate and unscientific, even if one privately lives by one's own conclusion or belief. It is necessary too, to allow the fact that in the light of new knowledge or experience, one's current understanding may need to change.

So over something like 5000 years, from the earliest attempts to explain why the way things are the way they are, humankind has gradually evolved means of acquiring knowledge. The trick, it seems to me, lies in accepting the uncertainty and limitations of all our knowledge and considering the implications for how individuals, groups and cultures interact and behave towards one another. How we accommodate and handle differences of opinion and understandings is the key to a more tolerant and peaceful future.

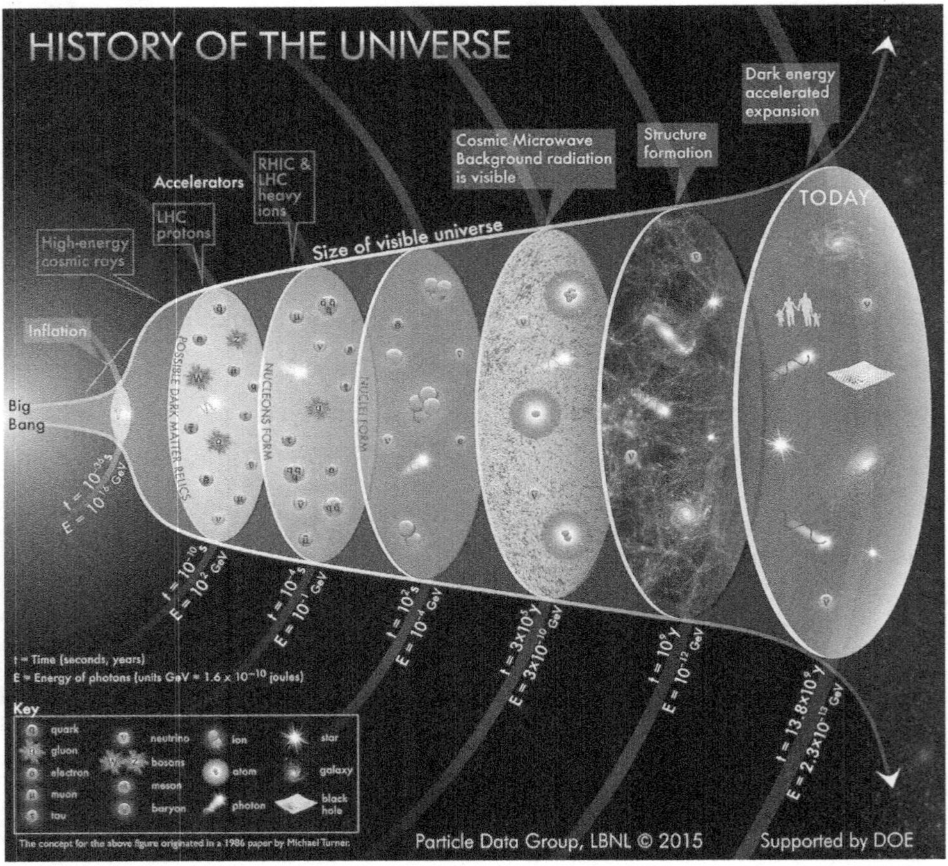

FIGURE 2 TIME–LINE OF THE HISTORY OF THE UNIVERSE

※∩ ※∩ ※∩※

CHAPTER 5

ORIGINS : EARTH HOMO SAPIENS

LANGUAGE SOCIETY AND COUNTRIES

Earth

In chapter 4 we dealt briefly with the idea of the "Big Bang", happening about 13.7 billion years ago. How, something called a singularity, a "point" that was infinitely dense and infinitely hot, "exploded" to form the beginnings of the universe. From this explosion, there was a massive release of energy and heat which spread out and expanded to form, what we call, the universe or space. At the same time, if that is the right phrase, this was also the beginning of space–time, as it is currently understood. Prior to this, so the current understanding goes, there was no space and there was no time. So any questions about what came before the singularity are regarded as having no meaning.

Not everyone accepts this notion of course, but for the time being (that word time again) this model/theory provides some explanation for many of the features and mathematical findings about the universe as we presently observe them. Along with the energy and the heat, there was also the formation of, what are known as, elementary particles. Quite a number of these have now been described and are categorised in The Standard Model of particle physics.[236] These particles have fascinating names, including quarks, electrons, photons, neutrinos, gluons and bosons. There are also anti–quarks and some of these associate with quarks and then we can get mesons and baryons and the electrons are

174

part of a group know as leptons. All these came into being during the first fractions of a second following the "Big Bang". Such tiny portions of time are known as units of Planck Time,[237] and each of which represents a 10^{-43} of a second. So within the first second of the "Big Bang" these particles came into being. Also during this time, the four forces previously mentioned (gravity, the strong and weak forces and the electromagnetic force) became distinguishable, and at this point, all these particles were moving around at vast speeds, colliding with each other. Then with the brief period of inflation, as previously mentioned in chapter 4, and the subsequent steady expansion, the universe began cooling down.

All things are relative of course and by cooling at this point, we are still talking about a trillion degrees. But this degree of cooling is sufficient for the elementary particles to start slowing down. Slowing down means their collisions don't result in annihilation of each other, rather they begin to coalesce. This results in the formation of photons and neutrons which in turn capture electrons and in this way form the first atoms – a nucleus of a proton and neutron with a circulating electron. These first atoms were principally hydrogen and helium. Thus, the building blocks of matter were born. Though now, even the definition of what may be called matter has evolved and become more complex, with some of the elementary particles themselves being regarded as matter in their own right.[238]

As the temperature of the universe continued to cool, the atoms joined up, forming molecules of hydrogen gas, and under the force of gravity, gradually coalesced to form gas clouds. These clouds expanded and could be millions of light years across.[239] Such clouds gradually collapsed inwardly on themselves and over millions of years, as they did so, their centre or core became increasingly dense and the temperature rose. This was the beginning of star formation. Eventually, a critical level was reached which precipitated a nuclear fusion reaction and changed the hydrogen into helium. This released vast amounts of energy and so radiated heat, causing the star to shine. Many stars may have formed within the same vicinity. Gravity then attracted more stars to move towards each other, thus forming the galaxies (collections of billions of stars, dust and planets). In turn, as these increased in size, their gravitational pull attracted other galaxies to join them, becoming known as superclusters.

In the core of the stars, the intense heat caused all sorts of chemical reactions, forming other elements and heavier atoms. Eventually, after tens of millions of years, the stars came to the end of their life, and exploded as supernovae. These spewed out the contents of the stars across the universe and the debris or cosmic dust reacted with other gas clouds and so seeded the development of more stars. Other debris coalesced and formed material of increasing size, such as grains, pebbles and small rocks. Some of these had their own gravitational fields which increased the attraction and so coalesced to gradually form the planets. Thus, our solar system was formed – first the sun as a star and then the planets. Under the influence of gravity, the planets gradually found their orbits around the sun. Similar processes continue today.

However, before the current planetary arrangements occurred, it is postulated that the earth was in collision with a hypothetical smaller planet called Theia. This allegedly knocked a chunk off the earth. This chunk careered into space and became our moon. The collision is said to have also caused the earth's current axial tilt of twenty three point five degrees towards the sun. It is this tilt that creates our seasons, with the earth rotating around the sun once per year, as well as rotating about its own axis every twenty four hours. So around 4.6 to 5 billion years ago our solar system came into being – the planets, including the earth, the sun, the moon and other rocky debris which make up the asteroid and Kuiper belts.[240] The oldest dateable rocks on earth being around 3.8 billion years old.

Thus, the earth formed as a rocky planet with outcrops and craters. The heavier material sank towards the centre and formed a dense molten metal inner core. This is surrounded by a cooler, less dense outer core which in its turn is surrounded by the mantle layer and outside of this is another layer forming the surface or crust of the planet. Superimposed on this is the surface water which covers about seventy percent of the earth at an average of about two to three miles deep, though at its greatest depth it is over seven miles. The water in these oceans may have been caused by condensation of water vapour from the atmosphere or by prolonged bombardment from ice containing asteroids and comets.

Whether all this has happened in quite this way or in this timescale is not precisely known. However, we do know we have an earth that is over seventy percent covered by water and that the rocky outcrops have formed land masses – islands essentially – that protrude above the level of these oceans. Some of these protrusions will be original whilst others have subsequently appeared as a result of rocks being forced to the surface because of increasing pressure developing within the hot inner core. Thus, the cratons (small continents) developed and coalesced to form the larger tracts of land that we have come to know as continents. However, their current configuration has only developed over millions of years (Ga = Giga anum – 1 billion years) as indicated by Figures 3 and 4. This movement has come to be called "continental drift." This idea of moving continents was first proposed by Abraham Ortelius in1596 CE and suggested by numerous other scientists, before the term was coined in the twentieth Century by a climatologist called Arthur Wegener. This movement occurs due to the relative movement of two further layers lying between and across the boundary between the Earth's mantle and its crust.

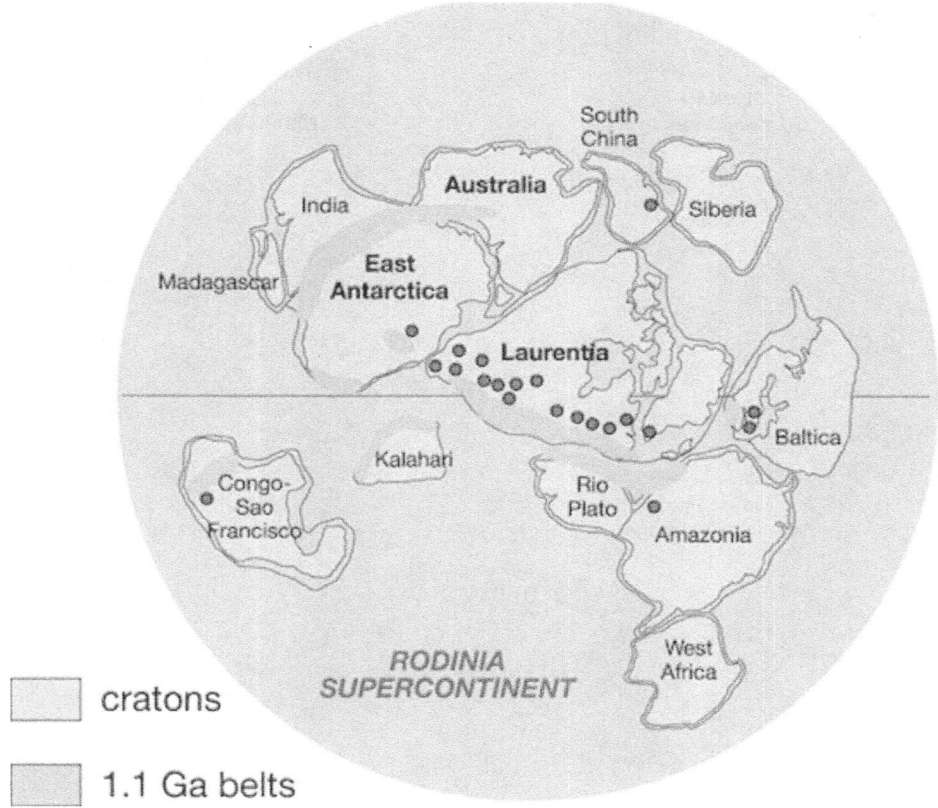

cratons

1.1 Ga belts

FIGURE 3 RODINIA SUPERCONTINENT – A COALITION OF SEVERAL CONTINENTAL LAND MASSES.

The outermost of these layers is called the lithosphere and below this is the asthenosphere. The lithosphere is separated into a number of large distinct "tectonic plates" and multiple much smaller ones. Depending on

178

the prevailing pressure and temperature between these two layers, the tectonic plates of the lithosphere can "float" on the asthenosphere.

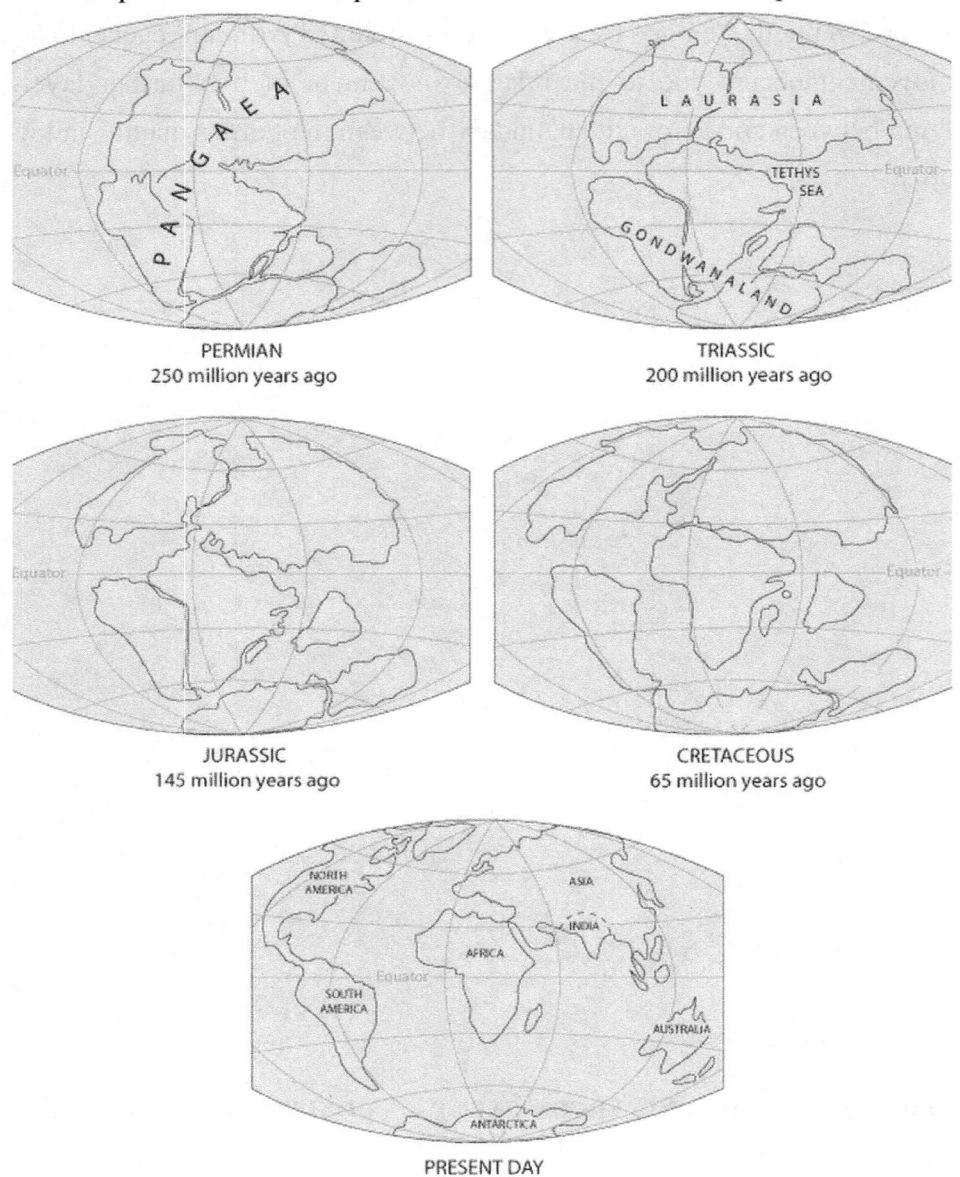

PERMIAN
250 million years ago

TRIASSIC
200 million years ago

JURASSIC
145 million years ago

CRETACEOUS
65 million years ago

PRESENT DAY

FIGURE 4 CHANGING CONFIGURATION OF THE CONTINENTS

The movements of these plates account both for the movement, relocation and separation of the landmasses and also the formation of mountain ranges, oceanic trenches, volcanic eruptions and earthquakes, as the edges of the plates move and collide with each other. The geography and position of the continents continue to change as a consequence of this movement of the tectonic plates. This gives rise to coastal erosion and deposition, through the occurrence of tsunamis, major earthquakes and volcanic eruptions. Within the layers of the earth's structure already mentioned, other layers or strata of rocks can be identified. These various layers were laid down, both at the time the earth was originally forming and also under the subsequent influence of changing temperatures, pressures, volcanic activity and changes in climate, over millions of years.

This activity has produced many different types of rock and sediment material and such layers can be identified, continuing across and between continents. The layers are dated using a set of principles known as "relative dating." For example, the layer above is younger than the layer below, material enclosed within a rock is older than the surrounding rock and so on.

More recently, over the last couple of hundred years, radioactive dating has enabled more precise figures to be attached to the timing of the categories shown in Figure 5. These layers of rock contain evidence of major changes in climate, and the fossils and artefacts they contain lend support to the theory of evolution of life forms – plants, birds and animals. They also provide estimates of the timing of these. So by these methods, geologists, palaeontologists, archeologists and other earth scientists build up a picture of how the earth and its environment and inhabitants have come to be the way they are.

Of course, this is still work in progress. We are currently experiencing changes in our climate and there are fierce arguments taking place about the role played by "global warming," secondary to carbon dioxide (CO_2)

EON	ERA	PERIOD	MILLIONS OF YEARS AGO
Phanerozoic	Cenozoic	Quaternary	--- 1.6 --
		Tertiary	--- 66 --
	Mesozoic	Cretaceous	---138 --
		Jurassic	-- 205 --
		Triassic	-- 240 --
	Paleozoic	Permian	-- 290 --
		Pennsylvanian	--- 330 --
		Mississippian	-- 360 --
		Devonian	---410 --
		Silurian	-- 435 --
		Ordovician	-- 500 --
		Cambrian	--- 570 --
Proterozoic	Late Proterozoic Middle Proterozoic Early Proterozoic		-- 2500 --
Archean	Late Archean Middle Archean Early Archean		-- 3800?-
	Pre-Archean		

FIGURE 5 DESIGNATED GEOLOGICAL TIMES IN MILLIONS OF YEARS AGO (MYA)

production and accumulation. The Gaia hypothesis, proposed by James Lovelock [241] argues that the whole of the earth, its living inhabitants and its geographical and oceanic configuration operates as a complete climate self–regularity system. However, there is great concern that human beings are now tipping this system out of kilter and many environmentalists have declared a climate emergency. Many people of faith are also concerned, in that they see this as abusing our responsibilities as stewards of God's creation.

LIFE

So it took about 9 billion years, from 13.7 billion to 4.7 billion years ago, for our solar system and for the earth, in terms of its composition, to form. Then over the next 3 billion years, earth becomes a water covered rocky planet and the continental configuration gradually takes shape as previously shown in Figures 3 and 4.

The next step is a bit of a jump – a leap of faith one might say – though the outcome is clear – life begins. The "bit" we are not sure about is how. What is the process from the formation of the earth to the start of life ? The basic requirements thought to be necessary for any form of identifiable life to arise are: liquid water, the six chemical elements – oxygen, hydrogen, carbon, nitrogen, sulphur and phosphorus plus some kind of energy source.[242] However, bacteria have been found that seem to thrive on arsenic or sulphuric acid and other chemicals thought to be toxic to life, so it may be that life may have started under conditions now considered incompatible with living organisms. We also know from the work of Watson and Crick, published in 1953 CE, that the genetic material required for life to continue and reproduce requires molecules known as DNA (deoxyribonucleic acid) and RNA (ribonucleic acid).[243] Their structures are shown in Figures 6 (A and B)

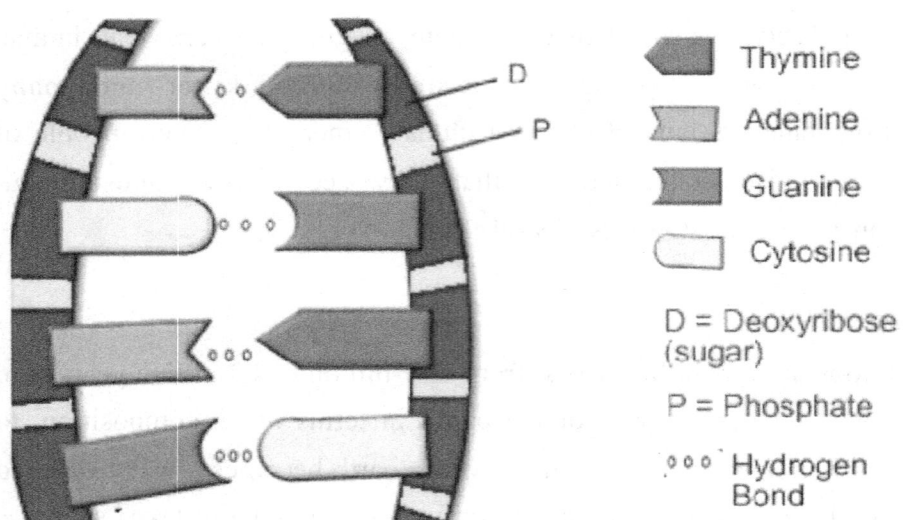

FIGURE 6 A DIAGRAMMATIC REPRESENTATIONS OF DOUBLE HELIX

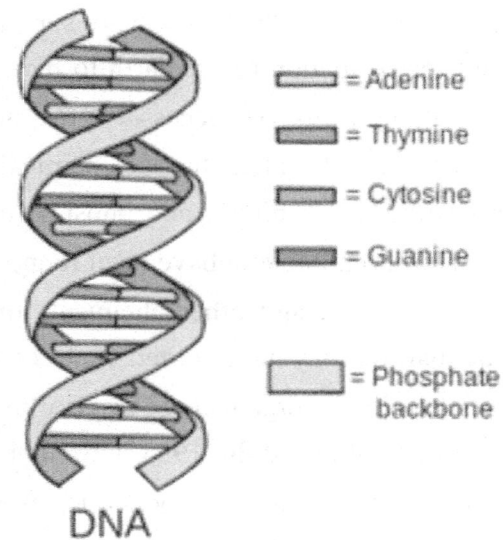

FIGURE 6B

Each gene is represented in turn by a portion or segment along the double helix, made up of a particular sequence of DNA components and each having a unique sequence and length. The DNA stores the genetic information whilst molecules of different RNAs are needed to transport (messenger m–RNA) that information, and make it operational (transfer t–RNA and ribosomal r–RNA, among others). It does this by organising amino acids to get together to make proteins. The chemical structures for these materials and their relationships to each other can be seen in Figure 7.

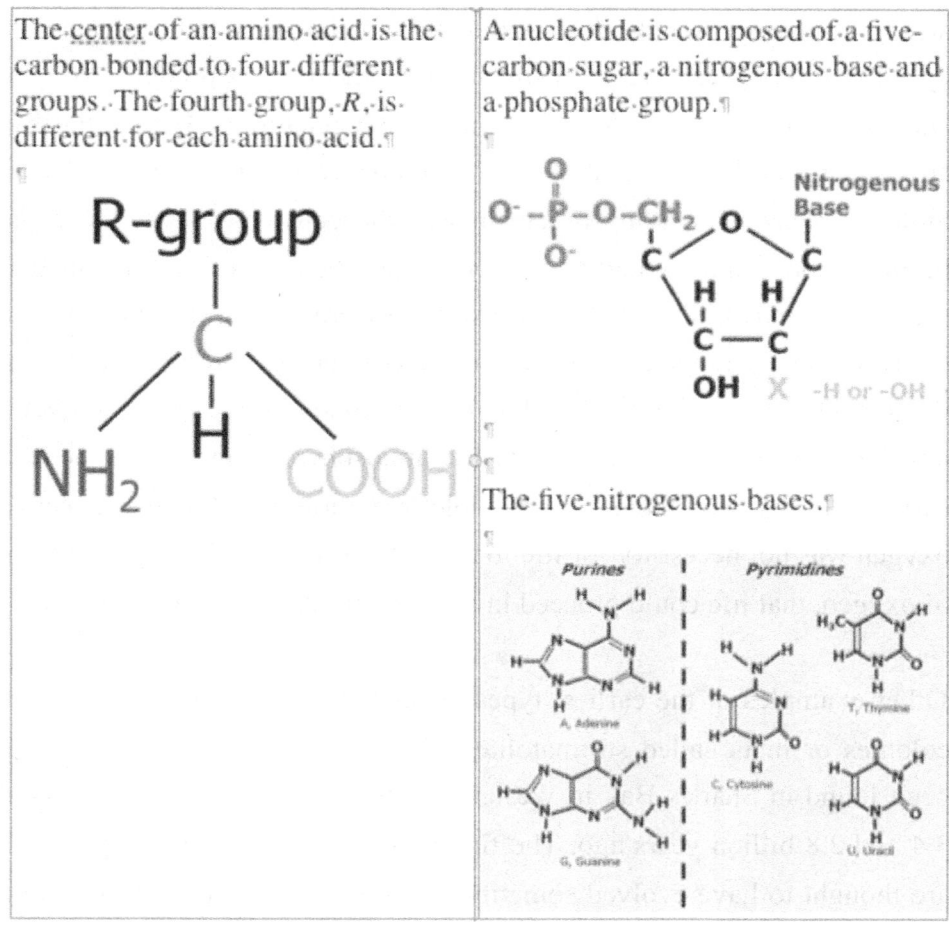

The center of an amino acid is the carbon bonded to four different groups. The fourth group, R, is different for each amino acid.

A nucleotide is composed of a five-carbon sugar, a nitrogenous base and a phosphate group.

The five nitrogenous bases.

FIGURE 7 THE CHEMICAL STRUCTURES OF NUCLEOTIDES NUCLEIC ACIDS AMINO ACIDS WITH PROTEINS BEING MADE UP OF A SEQUENCE OF AMINO ACIDS

It has been suggested that RNA was the initial repository of genetic information – The RNA World hypothesis – but was superseded, once DNA made an appearance. Each characteristic, that helps create each member of a species of plant, animal or human, is represented by a gene.

Anyhow, however this heady cocktail of chemicals came about and combined, they did, and the first unicellular (prokaryotic) forms of life arose. Amongst the first are thought to be blue–green algae – cyanobacteria. These early organisms were the first to produce oxygen and were responsible for the initial oxygenation of the Earth's atmosphere. Perhaps around 2.7 to 2.2 billion years ago. Thus, the basic building block of a living organism was the cell. Over time, the cells became more complex and began replicating themselves and organising themselves into multi– cellular (eukaryotic) organisms. Then as plants evolved later, they continued this oxygenation process, through photosynthesis, causing the build up of oxygen levels and the subsequent development of the ozone layer in the atmosphere. This then protected early lifeforms from the sun's ultraviolet radiation. So although it seems oxygen was not necessary for life to start, it was only with the production of oxygen, that life could proceed in a more abundant and varied manner.

Other examples of the earliest types of unicellular bacteria, formed into colonies or mats called stromatolites. Fossilised remains of these have been found in Shark's Bay in Western Australia, dating back to between 3.4 and 2.8 billion years ago. The first more complex cellular organisms are thought to have evolved sometime between 2.5 and 1.7 billion years ago, perhaps alongside the rise in atmospheric oxygen. The nucleus in these cells was able to hold and protect complex molecules such as RNA and DNA and the first multi–cellular life probably arose around 1.2 billion years ago.

Given this rudimentary background introduction, we can leap forward to Charles Darwin. Clearly he did not know anything of what we have just described about proteins and DNA, but he was aware of geological stratification. It seems he took with him on his voyage in HMS Beagle (1831 to 1836 CE), Charles Lyell's book – *Principles of Geology*, which had just been published in 1830 CE. In fact, much of Darwin's study on this trip related to the geology of the places visited, as much as it did to what has become much more well known – his study of the flora and fauna. This eventually contributing to his theories of natural selection and evolution. Strangely perhaps, the fossils he came across did not seem to impact very much on his theories of the origin of species.

Although Charles Darwin is credited with the theory of evolution, notions of species changing or arising from some common ancestor had been around since the pre–Socratic philosopher Anaximander. A short but useful summary of the evolution of some of these ideas relating to evolution is provided in Wikipedia.[244] Such ideas are not as clear as those of Darwin or based on the same degree of research, but do demonstrate that the creationist idea that God had created known species just as they are, from day one, was not a universally accepted one. What may have seemed like a reasonable or obvious idea, that God had created things in the form in which they were known, was questioned as far back as 500 BCE. Later, Origen, Augustine and Thomas Aquinas questioned it and proposed the idea of the description in the book of Genesis be seen as allegory rather than fact. Although it is not known whether Darwin was aware of any of this history, he would have been aware of the views of his grandfather, Erasmus Darwin, who in his book *Zoonomia*, wrote:

"that all warm-blooded animals have arisen from one living filament, which THE GREAT FIRST CAUSE endued with animality...and thus

possessing the faculty of continuing to improve by its own inherent activity, and of delivering down those improvements by generation to its posterity, world without end?."

Also around the time of Charles, others, such as Georges–Louis Leclerc, Comte de Buffon, James Burnett, Lord Monboddo, and Jean–Baptiste Lamarck were courting similar ideas. In his book – *Vestiges of the Natural History of Creation* - Robert Chambers expressed similar sentiments. and so did Alfred Russel Wallace, who also co–published with Darwin. Figure 8, shows in diagrammatic form, the relationships between species and their evolution from the beginnings of life. But Darwin not only contributed to the notion of evolution but also supplied a mechanism of operation, with his theory of natural selection.[245]

This theory describes the idea that the characteristics of an individual plant, animal or human are passed on as genes. The genes of those individuals that best adapt to their environment, continue to be passed on through reproduction. Those individuals least well adapted, will die out and take their genes with them. Darwin was very aware of the potential implications for how people might respond to his ideas, which went against the creationist view of the origin of species; that is, that God had created all creatures in their current form from the beginning, therefore this form was unchanging and unchangeable. Not withstanding the fact that some of the early Church fathers had centuries before, advocated treating the book of Genesis as allegory. He was also concerned about how people might respond to the notion that they had an ancestor in common with the great apes.

For these and other personal, health and work related reasons, it was not until 1859 CE that his book *On The Origin of Species* was published.

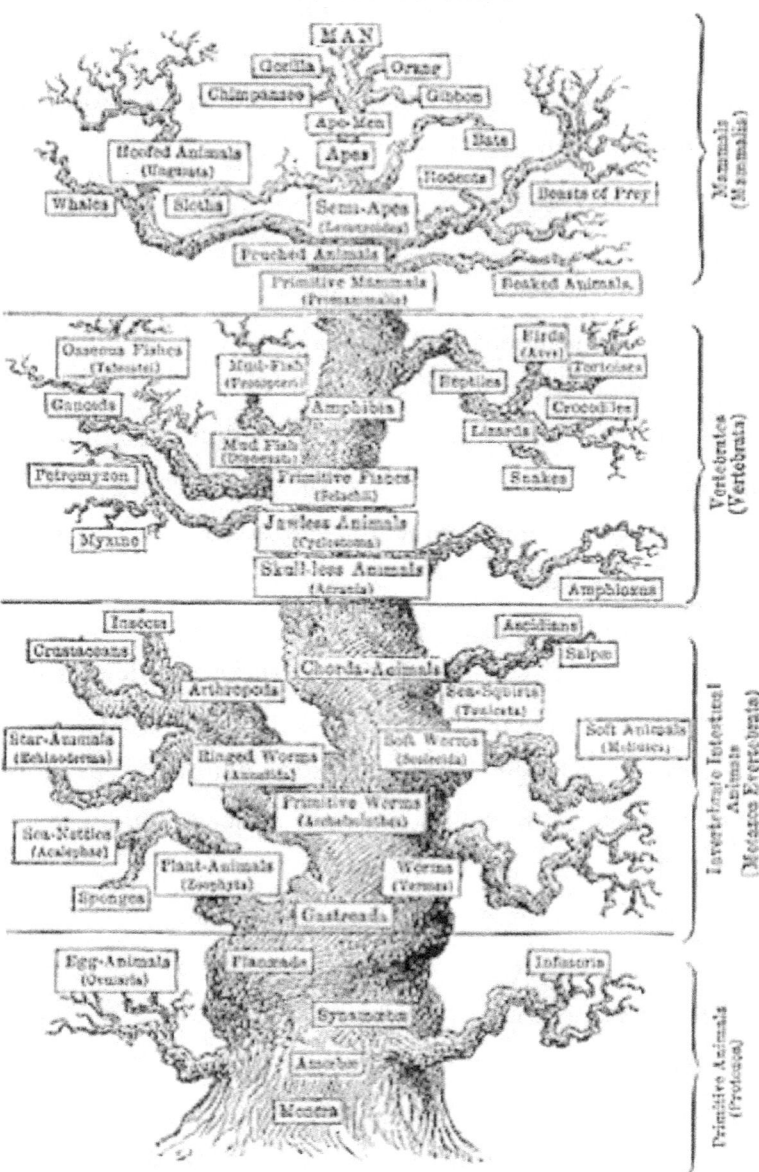

FIGURE 8 THE TREE OF LIFE

Subsequent research, the development of genetics, new fossil finds, advances in biology, natural history and scientific techniques of investigation, tended to support the general concept of an evolutionary process for species development. This mechanism being – natural selection. Although there are still many unanswered questions, for now, these ideas provide a useful working concept of how we, and the universe, have reached the point we have.

Both the interconnectedness and the diversity of evolution, under the influence of the environment and genetics, is well illustrated by the numerous different pathways of development. Resulting in a wide variety of endpoints. I say endpoints, but of course evolution continues and so the creature or plant or bird currently at the end of these branches will undergo further development as time goes on. At the top, a procession of human like characters leads nicely to the next section of the chapter.

HUMAN – EVOLUTION OF HOMO SAPIENS (modern humans)

The discovery of fossils, skulls and other skeletal fragments, in different parts of the world, has helped scientists build up a picture of the various stages along the pathway to the modern human race of Homo sapiens.

As evolution progressed, the various branches spread out with new types of living organisms. One of these branches became a pathway for the emergence of the primates. These had further branches for the lemurs, old and new world monkeys and branches for the great apes – Orangutan and Gorilla. The chimpanzee – human last common ancestor (CHLCA), or most recent common ancestor (MRCA) then branched into the Pan line of chimpanzee and bonobo and the Homo line of humans. Such progression is shown in diagrammatic form in Figure 9.

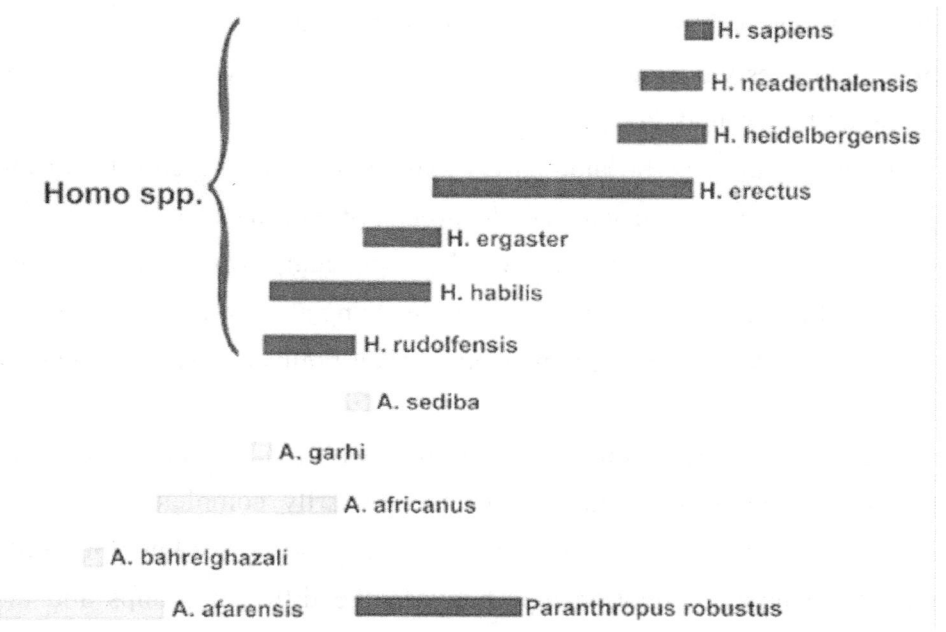

FIGURE 9 TIME LINE OF HUMAN EVOLUTION

This shows a bar chart of some of the early species of hominid that gradually evolved or died out but ultimately culminated in the development of Homo Sapiens. The estimated time line, for the species covered by the diagram, ranges from about 4mya to the time of the initial appearance of Homo Sapiens, some 200 thousand years ago. But naturally includes their continued evolution.

However, it must be remembered, that the picture portrayed in Figure 9 reflects the current state of knowledge, based mainly on fossil evidence, but also on finds of primitive stone tools and implements. If more fossil remains are found that match any of the species already identified, and can be dated earlier or later than any of the specimens currently known about, this will change the configuration of the chart and the time relationships between the different species. The characteristics of these

190

different incarnations would also have changed as they emerged from a presumed lifestyle based around living in trees to one more suited to living on the ground. For example:

- they may have became darker skinned to protect themselves from the sun particularly as they lost much of their hair.
- they began walking on two legs (bipedal), which was a more efficient way to get around, than by using all four limbs.
- Their skull shape also changed, as their brains increased in size.

The chart is not comprehensive in terms of all the human like remains that have been discovered. Nor is it necessarily completely accurate. However, it does give a basic structure to the story as we know it so far, with the suggested relationships between the different groups and the possible associated time periods. The dates for all this are somewhat uncertain but it is thought to cover somewhere between the last 5 to 10 million years. So the next part of our narrative is the story of how humans evolved and then managed to colonise the earth.

The current thinking, based on genetic evidence, recognises the possible evolution from a common female ancestor, affectionately known as "Mitochondrial Eve" and her counterpart "Y–Chromosomal Adam."[246][247] As already mentioned, all living organisms are made up of cells, the basic structure of which is the nucleus bounded by a membrane which is then surrounded by cytoplasm. The cytoplasm in turn contains smaller structures called organelles, of which the Mitochondria is one such. All of this is surrounded by a membrane structure keeping everything together. The nucleus contains the nuclear genome DNA carrying our genetic material; that is, those characteristics we inherit from our parents. These are configured in the form of twenty two pairs of chromosomes plus one pair of sex chromosomes. The DNA of the chromosomes is arranged in sections. Each section represents the information required for a particular characteristic or gene.

This information is "coded" through the arrangement and number of the nucleotides and nitrogen bases and every code is unique for each piece of genetic material. The pair of sex chromosomes are designated either as an X plus an X for a female or as an X plus a Y for a male. The mitochondria, unlike most organelles, have their own set of DNA and genes, although this is very small compared to that of the nuclear chromosomes. It codes for thirteen proteins involved in mitochondrial energy production and the production of specific proteins.[248] I include this detail by way of preparing the ground for the next bit of crucial information.

By retrospectively following mutations (naturally occurring changes) to the mitochondrial DNA and the Y chromosomes, using material extracted from the fossils of our ancestors illustrated in Figure 9, it seems that women alive today can be traced back to a single female carrying a specific mitochondrial DNA. Similarly, men can be traced back to a man carrying a specific Y chromosome. This is not to say that there was only one man and one woman – as in the biblical story of Adam and Eve, but that other lines of development have gradually died out and their DNA with them. Other male and female lines would have developed, but over time, for various reasons, these do not seem to have survived. Neither is it to say that "Y–Chromosomal Adam" and "Mitochondrial Eve" were ever personally connected, though it is thought their timelines would have been fairly close together.[249] The geographical setting for the start of these developments is thought to be the African continent. The story goes something like this:[250]

The first humans (homo) emerged around 2 – 3 million year ago from a type of ape known as Australopithecus, somewhere around the south east part of Africa. The first of these humans is known as Homo Habilis. As the groups grew in number, some members began spreading to other parts of Africa. Then in response to differing environmental conditions, several other types of human evolved, including for example, Homo Ergaster and Homo Erectus around 2 to 1.8 million years ago respectively. In this way, there would be several species of human around at the same time. Some of these would die out whilst some members of other species began to migrate, both across Africa and beyond.

The northern route, was through the Horn of Africa, Egypt, The Sinai Desert and into the Levante area (now considered to include: Cyprus, Egypt, Iraq, Israel, Jordan, Lebanon, Palestine, Syria, and Turkey). From here, there was access into Western Europe, or north, into Russia and then over to China or by turning east, gaining access into India and Asia. From Russia it was possible to cross a land bridge, (The Bering Strait, which existed somewhere between 70,000 and 60,000 years ago and then again somewhere between 30,000 and 11,000 years ago), into Alaska and down into North and then South America. The southern route was across the Red Sea at its narrowest point (Bab al Mandab Strait), into Saudi Arabia on into India, South East Asia, Indonesia and down into Australia.

Figure 10 illustrates potential routes, out of Africa, taken by some of our ancestors. In this way, it is thought that about 1.9 million years ago, Homo Erectus left Africa and gradually moved into western Europe (reaching Britain about 250,000 years ago) and travelling as far as China and into Asia. There is evidence of human habitation in south east Asia 1.7 mya, in China as early as 1.6 mya and in Europe 1.2 mya. But they not only migrated they also evolved further as a species. Fossil evidence and finds of stone implements suggest by about 800,000 years ago a new form – Homo Antecessor was found in Europe and about 600,000 years ago Homo Heidelbergensis had developed in Africa. This form, in turn, spread across East Africa as Homo Rhodesiensis and to Eurasia, where it gave rise to Neanderthals (named after the Neander Valley, just east of Düsseldorf, Germany where the first fossils were found) and Desnovians (fossils found in a cave in Siberia, on the borders of Kazakhstan).

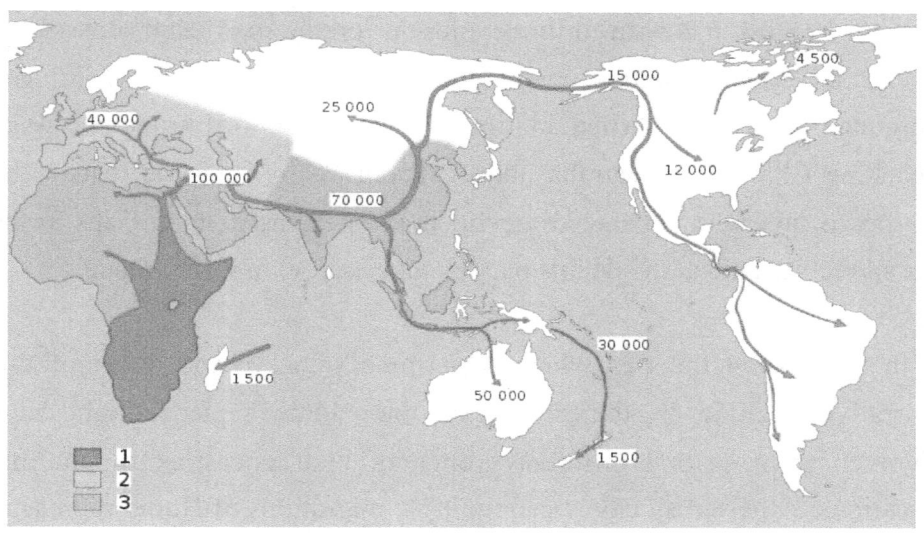

FIGURE 10 INDICATIVE MIGRATION ROUTES OF HOMO SAPIENS WITH ESTIMATED TIMINGS (THOUSANDS YEARS) 1(RED) HOMO SAPIENS, 2 (OCHRE) HOMO NEANDERTHALENSIS 3 (GREEN) HOMO ERECTUS.

Eventually Homo sapiens emerged as a species, from the Horn of Africa, about 300,000 to 200,000 years ago. It is assumed that this group was the only source of Homo sapiens and there were no other parallel evolutions. They too migrated, somewhere between 150,000 and 70,000 years ago. This migration included movement across and beyond Africa. Some crossing into Saudi Arabia, making their way to Indonesia and down into

194

Australia, around 65,000 year ago. Others entered Europe, around 50,000 to 40,000 year ago, as evidenced by fossils found in the Cro–Magnon rock shelter in southwestern France. Hence the name of Cro–Magnon for this European group of Homo sapiens. They expanded across Europe with evidence of their presence in Britain around 42,000 year ago.

So, the current generally favoured story is the development of early humans about 2–3 mya, their evolution and expansion out of Africa around 2 to 1.5 mya with further evolution "on the road" as it were. Then about 300,000 to 200,000 years, Homo sapiens evolved in Africa, and migrated both across Africa and into the continents of Europe and Asia, reaching China and Australia and eventually colonising the rest of the globe, as other human species declined and became extinct. Gaps in the fossil records, mean the details may change as new finds are made.

Other proposed theories which undermine some of the detail of this narrative, include: 1. the suggestion that Homo sapiens might have evolved from several different groups of their ancestors in multiple locations 2. there may have been multiple migrations of Homo sapiens at different times 3. the possibility of reaching Europe, having first crossed the Sahara area (which at different times has been a rich savannah type area) and then crossing the Straights of Gibraltar, even as far as Sicily.[251]

Also, the dating of these fossils and migrations is controversial. Some scientists suggest that Homo sapiens may have migrated to the Levant and Europe as early as 130,000 to 115,000 years ago and possibly even as early as 185,000. Such early migrations do not appear to have led to lasting colonisation and may have receded by about 80,000 years ago. Others suggest the possibility that this first wave of expansion may have reached China as early as 125,000 years ago. Yet other scientists argue that modern humans left Africa at least 125,000 years ago, using two different routes: the northern route as early as 120,000–100,000 years ago and the southern route as early as 125,000 years ago.

However, the point is, that although the time–lines of this narrative are only estimates and the relationships between the different human species are speculative, we are here! Homo sapiens has colonised the world, and our characteristics have evolved to help us adapt to living in a wide variety of environmental conditions. So we evolve as a species, alongside the evolution of our understanding of that evolution !

EVOLUTION OF HUMAN SOCIETY

Various reasons are proposed for why the human species migrated. They include such factors as: becoming bi–pedal and so finding it easier and more efficient to travel long distances, physical changes that facilitated movement and the ability to carry things, pressure on food supplies, climate change and curiosity.[252] Why Homo sapiens came to be the dominant human species is not clear, but suggestions include: an increasing brain capacity,[253][254][255][256] the development of more complex language and hence improved communication, greater use and creation of tools and weapons, more sophisticated social organisation [257][258][259] and the discovery and the use of fire. Evidence of this has been found in the Wonderwerk Cave in South Africa, from about one mya.[260]

Initially, the first humans would have been foragers and scavengers before developing their hunting skills. Such skills then allowed them to broaden their food sources and develop their lifestyle as hunter gatherers.[261] Life as a hunter – gatherer is thought to have been a somewhat precarious nomadic type of existence, with any settlements being of only a temporary nature. More permanent settlements have been linked to the introduction of horticulture (cultivation of small gardens with a variety of plants) and agriculture (cultivation of larger areas perhaps with a single crop) which clearly required a commitment to a more settled existence. However, where hunter gatherers had a more secure food source such as fish from rivers or the sea, it would make sense for them to develop a small settlement there. Thus, it is now acknowledged, that the development of agriculture and formation of settlements are independent drivers and are not always automatically linked. But clearly, the growing of crops needed the community to be more committed to staying put for longer periods than perhaps the quest for food by hunting and foraging required. Various theories place the introduction of agriculture between 14,000 and 5,000 BCE.

Intertwined with these transitions, there is evidence of the domestication and herding of animals and the development of an efficient nomadic and herding culture.[262] [263] This really begins with sheep and goats around 8000 BCE, chickens 6000 BCE, camels and horses around 4000 BCE and the Bactrian camel (two humps) around 2500 BCE. The dog is thought to be the first animal to be domesticated around 20,000 years ago. Just out of interest, the wheel is first seen around 3400 BCE.

How the sites for the first settlements were selected is not always clear. Factors such as fertile ground, access to water, proximity to trade routes, access to river transport, availability of food and strategic defence would all become important considerations, once the development of larger settlements became a feature. The establishment of such settlements should not however be seen as a linear transition or as progress from a primitive nomadic existence to a more sophisticated settled civilised existence. Some communities fluctuated between a nomadic and a settled existence, depending on circumstances.[264]

What particular lifestyle was adopted was more related to the development of a mixed food economy and adaptation to a specific environment, than to a sense of increasing civilisation. But once communities were established, this allowed tasks to be more widely shared and also encouraged a more efficient use of time and resources. Settlements then grew and a more diverse society gradually evolved. As mentioned in chapter two, some people might then be freed from the day to day business of survival and food production and given time to think and be creative and concentrate on other things. Thus, more intensive food production allowed communities to grow and freed others to trade and thereby increase the communities wealth. They were also able to sustain a leadership elite and standing armies. With community growth, came the need for more societal structures to keep a settled and ordered society for leaders, laws, social norms and rituals to gradually emerge.

One well preserved example of a set of laws is the Code of Hammurabi.[265] Hammurabis was the sixth Babylonian king, and produced his code around about 1754 BCE. It consists of 282 laws, with scaled punishments, adjusting "an eye for an eye, a tooth for a tooth" as graded depending on social status, of slave versus free, man or woman. The code deals with matters of contract, like establishing the level of wages to be paid to an ox driver or a surgeon or determining liability in cases where a house collapses, or property that is damaged while left in the care of another. A third of the code addresses issues concerning household and family relationships such as inheritance, divorce, paternity and reproductive behaviour. Earlier collections of laws include the Code of Ur–Nammu, King of Ur (around 2050 BCE), the Laws of Eshnunna (around 1930 BCE) and the codex of Lipit–Ishtar of Isin (around 1870 BCE), while later ones include the Hittite laws, the Assyrian laws, and the Mosaic (from Moses) Law. These codes come from similar cultures in a relatively small geographical area, and they have passages which resemble each other.

A description of the processes, pressures and drivers leading to the agricultural revolution and its subsequent impact on the development of settlements, cities and empires is described in some detail in *"Sapiens"* by Yuval Noah Harari.[266] He presents a narrative of how myths and fictions and imagined concepts became established, to hold communities together and help cement stable, societal structures. Thus, the majority of the population bought into a common understanding of the order of things that both sustained societal organisation and supported cooperative action to defend it. This common understanding was facilitated by the invention of money, imperial rule and religious belief.

There is much room to debate the actual timing of these developments, but the earliest five major settlements and agricultural sites are thought to be: 1. within the fertile crescent in the Levante and Mesopotamia regions, around 11,000 BCE – growing barley, wheat, and hemp, 2. India, in the Indus valley around 9,000 BCE – growing rice, barley, beans and other crops, 3. Egypt around 8,000 BCE – growing domesticated cotton, wheat and barley, 4. China around 9,000 BCE) – growing rice, wheat and lentils 5. Meso–America (Mexico, Belize, Guatemala, El Salvador, Honduras, Nicaragua, and northern Costa Rica) around 7,000 BCE – growing acorns, mesquite, prickly pear and beans. It is thought all these centres developed independently.[267]

EVOLUTION OF LANGUAGE AND WRITING

As we have said, one of the factors influencing the successful evolution of Homo sapiens is the development of language. Now, identifying the origins of language is tricky and a number of conflicting theories have been proposed. Note that we are considering here, a properly formed and constructed system of language not just the ability to communicate through sounds or gestures. It is generally accepted that the emergence of the first proper language must have happened after humans had split from the Latest Common Ancestor, otherwise chimpanzees would be able to speak.[268] This puts the earliest date at around 2mya. Also, it must have happened before the first Homo sapiens left Africa at around 70,000 ya. So it is possible that the early ancestors such as Homo Erectus may have had language, though it is generally thought more probable that language arose within Homo sapiens i.e., between 200,000 and 100,000 ya. This language is now regarded as the base from which all the other 6000 plus languages we know today, and those that have been lost or superseded, have evolved.

Several theories have been put forward to both explain why language may have started [269] [270] and what was its purpose. However, dating languages is an imprecise science and it is only possible to go back between 5 and 10,000 years. Many linguists (Linguistics, is the study of language(s), looking at how language changes over time and describing the differences between languages) see little point in spending a lot of time on how, when and where language began. They are more interested in how languages have evolved and changed, as they do quite rapidly. As mentioned in chapter 2, we will have experienced in our own lifetimes words that have changed their meanings. Thus, the changing nature of language has a bearing on how ancient documents are interpreted. This applies no less to the way we need to approach the understanding of scripture.

Although it seems self evident that language is all about communicating with other people, the paradox is that having started with a common language it seems counterproductive for so many new languages to emerge, some, within only a few miles of each other. One suggested reason relates to security. The thinking behind this idea is that it was more important to communicate with people in the same group than with strangers. It was more important to be able to identify strangers and one way of doing this was to have one's own language or dialect with words or pronunciations that strangers would not be familiar with. Thus, outsiders would be more easily identifiable and so levels of awareness regarding possible hostile intentions would be aroused and being forewarned was to be forearmed. The spread of languages is thought to have been facilitated by the movement of peoples, as a consequence both of agricultural expansion, trade and war.

But with expansion of settlements and increased food production and the establishment of trade and exchange of goods and services, there was a need to keep some sort of track of the transactions. Memory was no longer adequate. So some form of recording system was required. The earliest known written down recording system is by the Sumerians in Mesopotamia in about 3100 BCE. This is called a "partial script" in that it was purely a way of recording transactions of trade – numbers and names. However, by 2,500BCE, we see its transformation into what is now called – cuneiform. This is a complete script and so enables the representation of the whole range of spoken language to be expressed in written form.

Around the same time, the Egyptians had also created their own form of writing – hieroglyphics. From about 1200 BCE there is evidence of writing in China and from America round about 1000 – 500 BCE. With increasing contact between peoples through trade and war, the concept of writing expanded and the role of interpreters and translators became a sought after skill. Thus, the foundations of inter–community communications were laid. The Pyramid Texts from Ancient Egypt are one of the oldest known religious writings, dating from between 2400 to 2300 BCE.[271]

EVOLUTION OF NATIONAL BOUNDARIES

As settlements became the norm, the concept of boundaries emerged. Initially, these would just be understood and accepted as areas the group generally stuck to rather than in any sense of being owned or specifically staked out. At first, they would be defined by the outer limits of the settlement structures comprising just a few family dwellings. And there would be some "gaps" or "no–mans land" between one settlement and the next. But as the settlement groupings/societies became more complex and organised, they would begin to expand beyond these immediate boundaries. So these tracts of "no–mans land" would gradually become narrower until one settlement would take over or join with the next. Such expansion might be in response to pressures on the supply of food or fresh water or a perceived need to neutralise potential threats from other settlements.[272 273 274]

These pressures would encourage expeditions across the settlement boundaries and in so doing, bring them into contact with other groups. In this way, settlements grew larger – a few dwellings became a village, villages became towns and towns became cities, and so areas covered by common systems of governance increased. Some cities turned into city states, a collection of cities and towns became kingdoms and eventually, through conquest, empires were created. The boundaries of these territories may have been initially ill defined or based on tacit agreement between neighbouring groups. But more often, would follow naturally occurring geographical features, such as – rivers, mountains, valleys, islands and oceans. Over time, the definition of such borders changed, mirroring the changing fortunes of the populations under consideration.

The historical evidence for the delineation of these early boundaries is not often well documented, but by using multiple sources, scholars can derive plausible approximations. As we get nearer to the present day, the historical evidence becomes stronger with more precise accounts and the availability of more accurate mapping. But even today, where we have very accurate maps for much of the Earth's surface and well marked "lines" that define the limits of identified countries and the borders between them, such "lines" may still be contested. It is perhaps natural today for us to assume, without really thinking about it, that countries have always been countries and pretty much configured the way they are today. The land masses, that is the continents and islands, continually change due to the effect of earthquakes, volcanic eruptions and by erosion or land deposition around their coast lines. But to all intents and purposes, their overall topography has been much the same for the last few hundred thousand years.

So countries like Australia, New Zealand, Japan and Indonesia and other island nations like Great Britain, Ireland, and Sri Lanka in recent times, haven't much changed in their overall configuration. Though it is likely that we may see significant change in the near future with a number of the South Pacific Islands, and places like Biafra being at risk from flooding and even extinction, as a result of global warming. But thinking about continental countries, we realise that their current configurations result from long, complex and often violent and bloody activities by humans.

Starting from just a few hunter–gatherer Homo sapiens migrating out of Africa we now have a whole patchwork of communities spread across the world. How this came about is the story of mankind. How a few settlements, dotted over an increasingly wide area of the globe, morphed into the complex arrangement and demarcation of societies we have today, I think is quite remarkable. Although the precise details, certainly from the beginning, are uncertain, historians and archaeologists have been able to piece together evidence from many different sources to provide an insight in to the evolving processes by which this came about.

For those interested in the ebb and flow and evolution of geographical demarcations, a fascinating online tool (the basic tool is available for public use) produced as part of the GeaCron Project,[275] shows the changing nature of these boundaries from 3000BCE to the present day. This project uses information from many sources to indicate the original boundaries of the world's early civilisations and their changing geography, as empires and conquered territories came and went over time.

By way of an example, using the GeaCron tool, we can follow the changing occupation around a particular region, such as the area known as Mesopotamia. This is situated at the head of the Persian Gulf, lying between the rivers Euphrates and Tigris and incorporating what now corresponds to most of Iraq, Kuwait, parts of Northern Saudi Arabia, the eastern parts of Syria, southeastern Turkey, and regions along the Turkish–Syrian and Iran–Iraq borders.[276] Although we will focus our attention on this geographical region, I will make passing reference to parallel conquests of other areas going on at the same time.

Though before we do this, it is worth making reference to the work of James DeMeo. His PhD Thesis entitled Sarahasia,[277] discusses one of the drivers of the migration and warring pattern of take overs of one group of people over another. (see Ego Explosion in chapter 8) He notes the occurrence of an environmental disaster across North Africa, the Near East, and Central Asia he calls Saharasia. This turned a fertile hospitable region into a desert, resulting in the need for the inhabitants to move house. However, this incident is said not only to have required the peoples to move in order to survive, but also led to a loss of their sense of being an egalitarian community.

A much more pronounced sense of self identity (I/me – us/them) developed, leading to competition and a preparedness to defend what was theirs rather than sharing what they had and an altogether more combative attitude to others. Thus, a relatively peaceful matriarchal society (one formed around characteristics associated with females) was transformed into a war–like patriarchal one (a society formed around characteristics associated more with males). These peoples, particularly the Indo–Europeans and Semites were war–like as well as theistic, and over the following centuries, conquered large parts of the world.

The Indo–Europeans eventually occupied the whole of Europe, parts of the Middle East and India, while the Semites conquered most of the Middle East. Over time they split into different groups. The Indo–Europeans sub–divided into peoples like the Celts, the Greeks, the Romans and the ancient Hindus, with Europe sequentially invaded after c.4000 BCE by Battle–Axe peoples, Kurgans, Scythians, Sarmatians, Huns, Arabs, Mongols, and Turks. An indicative chronological time line for this occupation of Mesopotamia is shown on the following page in Figure 11.[278]

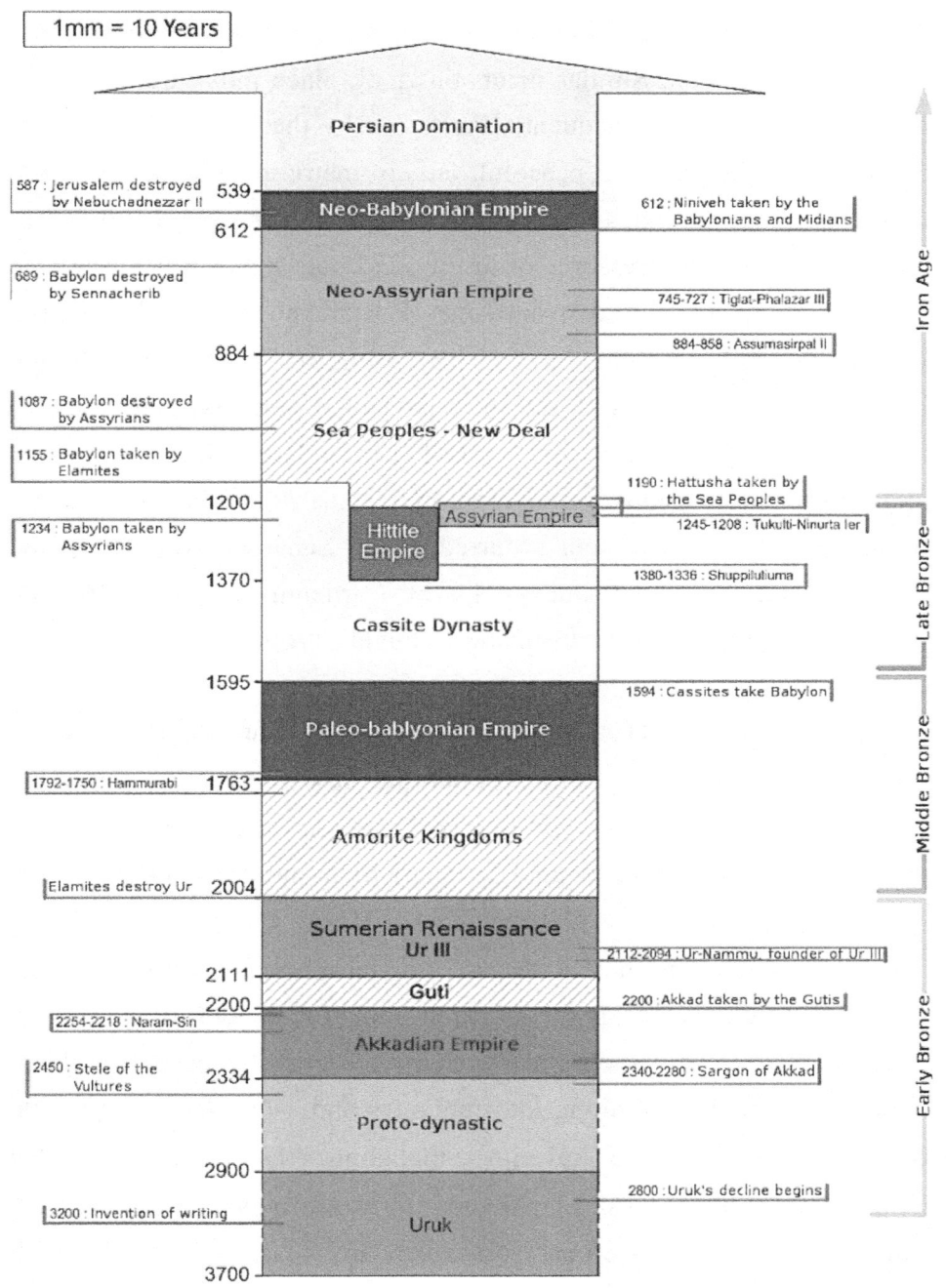

FIGURE 11 CHRONOLOGICAL TIME LINE FOR OCCUPATION OF MESOPOTAMIA

The Semites sub–divided into peoples like the Hebrews, the Philistines, the Arabs and so on. Similar incursions took place into parts of China, around 2000 BCE with sequential invasions by the Huns, Mongols and others. In South Asia, the peaceful, largely matriarchal settlements and trading states of the Indus River valley collapsed after around 1800 BCE, under the combined pressures of aridity and patriarchal warrior–nomad invaders from arid Central Asian lands. Patriarchalism spread thereafter into India, and consolidated later by Hunnish, Arab, and Mongol invasions, also from Central Asia. Matriarchalism similarly predominated in Southeast Asia until the learned patriarchal behaviour fuelled migrations and invasions from China, India, Africa, and Islamic regions. Similar patterns of behaviour occurred In sub–Saharan Africa, under the sequential influence of Pharaonic Egypt, Carthaginian, Greek, Roman, Byzantine, Bantu, Arab, Turkish, and Colonial Europeans.

Thus, as we use the GeaCron tool to chart the changing fortunes and geographical influences of different cultures, it is worth bearing in mind that the drivers and human war–like characteristics may have arisen as a result of the major environmental disaster a thousand years earlier.

The GeaCron graphic starts with the Sumerians, thought to be one of the earliest civilisations, occupying this area from around 3000 BCE. Though there is evidence of some human activity in the area even before them. Expanding northwards, along the Euphrates and Tigris Rivers, by 2300 BCE they established several cities, including – Ur, Kish, Umma and Uruk. Then, a new group, the Akkads, formed around the city of Kish and expanded to occupy Sumer, somewhere around 2270 BCE. They also extended their occupation northward, as far as the eastern Mediterranean coast, but their area of control contracted, as they in turn, were overrun by the Gutians in around 2200 BCE.

In parallel with these developments, the Minoan culture was establishing itself in Crete whilst the Egyptians were settling and expanding along the Nile Valley and by 2600 BCE, a civilisation was also evolving in the Indus Valley. From around 1600 BCE until about 1100 BCE, the Mycaenian Greeks came into being. It is a source of controversy as to why their civilisation collapsed, but following them was a period known as the "dark ages," before the star of the Ancient Greeks began to rise around 800 BCE.

So a pattern of changing occupation was established and repeated globally over the next 4000 years ! This process of expansion and dominance of a particular ethnic group, followed by its own decline and or disappearance as it is in turn dominated by another group, becomes the continuing norm as to how the geo–politics of any particular area of the world is established and evolves. In Peter Frankopan's book *The Silk Roads : a New History of the World* [279] there is an account of the changing face of global conquest as new cultures establish themselves and dominate their adjacent neighbours. They capture territory by a combination of terror, fear, violence and enlightened people management, before themselves being overthrown by the next dominating personality. Empires came about by deliberate conquest, rather than just expansion of the settlements and through trade.

It seems that little has been learnt as such activity and threats of such activity continue. Though it may be that some organisations of international collaboration – such as the European Union, NATO (North Atlantic Treaty Organisation), the UN (United Nations) or the deterrent effect of weapons of mass destruction such as nuclear bombs (though this strategy may have outlived its usefulness as more countries try to develop the capability for making their own), may exert some constraining effects.

Acceptance and toleration of other peoples' understanding of theology, religious faith or other philosophical belief systems, may also contribute positively to a reduction in tension and to calmer and more conciliatory relationships. Perhaps this book may contribute in some small way. But, getting back to the Mesopotamian region. The Gutian take over was followed by the rise of the city of Ur which in turn was usurped by the Amorites and who, in their turn were overthrown by the Babylonians. This ebb and flow is shown in Figure 11 which indicates a sequence of take overs by different cultures until the second period of Babylonian rule, around 612 BCE. By about 1030 BCE, an area known as Israel was established, on the East coast of the Mediterranean.

However, in 930 BCE, the inhabitants split into what was called the northern Kingdom of Israel and the southern house of Judah, centred around Jerusalem. This area was known as Judea. The northern territory was taken over by the Assyrians in about 720 BCE and the Babylonians sacked Jerusalem and the house of Judah in 586 BCE and took them into captivity in Babylon. This is situated just South of modern day Baghdad, about in the middle of Mesopotamia, on the East bank of the river Euphrates. When Babylon fell to the Persian Cyrus the Great, in 539 BCE, Judea became an administrative division within the Persian empire. Then Darius the Great seized the throne in about 521 BCE, and he allowed the Jews to return to Judea and rebuild Jerusalem in about 515 BCE. The sequential march of these changing areas of occupation can be followed on the GeaCron tool. This shows the retreat of the Persian empire as the star of Alexander the Great is on the rise.

By 334 BCE Alexander the Great was beginning his conquests and by 330 BCE had overrun the Persian Empire. He extended his own influence into Egypt and into Asia Minor through Afghanistan and as far as what we now call Tajikistan. When he died in 323 BCE, his followers – the Diadochi – carried on his legacy. These generals divided the empire between them but by 300 BCE they had been usurped by the Seleucids who took over all the land Alexander had previously occupied. At the same time, from about 320 BCE, Rome was becoming established as a thriving city and beginning to extend its sphere of influence. By 143 BCE the power and area of control of the Seleucids was on the wain as the Parthians came to prominence and were dominating the area by 93 BCE. They did so, apart from a brief incursion by the Armenians, until about 224 CE when they in turn were overrun by the Sassanids. The Sassanids remained in charge until 651 CE, holding the area from southern Syria into Asia Minor and up to the Caspian Sea.

Whilst all this activity was going on in Mesopotamia, there was at the same time another little side show taking place – the birth and evolution of the Roman Empire. It began to take shape from about 300 BCE and endured for about 800 years. At its height in 117 CE, it stretched from England, across Europe, included the North coast of Africa and down as far as the Persian Gulf. This incursion down to the Persian Gulf was only temporary, as the Parthians regrouped and pushed them back northwards to the northern part of Syria. From then on, the Parthians, and later the Sassanids, generally confined the Romans' southern border to this part of Syria, though in 299 CE, Diocletian campaigned successfully against the Sassanids and sacked their capital, Ctesiphon (situated about in the middle of Mesopotamia, on the East bank of the River Tigris).

One can imagine, that with the Roman Empire being so vast, managing it without telecommunications, the internet and air travel would be a major headache. And so it was. There were continual incursions at different parts of its borders by warriors of the various local cultures wanting a piece of the action. So in 285 CE Emperor Diocletian realised that a single emperor was inadequate in the face of internal pressures and military threats. So he split the empire in two, along a northwest axis just east of Italy, effectively creating what would eventually become two distinctive regimes – the Western and Eastern Roman Empires. He also realised that one centre of power was inadequate to maintain overall control of all the occupied territories and so by 293 CE he had sub–divided the administration into four parts, centred on Nicomedia (modern Izmit in Turkey), Mediolanum (modern Milan in northern Italy), Sirmium (modern Sremska Mitrovica in Serbia), and Augusta Treverorum (modern Trier in Germany) which were nearer to the frontlines, than Rome. This situation became known as the Tetrarchy, each with its own autonomous ruler.

This system continued for a number of years, though somewhat uneasily, with tensions between the four rulers. Over time, the ruling personnel changed through natural and unnatural deaths, until Constantine the 1st defeated Maxentius at the battle of Milvian Bridge [The legend surrounding the victory of Constantine I in the Battle of the Milvian Bridge (312 CE) relates his vision to seeing the Chi Rho (see reference) [280] in the sky, and so he reproduced this symbol on the shields of his troops.]This left himself and Licinius as sole rulers. By 324 CE Constantine finally defeated Licinius, and so reunited the two halves of the Empire and declared himself sole Augustus or Emperor. He moved his headquarters from Nicomedia to Byzantium which sometime between 330 CE and his death in 337 CE became known as Constantinople (now known as Istanbul). Constantine legalised Christianity definitively in 313 CE and in 325 CE convened and presided over the First Christian Ecumenical Council of Nicaea. The future of the Christian Church became intimately tied up with the fortunes of the Roman Empire, particularly the Western empire and we will touch on that in the next chapter.

The Empire was then intermittently separated and united until Theodosius became the last Emperor to rule over a united Empire. Following his death in 395 CE the empire could again be clearly distinguished as two separate parts. The Eastern Empire based on Constantinople and the Western Empire based in Milan and then in Ravenna (North East Italy). But, from around this period, the fate of the Roman Empire followed the established pattern – just as they had overrun vast expanses of land and subjugated numerous different ethnic groups, so the Western Roman Empire itself suffered the same fate.

There had been continual incursions across its borders from various groups, including Britons, Saxons and Jutes to the West and Burgundians, the Alamanni, Visigoths, Vandals, Ostrogoths, Huns, Bulgars and Franks from the North and East. The fortunes of these peoples waxed and waned over time and their areas of influence changed. In 410 CE Visigoths attacked and sacked Rome, and then in 455 CE Vandals did the honours and sacked it again. These were followed by the Ostrogoths, who by 476 CE had taken over the remains of the Western Roman Empire and ruled over what is now known as Italy and the western Baltic area. Thus, 476 CE came to be regarded by some historians, as the formal end of the Western Roman Empire.

What had emerged during this period of 150 years or so, was a Europe that had become dominated by three great powers. Firstly, the Franks (Merovingian dynasty) covering much of present day France and Germany during 481 to 843 CE; secondly, the Visigothic kingdom between 418 to 711 CE, covering modern day Spain and thirdly, the Ostrogothic kingdom between 493 to 53 CE covering modern day Italy and parts of the western Balkans. The Ostrogoths, in their turn, were later replaced by the Kingdom of the Lombards 568 to 774 CE.

This left the Eastern Roman Empire, in about 568 CE, mainly covering an area approximating to modern day Greece, Turkey, Syria and including northern parts of Egypt. This was designated as the Byzantine Empire by a historian in about 1555 CE. These boundaries expanded and contracted again over the next 200 years, by which time the Franks had also expanded their domain across Europe beyond France and Germany to Italy, by replacing the Lombards. In 800 CE, Pope Leo III crowned the Frankish King, Charles I, Emperor. This meant there were then two Roman Emperors – one for the East and now one for the West. The territory occupied by the subsequent Emperors, eventually became designated as The Holy Empire under Frederick I Barbarossa in 1157 CE and as the The Holy Roman Empire from about 1254 CE onward. In 1806 CE, Napoleon over ran this territory and the Emperor, Francis II abdicated. Napoleon then re–configured the borders of Germany, Prussia and Austria.

Meanwhile, back in Mesopotamia, from about 632 CE the Muslim faith had emerged through the prophet Mohammad and the Islamic forces were on the march. By 661 CE. they had overrun all the territory covered by the Sassanid Empire and so came into contact with the border of the Eastern Roman Empire. The Muslim forces, by 752 CE had spread west to Spain via Egypt and North Africa and east to Afghanistan and northern Kashmir. Then, following the conquest of Persia, Islam penetrated into the Caucasus region, parts of which later became part of Russia. Muslim forces were able to take further European territory, including Cyprus, Malta, Crete, and Sicily and parts of southern Italy. Their advance in the west, into the Iberia Peninsula (modern Spain) was halted by Charles Martel, the first of the Carolingian rulers, at the battle of Tours/Poitiers in 732 CE with a combination of Frankish and Burgundian forces. Though this was little more than a Muslim raiding party as opposed to an all–out invasion.

The Islamic territories were led by changing Muslim dynasties. Following the death of Mohammad, the Rashidun – "The Rightly–Guided Caliphs" – that is ; "those on the right path", provided the first four caliphs, before the Umayyad dynasty took over. These were replaced by the Abbasids, who also took over some of the southern part of Mesopotamia. Other Arab groups had a foot in this area until the current configuration of Arab countries was established during the twentieth Century. This whole region, including the Eastern part of Mesopotamia ,saw a number of changes in rule over the centuries, by different ethnic and Muslim groups. These included, the Seljuks in 1048 CE , the Mongols in the 13th Century. and the Timurids, descendants of the Mongol Amir Timur or Tamerlane, between 1370 and 1507 CE. The Timurids also established the Mughal Empire in Central Asia and Indian subcontinent between 1526 and 1857 CE. The Ottomans moved into Mesopotamia, in1535 CE.

Although, the preceding description is largely centred on the area of Mesopotamia, it is provided as an example of the similar processes that were taking place in China, the far east, and in North and South America.

Country boundaries are still evolving. As countries are invaded, their boundaries are also modified. We can see the changes and re–configuration of boundaries in Europe within recent history in the 1990s with the disappearance of Yugoslavia to be replaced by its previous constituent parts of Bosnia and Herzegovina, Croatia, Macedonia, Montenegro, Serbia, Slovenia, and Kosovo.

SUMMARY

This chapter summarises the evolution of humankind and our understanding of how, as a species, Homo sapiens has emerged and dominated the planet. We have looked at not only the physical and intellectual changes to our species, but also to the physical, geographical and cultural environment we inhabit. Our understanding of these processes is still incomplete and continues to change, as different theories are proposed to explain how this has all come about, and how the socio–political and economic drivers and differing interpretations by historians and so on have all played a part in contributing to the changes, and themselves influenced our understanding of how it is. This is in keeping with the general tenet of the book, in that evolution and our understanding of it, covers the whole of humanity, the world and the universe. So would we expect our religious understanding to be exempt ? This is the subject of the next chapter.

CHAPTER 6

DEFINITION, ORIGINS

AND CAUSES OF RELIGION I

The preceding chapters have looked at the evolutionary nature of science, philosophy, cosmology and human development, along with a discussion about the nature of knowledge itself. This is all by way of preparing the groundwork to support the idea that a similar process of evolution is therefore likely to apply to our ideas of religion, faith and God. So having got to this point in the book I thought, naively perhaps, that this chapter would be fairly uncomplicated. That I would find information about early religions and get a sense as to how they might be the forerunners of the major faiths, as we now designate them. That there would be some common stories and ideas between them and that it would be possible to identify the origins of the major faiths and follow how they have developed over the centuries. However, as we shall see, things are not going to be quite that straightforward. We have at least three major problems. The first is with the definition of religion, the second is with understanding and identifying the origins of religion and the third is with the causation of religion; that is, what it was in the first place that lead to the initiation and implementation of practices we now designate as religious.

None of this is to say that using the term "religion" is not useful. After all, we are where we are and the term "religion" is used the world over.

But to sound a note of caution. That whenever and wherever the word religion is mentioned, there is a need to understand the meaning and context in which it is being used. Whether it is a description of some cultural activity, whether it is in reference to the perceived impact or effect of "religion' on society or whether it relates to inter–cultural comparisons. This is all in the name of trying to ensure the basis of any discussion is factual and understood. Nevertheless, it does underline the fact that religion cannot be treated as a single, uniform concept.

But before we start our look back into history, just a little bit of contemporary context. Firstly, new religions continue to be formed, with Wikipedia listing nearly 300 new religions, formed since 1800 CE and most during the last one hundred years.[281] Secondly, new perspectives are developing such as Lived Religion, championed by David Hall[282] and Robert Orsi.[283] They have described feminist viewpoints, cultural materialism and secularisation. Lived religion describes a holistic framework for understanding the beliefs, practices, and everyday life of religious and spiritual persons, within a community. Encouraging the people themselves to narrate and interpret their own experiences.

Hall and Orsi argue that the study of Lived Religion shapes and is shaped by the way family life is organised. That religious practices and understandings, only have meaning in relation to the life experiences and actual circumstance of the people using them. This is not a million miles from the approach advocated by Evans–Pritchard, who we will meet later. However, it is in contrast, they say to Popular Religion,[284] which is only concerned with the practices of "common folk" and distinguishes their practices as separate from the "official religion" of the elite in society.

Thirdly, there is also more interest recently in the geographical location and spread of religion and how it is disseminated. Park gives an account of this, focusing on the distribution of religion and the role of sacred spaces, in his chapter in Hinnel's Companion to the Study of Religion.[285] Finally, despite the decline in Christian religious affiliation in western Europe, worldwide those aligning themselves with a religion seem to be increasing.[286] Based on data from 2020, Christianity has the greatest number (2.3 billion – twenty nine percent), Islam (1.9 billion – twenty four), Atheist, agnostic, secular (1.1billion – fourteen percent) Hindu 1.1 billion – fourteen percent) followed by Buddhism, folk religions and others.[287] A broad reflection of the degree of religious adherence, world wide, is shown in Chart 1. The more intense the colour, the greater the religiosity. The grey areas are unknown.

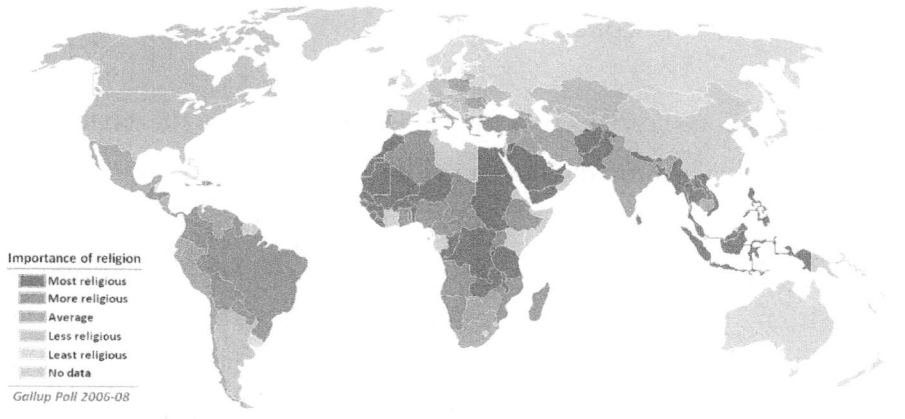

CHART 1 WORLD MAP OF THE DEGREE OF RELIGIOUS INTENSITY ACROSS THE WORLD

The emphasis in this chapter is on the historic aspects of religion in our search for definitions, origins and causes. There is however, much contemporary interest and development in both the study of religion and its practice, as it is popularly understood. Despite the fact, that from an academic point of view, there is no agreement on what it is that defines religion.

DEFINITION OF RELIGION

So to the first problem – defining what is meant by religion. This may seem to be a somewhat surprising problem to have, as I suspect that most of us have a sense of what we mean by religion. But when we begin to look for formal definitions, we find in the Definition of Religion entry in Wikipedia, the following:

"the definition of religion is a controversial subject with scholars failing to agree." [288]

Why should defining something, that most of us will feel we have some sort of handle on, be so controversial ?

It may come as a surprise, it certainly did to me, that the concept of religion, as we have come to understand it as a distinctive part of cultural life, did not emerge as a separate idea until the fifteenth and sixteenth centuries.[289] [290] [291] This occurred as part of the transition in European thought from Mediaeval Society through the Renaissance to the Enlightenment. Before this period, there appears to be no reference to "religion" or any words with similar meaning, in the Bible or in the sacred texts of other "religions," that describes what we now regard as religion or religious related activity.[292]

Even though traditions and practices have existed for tens of thousands of years, and sacred texts for around 2 – 3000 years, most cultures did not separate everyday life from the sacred or religious. They did not identify things as "religious" or "non–religious." Life was seen and experienced as a "whole" and was not compartmentalised into different areas in the way we do today. The rituals, practices and activities which we may now designate as religiously related, were just accepted as a natural part of everyday life. They were seen as a natural part of existence, without requiring a separate, generic name.[293]

Within western Europe these rituals, practices and activities were predominantly Christian. But from the fourteenth Century onwards, despite the dominance of Christianity up to this time, the cumulative effects of a number of factors, posed a challenge to the authority and teachings of the Roman Catholic Church.[294] These included, the Black Death, increasing personal wealth, greater exposure to other cultures through trade and education, the re–discovery of Greek and Arab scholarship, facilitating cross fertilisation of ideas and the development of printing. The cumulative effect of these contributed to the emergence of the Protestant Reformation in the mid sixteenth century. Within this changing environment, the notion grew that there were aspects of life that did not belong to the Church and that the Church and its practices was seen as something separate from other parts of day to day existence. This separation of Church from the political, economic and cultural aspects of society, helped create a new independent cultural category – religion.[295]

By the eighteenth and nineteenth centuries, with the advent of science and the scientific method, questions were being asked as to whether such methods might also be applied to the examination of religion. Consideration was given to whether religious activity could be found in other cultures. So with an increasing accumulation of information from tribes, cultures and peoples from all over the world, gained through trade, missionaries, explorers and war, systematic study of other cultures became possible. The founder of the scientific study of religion is generally considered to be Max Müller (1823 to 1900 CE). He advocated the idea of comparative religion – the comparison of religious practices across cultures. So began the discipline of Religious Studies, applying scientific methodologies to the study of religions. Applying such scientific methodologies and discipline inevitably meant concentrating on natural explanations and ideas that could be investigated and examined, rather than accepting supernatural explanations which could not be verified.

However, study of a subject, requires some sort of agreement or definition of what that subject is about. Thus, the conditions were right for creating a set of parameters to define what might or might not be regarded as religion. Various criteria and classifications were devised to do just that.[296] [297] [298] However, these parameters were derived from within the context of western european Christianity. So the structure and practices of Christianity were taken to be the "standard" or "criteria" by which to judge or designate whether something should be regarded as religious or not. The agreed designated criteria included evidence of : worship, a reverent regard for the memory of ancestors, human burials with artefacts from their lives and an element of sacrifice or having a means of communication with some sort of supernatural spiritual being or with the supposed spirit dwelling within a natural or man made object.

Having devised these criteria as defining what constituted religion, they were then applied retrospectively. They were imposed on long established cultural traditions that were very different and very much older then western Christendom. As we now know and have already said, these cultures did not see their rituals and practices as being in any way separate from day to day life and so would not have recognised or described a particular aspect of their lives as "religious." So the use of a set of pre–determined criteria inevitably coloured any classification and conclusions about the "religiosity or not" of other cultures. As the early societies or groups had no equivalent word to describe a concept of religion as a separate cultural idea, when researchers or explorers came across the activities of these early societies, they either "shoe–horned" them into these predetermined categories or disregarded those that did not fit as not having any religious connotations. Thus, any activities not included within the designated criteria, were not recognised as being of potentially religious significance, and so excluded from consideration.[299]

The reasoning behind applying these pre–determined criteria, was to enable the classification and study of, what could then be designated as the cultural religious aspects of various groups and societies, enabling inter–cultural comparisons.[300] But such a retrospective approach meant that particular words from ancient texts were erroneously translated as meaning religion. Whereas in fact, as we now know, there were no directly equivalent words for religion in either tribal religions or the ancient texts and scriptures of what we now know as the major religions.

ETYMOLOGY

A number of candidates have been nominated as being the etymological source of "religion" and were originally translated as "religion." They include – the Latin – religio, the Greek – threskeia and the Arabic – dïn. However, it is now realised, these words did not, in their original usage, properly convey the complex notion of religion as we understand it today, or even as it was conceived in the fifteenth or sixteenth centuries.[301] Thus, when we read about "religion" in ancient times, or of the ancient Greeks, or of the Hebrews and so on, we can easily be mislead into imagining that what is being referred to is something akin to the modern concept of religion. For example, people did/could not identify as a Hindu, Buddhist, Taoist or many of the other of the world religions until the eighteenth and nineteenth centuries, when these terms first entered the English language. [302] [303]

These points are underlined by Nongbri in his book *Before Religion – A History of a Modern Concept*.[304] In it he articulates the confusion caused by applying the word "religion" to ancient tribal cultures. He makes the distinction between a "description" of what an observer believes is happening within a particular tribe or group and a "re–description," whereby the observer applies a set of pre–determined criteria to activities that the group itself would not classify in those terms. That is to say, certain activities were "shoe–horned" into a category called "religion" that conformed to a western pre–determined definition, rather than fitting such activities into a description and interpretation of the cultural nature and context of that particular activity, in a way that reflected the tribe or group's understanding of what they were doing.

This retrospective – re–descriptive – approach has been criticised, with some scholars feeling it is inappropriate and not necessarily helpful. However, Nongbri goes on to say that he believes it is OK, provided the distinction is understood when one is applying the descriptive or the re–descriptive approach.[305] Otherwise, any conclusions drawn may be incomplete at best and inaccurate and wrong at worst, mistakenly leading to the opinion that the practices of the past have the same connotation of religion as we have today. However, Evans–Pritchard feels that even the descriptive approach is insufficient to actually understand what is really going on. He argues that much deeper enquiry is necessary, together with some informed interpretation, if a proper understanding is to be reached.[306]

Superficially at least, trying to discover a definition of religion is not unlike the problem we had with science. With science, the word and definition did not arise until the eighteenth century and then notions of science were applied retrospectively to the work and thought of people working and thinking nearly three thousand years earlier. So too with religion. The word and definition do not appear until around the sixteenth and seventeenth centuries and have then been applied retrospectively to the people and cultures of the past. However, in the case of religion, the situation is more complicated. The application of western–Christian–centric criteria leads to a circular argument. Using such criteria to identify religious practice means that any practices identified will, by definition, conform to a western Christian notion of religion. This approach excludes other activities that may perform a similar function to religious practices, but that do not belong to the list of pre–determined criteria. Also, there is no documented evidence to indicate the intentions and thoughts in the minds of early peoples as they initiated and implemented their rites and rituals and other potentially "religiously related" practices. So any conclusions obtained from the retrospective application of pre–determined criteria of religion, are likely to conform to a preconceived notion of religion, rather than provide a real understanding of how the societies themselves understood these practices and how they integrated them into their cultural world view.

Also, from a western Christian point of view, definitions of religion are often intimately bound up with ideas of faith and God (see chapter 8). However, this is not necessarily the case, particularly for the early religions. It is possible to think about each idea separately. In his book *The Evolution of the Idea of God – An Inquiry into the Origins of Religions*,[307] Grant Allen clearly makes a distinction between them. He suggests that the term "religion", refers to a series of practical responses, including the use of rites, altars, priests, sacrifices and offerings (rather like the pre–determined criteria for defining religion as already mentioned). He also suggests that any consideration of whether a particular set of beliefs or faith, including the idea of God or gods which may or may not underpin such religious activities, should come within the separate concepts of mythology or theology (although mythology has been used by some researchers as a defining criteria of religion or as a precursor in the causation of religion).[308]

Not all religions identify a God figure within their belief framework and it is possible to believe in a God without necessarily engaging in any sort of structured religious activities. There may also be a much wider spectrum of activities and ideas that too have helped some societies connect with "the something other than or outside themselves." These may fulfil a similar function to religion, within a particular cultural context, but may not be recognised as such, if they fall outside the pre–determined list of activities designated by western scholars as religious.[309] However, despite this distinction made by Allen, most modern dictionary definitions still usually include reference to a deity and or belief system. For example, the Oxford English Dictionary defines religion as "the belief in and worship of a superhuman controlling power, especially a personal God or gods."[310]

In addition, the distinction drawn in the preface, between religion and faith, primarily considers religion from a modern organisational point of view (though also including the doctrines and practices of the national institution). Such an organisational perspective relates well to the three main Abrahamic religions – Judaism, Islam and Christianity with its denominations and offshoots— but is not transferable to the early and tribal or indigenous religions. Nor, as a concept, is it really applicable to the other major faiths.

Thus, rather than trying to identify a "universal" definition of religion and so including some practices and early societies and excluding others, it is perhaps more productive to understand the significance of the rites, rituals and practices within specified cultural contexts. Such practices do suggest a sense of there being "something other than or outside themselves" that needs to be acknowledged in some way. The customs, rites and rituals that are practised are a way of doing that.

Nevertheless, it is the insights from the methodologies of Religious Studies that have allowed the construction of a classification of religions – both ethnic (tribal and folklore) and the modern major religions. So despite the drawbacks previously mentioned, the information gathered by this approach can still be instructive and valuable in its own right, and as a framework for our exploration of the origins and causation of the concept of religion.

Origins

Now to the second problem – identifying the origins of religion. In Steven Weitzman's – *"The Origin of the Jews,"*[311] we find, a broad discussion about origins, and the notion that even the very idea of origins may be illusory. He notes that "origin" may be used to define the beginning of something, or refer to the roots of something – that is, the things or conditions out of which something grows or emerges. Or as a sense of something breaking free from the past. The sociologist Émile Durkheim (1858 to 1917 CE) in The Elementary Forms of Religious Life, published in 1912 CE, was equally suspicious of the value of even inquiring into the origins of religion and dismissed the search for the origin of religion as a non– scientific pursuit that "should be resolutely discarded" – "There was no given moment when religion began to exist." To quote Weitzman :

"Since everything is rooted in what precedes it, going back to the beginning of the universe, it is always possible to push an origin further and further in the past, and there is no clear–cut rule for how to distinguish an origin from a mere transitional moment—that decision is always going to depend on our starting point as interpreters of the past.

Other scholars treat origin as an act of imagination, posing questions drawn from literary studies or folklore analysis: What cultural and social functions do origin stories play? What meanings do they super– impose on reality? From these kinds of perspective, it makes no sense to investigate origins empirically, to look for them as if they were events that happened in the past: the subject requires inquiry into the workings of consciousness, reason, or imagination."

Pals makes similar points in the introduction to his book *Seven Theories of Religion*.[312] Weitzman concedes it is natural for human kind to want to know the origins of things, for example: Darwin's theory of evolution, Wegener's search for the origin of continents and Lemaître's theory of the "Big Bang", and that the search to understand the world by moments of origin goes back to the ancient Greeks. Nevertheless, he suggests:

"there is no way to reach a definitive understanding of origin because the concept itself is so hazy, so hard to pin down, and has been understood in so many different ways."

This may be thought to be going too far as it may seem self evident that, particularly for Christianity and Islam we should easily be able to identify their origins. After all, their starting dates are fairly well known, as we have documented evidence, certainly to within a few years, of the births of their founders. However, Weitzman would argue that even for these, the true origins would still go much further back than the dates of their founders. Any new movement emerges from and is influenced by, the practices and beliefs that have already been part of the existing culture. This is not to say that there is a natural or inevitable progression stretching from prehistoric times to the present day of ideas that go from simple to complex, primitive to more sophisticated or worse to better. Although the early theorists like Tylor (see later) did view the evolution of religion in this way.

By the time Schmidt (1868 to 1954 CE) was writing in the early twentieth century, this idea had little currency.[313] Rather, that new ideas are not just parachuted in from nowhere but arise out of a long history of ideas and thoughts. Progress or development should not be seen as automatically being better or cleverer but as more encompassing, because of the insights and knowledge gained by previous generations. Earlier cultures were probably just as clever and sophisticated in their own ways as current generations, but were naturally limited by their knowledge and brain processing capacity at their particular moments of time. But each generation pushed the boundaries and so helped the next generation to start from further on.

Whilst we may not be able to identify the true origin of religion, I think it is still worthwhile to look as far back as we can, to find the earliest information available. Whether this relates to physical evidence of religiously related practice or the speculative interpretation of such evidence, we may get a sense of the emergence of the "religious" idea.

Early Physical Evidence from Anthropology.

If you remember, evidence of a belief in life after death was one of the criteria used to try and define religion. So findings from presumed burial sites with corpses in shallow graves is taken as early evidence of religious practice. Where there is evidence of stone tools and animal bones within the graves, this is thought to indicate a belief in the afterlife. Such sites have been discovered by evolutionary archaeologists from around the Middle Palaeolithic era (45 to 200 thousand years ago), in Atapuerca in Spain. Here, the bones of thirty individuals believed to be Homo heidelbergensis, whom we met in chapter 5, have been found in a pit. Also, Neanderthal burial sites have been found at Shanidar in Iraq and Krapina in Croatia and Kebara Cave in Israel. The earliest known burial of modern humans (Homo sapiens), is from a cave in Israel located at Qafzeh, (a prehistoric archaeological site located at the bottom of Mount Precipice in the Yizrael Valley of the Lower Galilee, south of Nazareth), dating from around 100,000 years ago.[314][315]

Also, the use of symbolism, as seen in some early cave paintings, may also be evidence of religious activity. The use of art is thought to be a more meaningful way of communicating supernatural ideas than can be accomplished simply by the use of language. Such use of symbolism is thought to demonstrate the capacity for abstract thought and imagination necessary to construct religious ideas. Some of the earliest evidence of symbolic behaviour is associated with Middle Stone Age sites in Africa, cave paintings at Chauvet (south east France) and other parts of Europe and the Aboriginal rock art, in caves across Australia, including Uluru (Central Australia) and the Northern Territories. Dating rock art is

229

difficult, as the pigments do not contain carbon, but there is evidence of human habitation both around Chauvet and Uluru from at least 30,000 years ago. Although the purposes of rock art are uncertain, a number of theories have been proposed, including art for arts sake, boundary demarcation, shamanism and religious purposes. But as Violetti [316] puts it:

"Since almost all cultural developments have multiple causes, it seems reasonable to suppose that the development of Upper Palaeolithic rock art has a multi–causal explanation rather than a single cause....If the assumption that at least some European rock art was created for religious reasons can be accepted, then it is safe to suppose that rock art is just the most archaeologically visible evidence of prehistoric ritual and belief, and unless rock art was the only and exclusive material expression of the religious life of prehistoric communities, we can assume that there is an entire range of religious material that has not survived. Some of the Upper Palaeolithic portable art could also be connected to religious aspects and be part of the 'material package' of prehistoric ritual. Our knowledge of the meaning of Upper Palaeolithic rock and portable art should not be considered either correct or incorrect, only fragmentary. The element of uncertainty, which involves the rejection of any form of dogmatic or simplistic explanation, is likely to always be present in this field of study. This should lead to flexible models complementing each other and the willingness to accept that, as more evidence is revealed, arguments will have to be adjusted."

These sentiments, of what we know being "fragmentary" rather than correct or incorrect and also of being "uncertain" until more evidence is revealed, leading to the "rejection of any form of dogmatic or simplistic explanation," can be seen in part as a summation of some of the themes and rationale of this book.

What is thought to be the oldest constructed religious site yet discovered is at Göbekli Tepe, in the southeastern Anatolia region of Turkey. It includes circles of erected massive T–shaped stone pillars, decorated with abstract pictograms and carved animal reliefs. It is thought to have been created around 9000 BCE, prior to the Neolithic period. It is conjectured that it may even have been built by hunter–gatherers, before the establishment of agricultural based settlements. If so, this would change some of the ideas associated with the lifestyles of hunter–gatherer communities.

In a study of a number of hunter–gatherer communities, peoples and colleagues,[317] explored the evolutionary nature of religion based on information about the rituals and practices of these groups, using some of the pre–determined religious criteria mentioned earlier. The information was taken from other studies and observations of hunter–gatherer communities from North and South America, Africa, the Far East and Australia. From analysis of these cultures, animism (the belief that both animate and inanimate objects are inhabited by spirits that can be active in the world, and a belief in souls) was found in all of the groups.

Shamanism (there is no universal agreed definition but the term includes the idea that a Shaman – man or woman – is a socially recognised ritual intercessor, healer, and problem solver who uses their power over spirit helpers during performances involving altered states of consciousness) was found in seventy nine percent, belief in the after–life in seventy nine percent, ancestor worship in forty five percent (with half believing these ancestors could be active in the world) belief in high gods (i.e., a supernatural creator) thirty seven percent and belief in active high gods (supernatural creators that are still active in the world) in fifteen percent.

Through various statistical and phylogenetic techniques,[318] demonstrating genetic relationships between ancestors, they suggest that the oldest trait of religion, present in the most recent common ancestor of present–day hunter–gatherers, was animism. Belief in an afterlife emerged later, followed by shamanism and ancestor worship. Ancestral spirits or high gods who are active in human affairs were absent in early human societies, but the trait "high gods" seemed to stand separately, suggesting that belief in a single creator deity can emerge in a society regardless of other aspects of its religion. Not all the practices were found in every place, but there is clearly evidence that such activity is wide spread across continents. However, it is important not to assume that similar activities indicate similar underlying beliefs or that they signify the same for each group or culture.

They conclude that some type of indigenous or folk religious activity has been part of human societies for thousands of years and may even have started just before or at the time of the exit from Africa by the first humanoids. This may explain the widespread evidence of religious practices across the globe and some of the common elements that have been found. It is postulated that such activities evolved as the groups developed, in addition to any syncretism (incorporation of religious ideas from one group to another) that occurred as groups met and interacted. These developments became more evident as organised religions emerged and with increasing travel for trade between continents, intermittent war, the slave trade between Africa, Europe and America westwards and exploration from Europe eastwards.

The importance of such evidence, even if somewhat speculative and incomplete, suggests that from the earliest times religious type activities have been part of human culture and experience. Though it should not be assumed that the information we have now represents precisely the nature of the activities at their inception nor that they are not still continuing to change. Brown and others regard belief in religion as "Universal," that is; one of a whole range of things that occur in all societies and cultures.[319] However, according to Fitzgerald, religion per se is not a universal feature of all cultures, but rather a particular idea that first developed in Europe under the influence of Christianity.[320] These two view points illustrate the earlier discussion around defining religion and so both may actually be correct. Though an added twist is provided by Henning [321]who's economic theory (see later) meant that religion did not begin until a society was stable in terms of its survival needs.

Causation

And to the third problem – the causation of religion. This is even more problematic than looking at definition or origin, as it involves much more speculation. Despite the difficulties and limitations already mentioned in relation to finding a definition or identifying the origins of religion, there were objective elements that provided some sort of frame of reference. However, with causation, we are considering the reasons behind why certain practices, rituals and beliefs may have come into being. Without written explanations from the people who initiated them, and of course there aren't any for early religions, it is difficult to know the reasons behind why such practices began. Even for the later organised religions we know more about the what and the how things happened rather than the "why did" aspect of their practices and beliefs.

So, the way researchers have approached this question falls broadly into three categories. The first is to speculate on certain factors or conditions thought likely to be necessary for the development of religion, together with a proposed explanatory narrative. The second is to speculate more broadly and devise comprehensive theoretical frameworks to explain the emergence of a religious dimension to cultural life. The third category is a combination of these.

A variety of factors and conditions have been proposed as being potentially necessary for the idea of religion to emerge. Such factors and conditions are not mutually exclusive and indeed some are themselves probably linked. So it is likely that all of them may have some part to play. These factors are linked to the state of development or evolution of firstly, our ancestral humanoids and then later ourselves, as Homo sapiens. Their state of development would both influence and reflect their capacity and capabilities to initiate complex and abstract ideas and process the significance of any "supernatural" communication – if there was any. (see below under supernatural). I will only deal with these factors briefly, delineating each as a heading in bold italics. They include: 322

Factors and Conditions

Increased brain size – It is proposed that the human brain must have grown to a certain size before it was capable of abstract thought and reflection (perhaps by about 500,000 years ago). *Tool use* (perhaps linked to brain size and the capacity for imagination and ideas about causation and problem solving). *Development of language* – enabling the formulation and communication of religious, ethical and philosophical ideas. With increasing brain size, the possibility of *developing a morality* along side the increasing sizes of societal groups.

This may have been necessary for societal cohesion, by developing systems of reward and punishment to maintain social control, conflict resolution and group solidarity. For this, a degree of judgement and reason would have been necessary. The degree of scrutiny and social control may have been reinforced by invoking the idea of supernatural beings (gods) and the notion of ancestors watching over the comings and goings of daily life.

In this regard, religion may also be seen as advantageous in terms of natural selection as it could have contributed to societal stability. This may relate to a sense that religion leads to people feeling better and fitter and calmer and so perhaps performing better under stress and so could have survival benefits. *Dopamine,* a chemical transmitter in the brain, may promote an emphasis on distant space and time which is thought to be a necessary condition for development of a religious perspective. Another idea is that of the so called *"God gene,"* proposed by Hamer,[323] that may pre–dispose towards developing a personal spirituality. He claimed to have identified a gene he thought may be influential in determining whether one was religious or not. A non–biological but related idea that may evolve by "natural selection", is that of the *"meme"*, first coined by Richard Dawkins.[324][325] A meme is thought of as a package of information that represents an idea or collection of ideas that can be propagated between people, societies and cultures. In this context, it is postulated that religion could be regarded as a meme. Its propagation being regarded as either having a positive or negative effect on humankind. The invention of writing played a major role in sustaining and spreading organised religion, as did printing.

It would seem self evident that the evolution of brain size and the development of language together with the invention of writing and

235

printing will have facilitated the emergence of new ideas and their transmission to others. But at what point along the religious timeline each kicked in is uncertain. The other factors mentioned above are fascinating but still only speculative. It will be interesting to see if they are developed further. For the purposes of this book, it is not necessary to go into any greater detail of these ideas, but they have been included to indicate the variety of "natural explanations" that have been suggested for the idea of religion to emerge.

There is a supernatural explanation of course, predicated on the idea of the actual existence of "God". In this scenario, "God" may have initiated or implanted thoughts about religion and the supernatural elements of that, or may have used the factors mentioned above to bring about ideas of religion and the implementation of various practices that express these thoughts. He may have used some combination of these or even some other way completely. In keeping with the evolutionary thesis, the level at which these ideas could have been initiated would have to have been commensurate with the capacity of people to comprehend, at whichever point in our humanoid evolutionary history he chose to initiate these ideas. In keeping with an open scientific approach, such a possibility, or something similar, given our now vast accumulated religious related literature, has at least to be on the table.

Theories of Religion

The following theories are speculations as to how the notion of "religion" may have come about. But of course, these theories were all formed in a context in which the concept of god(s) and the supernatural already existed. This awareness of itself affects the way in which a theory of causation may be crafted.

In recent times, anthropologists, sociologists, scientists and philosophers have produced theoretical models of how the concept of religion and any activities following its inception, may have come about in the first place.[326] [327] [328] [329] [330] In his book, *Seven Theories of Religion* (and the later updated versions of eight theories and nine theories), Pals[331] pulls together some of the theories regarded as having a "shaping influence on religion and the intellectual culture of our century." He provides details of the theories themselves with commentary and some of the criticisms levelled against them and sets them within a wider approach to the scientific research of religion.

The early modern theorists were driven by the idea that there would be a common basis or fundamental explanation underpinning the formation of all religions. A sort of "universal theory for everything." Much of their work was based on second hand information from missionaries, explorers and other researchers, rather than on their own observations and fieldwork. They also generalised on the basis of information from only a few populations. This led to accusations of them being "armchair anthropologists" from the likes of the English social anthropologist Sir Edward Evans-Pritchard (1902 to 1973 CE). According to him, these methods led to superficial interpretations of the significance of certain activities and to assumptions that similar practices in different groups had similar significance and meaning. He argued that by being immersed in a western–Christian–centric environment and using second–hand accounts from people who would only have had a brief superficial acquaintance with another culture, any conclusions misrepresented the reality. He claimed that anthropologists rarely succeeded in entering the minds of the people they studied. Thus, there was a danger of ascribing motivations to the studied peoples, which more closely matched those of the researchers' own culture, rather than to the one they were studying.

He suggested that any rituals and practices need to be interpreted within the context of the particular culture under study and that generalisations were therefore in error.

Evans–Pritchard did his own research. He undertook two particularly in depth field studies of the Azande people of north central Africa and the neighbouring Nuer people from the Nile Valley in south Sudan. His view was that it was only possible to make sense of the rituals and practices of another culture if one took the time to understand what was going on below the surface of any particular activity. Such understanding was then often specific to that particular culture and not generalisable to all cultures, even those with similar practices. However, he continued to believe that a unifying underpinning idea of causation of religion might be found.

Whereas, the anthropologist Clifford Geertz (1926 to 2006 CE), who also undertook his own field studies amongst the Javanese people, thought that such a common core idea was not possible and that all religions, though they may exhibit similar features, were unique to their particular situation and had arisen independently. Even though there may have been cross–fertilisation of ideas and practices.

Geertz held the view that explanations that merely described religions and cultures were not sufficient. Like Evans–Pritchard, he believed that interpretations (thick descriptions) of tribal customs and practices was necessary to make sense of what was really happening. Rather than taking some practices at face value and making superficial judgements about primitiveness and savagery and so on, he believed that once one got underneath the ritual, practices made logical sense and their benefits could be found. He advocated that these thick descriptions were necessary to interpret symbols, by observing them in use. He thus became known as a founder of symbolic anthropology. Symbols may be words, actions or objects that convey a concept, meaning, feeling or emotion beyond its own superficial description. For example, a fish is used by Christians to symbolise Christianity.

Evans–Pritchard also felt that the theories tended to be coloured by the beliefs of the proposer of the theory. He argued that believers and non–believers approached the study of religion in very different ways, with non–believers being quicker to come up with biological, sociological, or psychological theories to explain religion as an illusion, and believers being more likely to come up with theories explaining religion as a method of conceptualising and relating to reality.[332] Schmidt actually suggests that those researchers with a religious faith are better able to understand religion as a whole because they will grasp the "supernatural" dimension that those without faith will ignore or exclude.[333] The theologian Rudolf Otto (1869 to 1937 CE) also emphasised the importance of religious experiences.

These are regarded as experiences of a different quality to those we encounter on day to day basis. When explained within a religious framework they are viewed as being a spiritual encounter with God or the supernatural. It is out of such experiences that religion emerges. He asserted that these experiences arise from a special, non–rational faculty of the human mind, largely unrelated to other faculties, so religion cannot be reduced to culture or society. Clearly such an explanation is untestable and so regarded as unscientific.

Theorists, like Karl Marx, Sigmund Freud and Emile Durkheim, who do not have a faith in God, produce explanations that exclude or deny the need for any god or supernaturalist component, as this is an aspect that is not part of their own personal experience. For those who do have a faith, some, like Evans– Pritchard and Mircea Eliade use their experiences and the same or similar knowledge but include the possibility of god or a supernaturalist component somewhere in the narrative. Others, like Tylor and Frazer set aside their personal faith beliefs and specifically exclude the roll of the supernatural as being un–scientific and so seek for explanations from within nature and the world as we experience it.

A common way of classifying the approaches and content of these theories, is into substantive (essentialist) theories and functional or reductionist theories. The substantive approach focuses on what religion is. That is, its content, rituals and practices and what it means for people. This approach asserts that people have faith because certain beliefs make sense to them, to the extent that they see them as important enough to guide their lives. However, this substantive approach has been accused of neglecting the sociological aspects of religion. The functionalist, approach seeks explanations of religion that are outside of religion, and focus on what religion does. The theorists themselves are generally (but not necessarily) atheists or agnostics. They reject divine or supernatural explanations for the status or origins of religions because they are not scientifically testable.

Substantive Theories
Edward Burnett Tylor (1832 to 1917 CE) defined religion as "belief in spiritual beings". This was accepted by many later researchers as a definition that was simple but also potentially wide ranging and so useful for encompassing a wide range of concepts. He postulated that the origin of this concept of spirits emanated from within peoples dreams and formed as a rational process of trying to explain why life was the way it

240

was. Thus, in considering the differences between life and death and the nature of dreams, the idea arose that a person had two things that belonged to them. Both separate from the body – life (in contrast to death) and a phantom; that is, personages found within dreams. If these two ideas were combined, the outcome is a notion of a spirit or soul. It then followed that if people have spirits or souls then why would all aspects of nature, both animate and inanimate not have them? So everything is seen as having a spirit; that is, Animism. It is then a simple step to see these spirits as being able to exist independently and act independently in the world and so such spirits come to be regarded as supernatural, sky gods. In this way, the pantheon of Greek and Roman gods, with their anthropomorphic (human–like) characteristics might evolve. Similar polytheistic ideas were observed in the Aztec and Hindu cultures.

Once the concept of gods had come about, ideas such as prayer or ritual sacrifice evolved as a means of influencing them. It is then only a small step to view one of the gods as being superior and becoming "The High God." These religious concepts would then supersede earlier ideas, designated as "magic." Magic was the word used to describe how rituals and human actions were performed to influence natural phenomena. It was deemed to work through imitation or contact. So for example, imitating the desired action; that is, sprinkling water imitates falling rain, and so a rain storm can be induced. Or by sticking a pin into a small doll representing an enemy, harm to that person can be brought about. Such rituals operated as if they were laws and so it was believed that if the ritual was performed correctly then it was inevitably followed by the desired action. With the advent of the ideas of independent gods, other rituals, sacrifices and prayers took over from the magic.

In *Homo Deus – A Brief History of Tomorrow,* Yuval Noah Harari [334] describes the progression of religion as follows. He suggests that the origin of religion stems from animism as a first concept of the interconnectedness of animals, humans, plants and the environment. However, this relationship becomes disconnected with the agricultural revolution and the development of a theist (god based) belief. Homo sapiens then comes to dominate the environment, though constrained through a relationship with god. With the advent of modern science and technology theism is pushed aside and Homo sapiens comes to dominate the world but without any constraint or obligation.

James George Frazer (1854 to 1941 CE), in *The Golden Bough* [335] (first published 1890 CE) took some of Tylor's ideas further, and postulated the evolution of the medicine man, shaman or witch doctor who acquired the skills and knowledge of magic, to ensure a likelihood of success. Such men may then become the Magician – King and head of the clan, group or tribe. So, with the concept of the supernatural god, the Magician – King is replaced by a Priest – King and religious thoughts gradually replace the magic; that is, prayer to the god causes the effects that once magic controlled. Thus, some historical religions moved from the role of spirituality of nature with magic being the interface of human effecting nature to the role of god acting within peoples lives.

Tylor's theory assumed that the psyches of all peoples of all times are more or less the same and therefore that his theory illustrated a progression in sophisticated thought. That is, that cultures and religions become more sophisticated over time. They progress from animism to polytheism (many gods) and then to monotheist (one god) religions, such as Christianity and eventually to science. Science then replaces religion by demonstrating that animism doesn't exist. However, this progression does not seem to have quite panned out, as we are now in an age of science, and yet religion with its beliefs and rituals still persists. Here, other theories (see below) by Marx (his opium of the people) and Durkheim may shed light on why this is, by indicating the usefulness of religion as a social construct. The "glue" which helps keep societies together. Even so, the idea of a uniform progression from a "primitive to a sophisticated" form of religion and the view that monotheism is more evolved than polytheism has been criticised as unverifiable.

Herbert Spencer (1820 to 1903 CE) was of the view that ancestral worship or ghost worship (Manism) formed the basis of all religious developments. This was associated with the notion, found in many cultures that people had "a double or ghost" that was able to exist independently and could leave the body during sleep, unconsciousness and at death. In other cultures this idea was equivalent to the soul or spirit which could enter other living people as an enemy or might become an independent god or supernatural being to be worshipped.[336][337]

Henning (1897 to 1955 CE) [338] rejected these earlier theories and suggested that economics was a prime cause of the development of religion. In earliest times before any religiously related activity existed, the principle concerns of early humans was survival. They would live in groups and their common survival instincts would have provided some sort of cooperation and social cohesion in order to maximise their survival chances. He states that *"In every man lives an inclination for the improvement of his condition."* Within society, there would be people who would perhaps stand out in their commitment to the group, looking after group members and generally having the group's needs at heart. In death, such people would be revered as someone who had "been there and got the T shirt". When some sort of stability was achieved and there was perhaps a little more time to think about other things than just survival, this reverence of former leaders might translate into formal burials and so provide a place to go and "talk" to them and seek advice about how to survive, hunt, grow crops, climatic conditions, respond to threats and so on. In this way, gradually rituals and practices that we later recognise as being religious, would evolve. Their spirits might then take on the role, as we now see it, as god or gods and so the relationship moves from one of reverence and a source of advice to one of actual worship.

Mircea Eliade (1907 to 1986 CE), describes early humankind as living with two concepts – the sacred and the profane. The profane refers to the whole of material life as we know it – the ordinary aspects of nature, meeting the needs of day to day living. The sacred refers to the supernatural or the extraordinary, mysterious and momentous. Like Otto, he sees religion as something special and autonomous, that cannot be reduced to the social, economical or psychological alone. So within the orbit of the sacred are the sky gods, the symbolism of the sun and moon and the cycle of life. Anything ordinary can thus become sacred if one so decides. He suggests there is a desire of humans to get back to a historic perfection of life, away from the drudgery of day to day existence, to escape meaninglessness. Another theory is that primitive man could not cope with the notion that his struggle to survive had no meaning, so religion emerges as a counter to this despair. Eliade used some personal knowledge of other societies and cultures for his theories, amongst which was his knowledge of Hindu folk religion. But Eliade, like Frazer and Tylor has also been accused of out–of–context comparisons of the religious beliefs of very different societies and cultures.

Schmidt on the other hand eschews all intermediary ideas and claims that even in the earliest societies, religion begins with the notion of a monotheist deity.[339] He argues that there is evidence of the supreme being – high god(s) – in the earliest cultures identified through ethnographical methods. This idea is to be found across all the major continents from Australia to North and South America. These gods may have either started off on earth and left because of human sin or have come down from the sky to earth. He claims there is evidence of such beliefs being present before those of magic or animism, totemism or the nature myths.

Functional Theories

The functional or reductionist theories focus on what religion does. That is, how it operates and influences the culture of its adherents. The actual beliefs are not their prime consideration. They contend that there are underlying structural and cultural or psychological reasons underpinning religious behaviour. The reductionist approach aims to explain particular complex phenomena by reducing them to the simplest common denominator or to its component pieces. Applying this to the foundations of religion means looking for natural explanations outside of the religious framework and not within the religion itself. Therefore references to God or supernatural phenomena as a potential cause or origin are specifically excluded. Functionalists see religion as performing certain functions for society.

The social philosopher Karl Marx (1818 to 1883 CE) held that the dynamics of society were determined by the relations of production – the alienation between the owners and the workers. Religion emerged as a product of this alienation as a means of relieving people's immediate suffering (hence the "opium of the people"), but by so doing masked the real nature of social relations. In this regard,[340] religion was not helpful in trying to tackle the inequalities and repression of society. Such ideas have a certain resonance but are not universally applicable to all religions across the world and across the centuries.

Sigmund Freud (1856 to 1939 CE) saw religion as an illusion, a belief that people very much wanted to be true, originating from unfulfilled psychological needs. He asserted that religion is a largely unconscious ,neurotic response to repression. By repression Freud meant that civilised society demands that we do not fulfil all our desires immediately, but that they have to be buried, repressed or held in check. Rational arguments to a person holding a religious conviction will not change the neurotic response of a person. These observations did not take account of the wide spectrum of early and organised religions across the world. This is in contrast to Tylor and Frazer, who saw religion as a rational and conscious, though primitive and mistaken, attempt to explain the natural world.

Émile Durkheim (1858 to 1917 CE) saw the concept of the sacred as the defining characteristic of religion, not faith in the supernatural. He saw religion as a reflection of the concern for society. He based his view on recent research regarding totemism among the Australian aboriginals. With totemism he meant that each of the many clans had a different object, plant, or animal that they held sacred and that symbolised the clan. Durkheim saw totemism as the original and simplest form of religion. According to him, the analysis of this simple form of religion could provide the building blocks for more complex religions. Durkheim held the view that the function of religion is group cohesion, often performed by collectively attended rituals. He asserted that these group meetings provided a special kind of energy, which he called effervescence, that made group members lose their individuality and feel united with the gods and thus with the group. This view that religion has a social aspect, became very influential.

The anthropologist Bronisław Malinowski (1884 to 1942 CE) was strongly influenced by the functionalist school and argued that religion originated from coping with death.

Max Weber (1864 to 1920 CE) one of the founders of sociology, thought that the truth claims of religious movements were irrelevant for the scientific study of the movements. He portrayed each religion as rational and consistent in their respective societies. Weber acknowledged that religion had a strong social component, but diverged from Durkheim by arguing, that religion can be a force of change in society. Weber's main focus was not on developing a theory of religion but on the interaction between society and religion.

Other social theories of religion include the rational choice theory. This has been applied to religions, particularly by the sociologists Rodney Stark (1934 CE –) and William Sims Bainbridge (1940 CE –) and described in their book – *Theory of Religion*.[341] In this model, religions are seen as systems of "compensators", with human beings viewed as "rational actors". Compensators are a body of language and practices that make up for or compensate for or take the place of the lack or loss or failure to achieve something else. The rational actor part assumes that people make "rational" choices by calculating the costs and benefits of actions that compensate for the lack, loss or failure. Religion is defined as a system of compensation relying on the supernatural. In this scenario, the rational choices are focussed on the satisfaction of wants. It is asserted that only a supernatural compensator can explain death or the

247

meaning of life or promise salvation and rewards in the after life. They suggest that most religions start out as cults or sects.[342] These are groupings which are in opposition with the generally accepted or tolerated beliefs of society. Over time, they tend to either die out, or become more established, mainstream and in less tension with society.

Cults are new groups with a new novel theology, while sects are attempts to return mainstream religions to (what the sect views as) their original purity. Mainstream established groups are called denominations.The theory of religious economy sees different religious organisations competing for followers much like the way businesses compete for consumers in a commercial economy. Theorists assert that a true religious economy is the result of religious pluralism, giving the population a wider variety of choices in religion. According to the theory, the more religions there are, the more likely the population is to be religious. Bruce [343] criticises this economic, market approach and argues that by ignoring social culture, it fails to properly answer questions of why people believe the things they believe, and why religions rise and fail.

Evolutionary theories view religion as either an evolutionary adaptation to environmental and cultural changes or as a byproduct of other evolutionary adaptations.[344] A primary adaptation occurs because there are survival advantages as a result of the adaptation. The concept of a byproduct is that these are produced incidentally alongside some other primary adaptation. Adaptationist theories view religion as a byproduct or spandrel that is of adaptive value to the survival of humans. Within evolutionary biology, a spandrel [345] is a phenotypic (physical) characteristic that is a byproduct of the evolution of some other characteristic, rather than a direct product of adaptive selection. They allow that there may be some benefits from this byproduct and so it is maintained, though it was not necessary as a survival mechanism. Chaterjee and Singh [346] conclude that at present the majority view favours the by–product theory.

In his critique of Loyal Rue's *"Religion is not about God,"*[347] Hubert Seiwert [348] notes that Rue belongs to the adaptive group. For Rue, religion is of the utmost importance for the survival and well–being of the human species and so may be seen as being on the other end of the spectrum to Richard Dawkins's biological interpretation of religion.[349] Dawkins, using the same evolutionary and cognitive psychology theories, concludes the opposite – that religion is one of the major obstacles to human well–being. Rue proposes a general and naturalistic theory of religion whose core argument is that there is a universal human nature that can be known by examining the evolutionary history of humankind. In this context, religion is seen as fulfilling vital functions in influencing the cognitive and emotional systems of humans that allows achievement of personal and social well–being.

Rue's thesis is that all religions have a common structure of narrative traditions or myths that collectively shape the religious life and links the cosmology to morality, integrating facts and values. As Rue sees it, every myth is centred on a 'root metaphor' – an idea that underpins and runs as a thread through the narrative. For example he identifies the root metaphor of the Abrahamic traditions is "God–as–person". The root metaphor keeps the mythic narrative together giving meaning both to cosmology and morality.

The theory is that Myths were born out of the narratives describing the nature and actions of the gods which provided an explanation for the relationship between humans and the natural phenomena observed in the world. So out of such concepts, the rituals and first religious ideas would emerge. Religion then emerged when storytellers imagined the first myths that brought together cosmological and moral ideas in a coherent narrative of gods and spirits. Thus the first religious traditions arose as these stories found a place within the continuing culture of the society.

Other Ideas

Other ideas have been expressed as possible causes of religion, including Brinton's[350] idea that humans have an innate sense that all they see around them is a manifestation of a spiritual power beyond or behind reality.[351] The Cognitive Science of Religion (CSR) seeks to explain how human minds acquire, generate, and transmit religious thoughts, practices and rituals by means of ordinary cognitive capacities.[352] In his paper "The Naturalness of Religion"[353] Lauonen discusses the notion of whether religion should be regarded as "natural"; that is, whether it is "innate" or in some way acquired. Contained within this methodology is the idea of Agent detection. This is the tendency for humans to presume the purposeful intervention of a sentient or intelligent agent (a supernatural being; that is, God, when applied within a religious context) in situations that may or may not involve one (see Gray and Wegner).[354]

There is also the theory of mind (ToM) which is the ability to attribute mental states—beliefs, intents, desires, emotions, knowledge, etcetera — to oneself, and to others, and to understand that others have beliefs, desires, intentions, and perspectives that are different from one's own. Having theory of mind allows one to attribute thoughts, desires, and intentions to others, to predict or explain their actions, and to posit their intentions. It is suggested that such an ability is necessary to conceive the idea of the supernatural. At what point such an ability occurs in human evolution is unclear, but it could be innate and or genetic. Another suggestion is that it is feelings that underly the development of religion – particularly the emotions of fear and love.[355] Fear of punishment by a supernatural agent leads to the development of rituals to propitiate this danger but the need for love or the experience of love generates the idea of a supernatural agent to take care of us. Tied in with all these ideas is the readiness to anthropomorphise (endow with human characteristics) the supernatural being so envisaged.[356]

Early Organised Religion

We have referred briefly to the early or indigenous religions and we will meet them again in a little more detail in the next chapter. We noted that indigenous peoples and tribes had a fundamental sense that religion was not identified as something separate from the way they ran their day to day lives but was rather an integral part of group or tribal community life. Thus, there would not have been the need for any separate complex organisational apparatus. Even so, it is likely there would have been some sort of structure or framework in which the various practices and rituals operated, so that people knew what was happening and how to fit in, may be together with someone to oversee the operation. This role being fulfilled in some cultures by the Shaman or Medicine Man, by virtue of the power of their perceived magic. With the transition of hunter–gatherers into more settled communities, in an agricultural environment, around the time of the neolithic revolution 11,000 years ago, individual communities grew in size.

To sustain such growth, more organisation of society was required and the Shaman could use their position of influence to rise and become the leader or chief of the group. This would have been in the role of magician–king or priest (divine)–king depending on the cultural ideas

251

prevailing at the time. Anthropologists have found evidence that nearly all state societies and chiefdoms in their history, have operated political power justified through divine authority. For example, in Egypt the Pharos were regarded as divine beings or as in Mayan culture, the leader was regarded as a high priest in communion with the gods. Within western Europe the notion of the Divine Right of Kings evolved in which monarchs were seen to reign at God's behest.

By this combination of the political and religious, leaders were able to exert control over their people by being regarded as divine, and threatening the wrath of the gods if the people did not fall into line. Thus, organised religion emerged as a means of providing both social and economic stability. They did so, by creating a central authority and organising the collection of taxes. In return, they provided social and security services by creating bonds between unrelated individuals, facilitating the stability and cooperation of groups which might otherwise be more prone to fighting. Thus, all sorts of systems of governance have arisen in which the ruler used a combination of human political and "religiously" related powers to exercise control over the population.

Such arrangements have evolved over the centuries, though the political and religious ties tend to weaken over time. Israel and some Muslim countries have religious involvement in government to varying degrees, whereas, in the West, religion is more generally confined to the personal faith of some of a country's leaders. Though, in the UK, Bishops of the Church of England still sit in the House of Lords as part of the parliamentary governing system.

Whilst this does not demonstrate an evolution of religion as a progression from early or folk religions into more "sophisticated" versions of religions, it does indicate that as societies evolve, there is a change in the nature of their relationship with religion that often occurs alongside that of the development of the political and social structures.

The importance of these theories, as Pals says in his conclusion, is not whether they are right or wrong, but that they lead on to other insights or new ways of thinking about the issues. All of them contain elements that can be generally accepted even if the thesis in its entirety cannot.[357]

SUMMARY

I will come to the development of specific religious movements in the next chapter, but for now, where are we ? Having uncovered and briefly considered some of the complexities associated with even discussing the concept of religion, let alone its content, where does that leave us in terms of the objective of this book ?

Clearly we do not have an agreed definition of what is meant by or what constitutes religion. We do have ideas of practices that may be included, though not an exhaustive list. There may be other activities performed that serve a religious like function that are not included. It is now generally accepted that practices and rituals need to be understood within their cultural context rather than being "shoe–horned" into a pre–determined western–Christian–academic framework. We have much information about the rituals and practices that take place in many cultures across the world as well as vast literatures on the so called major religions. But we do well to remember that similar activities performed by different groups in other parts of the world, do not necessarily indicate similar understandings or meanings for that particular activity.

Clearly we do not have a specific point in time or place when the idea of religion emerged. We do have a sense that most, but not all, societies

253

express activities, rituals and beliefs that can be considered "religious" and such clear practices have been around from at least the time of the first humanoid exit from Africa around 300,000 year ago. It is agreed that there is no simple evolutionary progression from some sort of primitive religion to a more sophisticated version. We do have physical datable evidence of when some of the practices and rituals associated with religion existed in antiquity. We do have much information about the earliest times of the major faiths which we address in the next chapter.

Clearly we do not have a common cause for the initiation of religion. We do have much research, using combinations of evidence, information and speculation as to the possible factors and theories that contribute to explanations and understandings of how and why practices and ideas we now label as religious, may have arisen. Also, how these practices are maintained and embedded in the day to day lives of peoples across the world. Ideas of the supernatural may have been implanted by God from the beginning or "drip fed" over time as the human capacity to comprehend expanded. They might have "just emerged" in response to the evolution of existence and had nothing to do with God or gods. Whether God is the creator of these ideas is likely to remain unproven. It is rather like the chicken and egg – which came first ? For those who have developed a faith, they are likely to see God somewhere in the process of the creation of religious ideas whilst those without, exclude this dimension altogether.

What is clear, is that whether one believes in a spiritual or supernatural dimension to religion or not, the rituals and practices of designated religious practice and the belief systems that perpetuate them, are an important part of the social fabric and structure of the cultural life of most societies. Religion has and continues to influence many aspects of both the lives of individuals and communities. It provides the motivation or excuse for a whole range of activities – both good and ill. So the fact of religion cannot be ignored.

The matter of whether one believes in a spiritual or supernatural dimension to religion or not is the real stumbling block for all the researchers, be they historians, anthropologists, sociologist, theologians, philosophers or whoever. Understanding many of the social dimensions that have influenced and been influenced by religion is important if we are to mitigate some of the negative effects of religion and perhaps support and encourage some of the positive aspects. But if this is done by ignoring or dismissing the basis of the belief system, namely the supernatural dimension of God, in those religions where God is the focus, then a whole dimension of understanding is missing and so is likely to lead only to a partial understanding. Rational decisions and practices within a framework of the supernatural may appear totally incomprehensible without taking that into account.

CONCLUSION
My intention in this chapter is not to have fully critiqued all the various ideas concerning religion. Rather, to indicate the breadth and variation of thought that has been devoted to attempting to explain the nature of religion and demonstrating the wide range of understandings that have emerged, often I suspect, outside the orbit of most of the general population. By exposing the wide ranging thinking that is going on by both religiously orientated people and others, then surely it is OK and legitimate for "ordinary folk" to question aspects of their religious understanding. Such questioning should be connected to this wider world in order to give support – not necessarily agreement, but support for the process of questioning and for connecting them to people who are thinking and questioning these matters

CHAPTER 7

ORIGINS OF RELIGION II

This chapter falls into two main parts. The first covers the ethnic and folk religions and some of the similarities between them and the major faiths, using selected stories from both the Old and New Testaments of the Bible. The second part explores the developments of each of the four major faiths Hinduism, Judaism, Christianity and Islam. We explore their origins and how their scriptures were written, adopted and understood. The structure of these faiths and their scriptures, did not begin in the forms we know today. They underwent periods of structural change and the scriptures went through various processes before being settled upon.

In the case of Christianity for example, a process of intense discussion, lasting several hundred years, preceded agreement of the current biblical format. Even now there is not one universally agreed form and new translations and commentaries continue to be produced. The manner of the structural changes and the processes for producing the scriptures may be different for each faith, but the complexity of their evolution is evident for each. So in order to do justice to these complexities it is necessary to provide rather a lot of detail. Such detail is important in supporting my assertion that in the light of God's communication with all his people (see chapter 10), that the, complexity and ongoing change makes it inappropriate for any single faith group to claim a monopoly of understanding of God.

I found this interesting to do, as I was not aware of much of this background. However, I realise that not all readers will necessarily find it so and may find the degree of detail detracts from the continuity of the main narrative of the book. It is possible therefore to skim through or omit this detail by going straight to the section summaries and the final conclusion, without missing any of the main arguments. For those whose interest is sparked, I hope you find it as fascinating as I did.

EARLY ETHNIC / FOLK RELIGIONS

In general, these refer to the rites, rituals, beliefs and practices associated with small tribes or larger ethnic groups.[358] [359] [360] They often stretch back into antiquity and reflect the earliest beginnings of religion, as far as is currently understood. These have developed along with the evolution of the tribal groups, and are distinct from the major or organised religions we identify today.

I include the following websites for those interested in finding a ready source of information on a wide range of religions, their practices and beliefs and the comparisons between them. They are: ✣ significant events in the development of religion from 300,000 years ago onwards, ✯ ethnic religions, ◉ all religions, ✪ new religious movements (NRM) and ✤ a chart comparing the characteristics of many of the more well known faiths. There are hundreds ! They are generally classified by geographical location by continent, for ease of reference. Also refer back to Chart 1 at the beginning of chapter 6 for a World Map of the degree of religious intensity across the world

✣ https://en.wikipedia.org/wiki/Timeline_of_religion

✯ https://en.wikipedia.org/wiki/List_of_ethnic_religions

⊙ http://www.humanreligions.info/religions.html

✪ https://en.wikipedia.org/wiki/
List_of_new_religious_movements

✡ http://www.religionfacts.com/big–religion–chart

From examining these lists, we can see that some religions are thousands of years old, others have died out, others have begun in the last few years and there is a whole bunch in between. By looking at examples of beliefs found in different early, ethnic cultures across the world, similarities and parallels can be seen within the early religions of Africa, America, Asia and Australia. Not all practices are present in every community, but the following list gives a flavour of some that are common to many religions. However, it should be remembered that similar actions do not necessarily mean a similar understanding of the purposes for which such actions may be being carried out:

- ancestor worship
- sacrifice
- not eating a designated sacred animal except on special occasions
- totemism
- belief in a resurrection and some sort of after life
- animism – belief in spirits
- shamanism
- multiple gods – sometimes hierarchical
- high god – one supreme being
- ritual worship
- ideas of appeasement

- magic
- sense of gods acting in people's lives or watching over them or "policing" them

In addition to these common ideas, there are some more specific stories and beliefs that are common to both the major and ancient religions. To highlight some of these in more depth, Thomas Doane's, – *Bible myths and their parallels in other religions,*[361] seemed initially to be a good starting point and a promising source of material. In his preface he makes a point of saying that he has put the material together in order to save future students of the subject a lot of personal effort in researching original sources for this information. Naturally, I was attracted to such an offer of "free help" and the idea of using some of his material as a basis for this section was very appealing. However, on closer inspection there is a need for caution when considering both his evidence and conclusions. We will come to these shortly.

His book is divided into two parts. Part I is concerned with stories relating to events in the Old Testament such as : 'the deluge (flood), the tower of Babel, the receipt of the ten commandments' and parallels found within other religions. Part II comprises a similar approach, but concerns events from the New Testament. These relate both to the life of Jesus and the beliefs and rituals of Christianity. Included are references to: virgin births, baptism, the concept of atonement (in Christianity, the forgiveness or pardoning of sin through the suffering, death and resurrection of Jesus), crucifixion, the crucifix (a person suspended on a cross), resurrection, the trinity [in Christianity refers to the three persons of God as Father, Son (Jesus) and Holy Spirit] and the eucharist (Lord's supper or communion). His general approach is to cite a biblical story or event and then give examples of similar stories and ideas found in different parts of the world. These are based on a range of sources he has managed to find in his researches. A long list of these can be found at the beginning of his book. The examples used are indeed widespread in religions and cultures across the world. They may be associated with gods or linked to historical figures such as Buddha, Zoroaster as well as national rulers such as Caesars, Pharos, Alexander the Great and others. Some of these references have been been gathered together and summarised into a more readily accessible list:[362] [363]

- Born of a virgin: Dionysius, Horus, Tammuz, Krishna, Zarathustra, Buddha, Lao–Tzu, Attis, Heracles, Jesus Christ

- Son of the Supreme God: Dionysius, Krishna, Mithras, Heracles, Jesus Christ

- Death or torture by crucifixion (including bound to or embedded within a tree or stone): Dionysius, Osiris, Krishna, Prometheus, Jesus Christ

- Resurrection and ascension: Osiris, Tummuz, Krishna, Mithras, Adonis, Jesus Christ

Points of Caution :

Some of these examples have been cited by prominent atheists such as Christopher Hitchens [364] and D. M. Bennet [365] in their arguments against Christianity. They suggest that finding similar references in widespread religions implies that these stories are just myth and so therefore Christianity is myth. Indeed Doane uses the same argument to conclude that the narratives of life events relating to Jesus Christ must therefore also be mythical. These arguments he saw as undermining the foundations of Christianity and thus denying its authenticity.

Partly as a counter to this, some of the examples used by Doane, specifically in relation to the stories of virgin births, crucifixions and resurrections, have been shown to contain fundamental inaccuracies. These particularly relate to the stories of Krishna in the Hindu faith and to Buddha.[366] [367] [368] [369] [370] [371] In his blogs, Ronald V Huggins [372] argues that even though some of these examples are easily shown to be historically inaccurate, nevertheless they have been regurgitated without critical analysis, as they suit the arguments of some atheistic writers and others.[373] He argues that some of the crucifixion references actually originated in the eighteenth or nineteenth centuries.

This presents us with two dilemmas. Firstly, Doane's belief that the finding of mythical stories in other religions and cultures, similar to events associated with the life of Jesus, undermine the foundations of Christianity. Secondly, do the inaccuracies of some of the examples used by Doane, specifically in relation to the examples of crucifixion and resurrection stories, render all the other information in his book, suspect ?

Not withstanding the forgoing, I believe Doane's book still has something useful to contribute. I will therefore be quoting quite extensively from it, but in the light of what I have just said in the previous paragraph, I need to make some brief comments. The purpose of my book is not to prove or justify Christianity and so I do not need to enter into a long rebuttal of the first dilemma other than to say, as you would expect no doubt, that I disagree with Doane's conclusion. The principle reasons being that :

- finding parallel stories associated with mythical gods is insufficient evidence, even though it may on the face of it seem reasonable, to support a conclusion that similar stories associated with Jesus are themselves mythical

- the argument itself does not logically lead to such a conclusion (that horse is brown, that animal is brown therefore that animal is a horse !)
- these arguments are illogical and so unscientific
- it may in fact suggest that many groups have found similar explanations for life's ultimate questions, which in itself, does not automatically make them untrue (may be quite the opposite – who knows ?)

The second dilemma is a bigger problem in that it does cast a shadow over the other aspects of the book. However, with the advantage of further scholarship and the uncovering of further written evidence, the inaccuracies seem to be mainly either misunderstandings or misreadings of the source texts. Or the source texts themselves are based on observations which the original author has unwittingly misinterpreted or misrepresented. I have not come across any other vast body of literature critical of other aspects of Doane's material, other than the references already cited. Therefore, for the purposes of this book, I present the material in good faith, as exemplars of the presence of ideas such as virgin births, crucifixions and resurrections being recognisable in other religions and cultures, even if the detail of their meaning has not always been accurately understood.

It is not necessary to use all Doane's examples, or refer to all characters mentioned in the list, but reference to just a few events will suffice to illustrate the point, that key Christian Bible stories often have antecedent or similar stories in religions from an earlier time. This, I believe, does not deny the truth or its particular relevance for Christian belief, but rather supports the idea that different groups have found similar explanations for life's ultimate questions. They may well have shared these ideas over thousands of years and incorporated new ideas into their own understanding (syncretism). They have then particularised such ideas to their own circumstances and experiences. Moreover, it lends support for the notion that not only has our understanding of religious ideas evolved, but the ideas themselves have also evolved, perhaps along more similar lines than has previously been envisaged. Such processes may have even be divinely inspired – who knows ?

We have dealt with ideas of cosmology and the "creation stories" in different cultures in chapter 4, so excluding these, I have chosen the following examples. A couple from the Old Testament – the deluge or flood and the receiving of the ten commandments and, what will perhaps be of particular interest from the New Testament, several key moments from Jesus' life: birth, crucifixion and resurrection. Rather than putting in lots of references, all the information, quotes and opinion statements are taken from Doane. The passages in italics and parentheses are taken directly from his book.

Before we commence, here is a brief timeline of the various scriptures, taken from the first of the websites mentioned earlier. We will deal with the development of scripture in more detail later in the chapter, but for now, this provides a historical framework for the writings when comparing examples and stories from the different cultures:

- 3000 BCE: Sumerian Cuneiform emerged from the proto–literate Uruk period, allowing the codification of beliefs and creation of detailed historical religious records.

- 2494 to 2345 BCE: The first of the oldest surviving religious texts, the Pyramid Texts, was composed in Ancient Egypt.

- 1700 to 1100 BCE : The oldest of the Hindu Vedas (scriptures), the Rig Veda was composed.

- 1250 to 600 BCE: The Upanishads (Vedic texts) were composed, containing the earliest emergence of some of the central religious concepts of Hinduism, Buddhism and Jainism.

- 6th/5th century BCE: The first five books of the Jewish Tanakh, the Torah are probably compiled.

- 6th century BCE: Possible start of Zoroastrianism

- 600 to 500 BCE: The earliest known Confucian writings

- 300 BCE: The oldest known version of the Tao Te Ching was written on bamboo tablets.

- 2nd century BCEHebrew Bible – The Tanakh in settled form

- 500 to130 BCE Tanakh translated into Greek – Septuagint

- 140 BCE: Possibly the earliest grammar of Sanskrit literature

- 100 BCE to 500 CE: The Yoga Sūtras of Patañjali, constituting the foundational texts of Yoga, were composed.

- 50 to 62 CE: The first Christian Council was convened in Jerusalem.

- 60 to110 CE: The four Gospels of the New Testament were written

- 1st to 3rd century CE: The oldest surviving Buddhist manuscripts

- c.✚ 383 CE: The Latin Vulgate version of the complete biblical canon commissioned. Full canon probably known by 367 CE and full Roman Catholic canon accepted by 382 CE.

- c. 647 CE: Definitive copy of Qur'an compiled in the form of a book

- 1429 CE: Guru Nanak (1469 to 1539 the first Sikh Guru

- 1708 CE: Gurū Granth Sāhib – the Sikh scripture designated as the final Sikh Guru

✚ c. = circa - around/approximately

OLD TESTAMENT STORIES

The Deluge or Flood

The story, as written in Genesis the first book of the Bible, gives an account of a flood covering the whole world. It is caused by God in response to the wickedness of the people. God tells Noah to build an ark (boat), to house his family and to take one pair (male and female) of all the animals and birds. Noah complies and after 40 days sends a dove out to see if it can find land. It returns "empty handed". On the second attempt it brings back a twig and on the third attempt it does not return. The ark then grounds on the top of a mountain and life is restored. Noah offers sacrifices to God in thanks for his safety. God sends a rainbow as a promise that he will never do this again.

The first written references to a flood are in the Sumerian Gilgamesh poems, around 2700 BCE.[374] A story of a flood can also be found in the history of the Chaldeans, written by the Babylonian writer Berosus in the early third century BCE. This apparently suggests that this history and the Hebrew Bible may have had a common source. In Hindu scriptures, there is also reference to a flood and also in writings from China, Persia, Scandinavia, the Celts, Mexico and the ancient Greeks. It is possible that some of these referred to local flooding rather than a world–wide catastrophe. Not all the references contain all the elements of the Genesis story. In Egyptian writings, older than these, there is no reference to a flood. However, it would appear that the idea of life being destroyed in order to start afresh is widespread and likely to have risen independently in different parts of the world.

Acquisition of the Ten Commandments

The story in Exodus, the second book of the Bible, has Moses going up a mountain, shrouded in a cloud with thunder and lightening. He meets God and they talk for forty days and nights. He is then given the ten commandments on two tablets of stone. Moses takes them down the mountain to the people, where he finds them worshipping a golden calf statue they have made. In a fit of anger he throws the stone tablets on to the floor and smashes them. So he has to go back up the mountain to get replacements. When he comes down the people comply with the wishes of God and prepare a tabernacle (tent or small moveable covered construction), an altar and an ark (box to house the ten commandments) and other things God told Moses he required. *[Scholars have questioned whether Moses was an actual historical figure, even though he appears through both testaments of the Christian Bible, the Jewish literature and the Qur'an and other Islamic literature. The details of his birth (being hidden in reeds to escape death and being found and raised by a royal personage and rising to become leader of the nation) are also found in relation to Sargon the Great, the king of Akkad, who lived about 2400 BCE (though the documented evidence of his life and reign is sparse). Moses is thought to have lived around 1600 to 1500 BCE].*[375]

Similar attributes to those of Moses, are ascribed to Bacchus, the Roman God of Agriculture and fertility and also known as the Law Giver. He delivered laws on two tablets of stone, he parted a river, performed miracles with his staff and struck water out of a rock. It is thought that the provision of the law to Moses is modelled on the story of Bacchus, but the inclusion of meeting God on the mountain is possibly taken from the story of Zoroaster in Persia, who, according to legend, prayed on a high mountain, in the midst of thunders and lightnings and the Lord himself appeared to deliver the "Book of the Law" which is called the Zend– Avesta, signifying "the Living Word."

Statements similar to the content of the ten commandments appear in association with Buddha. Also, according to the religion of the Cretans, Minos, their Law–giver, ascended a mountain and received the sacred laws from the Supreme Lord (Zeus). The Supreme God of the ancient Mexicans was Tezcatlipoca who occupied a position corresponding to the Jehovah of the Jews, the Brahma of India, the Zeus of the Greeks, and the Odin of the Scandinavians. His name is compounded of Tezcatepec, the name of a mountain (upon which he is said to have revealed himself to man) amidst the dark and smoke.

"According to Egyptian belief, it is Thoth, the Deity himself, who speaks and reveals to his elect among men, the will of God, portions of which are said to have been written by the very finger of Thoth himself. Diodorus, the Grecian historian, says :

The idea promulgated by the ancient Egyptians that their laws were received directly from the Most High God, have been adopted with success by many other law–givers, who have thus insured respect for their institutions"

And Doane says:

"that almost all nations of antiquity have legends of their holy men ascending a mountain to ask counsel of the gods."

Thus, we see that other nations, beside the Hebrews, believed that their laws were actually received from God, that they had legends to that effect, and that a mountain figures conspicuously in the stories.

Professor Oort, speaking on this subject, says :

" No one who has any knowledge of antiquity will be surprised at this, for similar beliefs were very common. All peoples who had issued from a life of barbarism and acquired regular political institutions, more or less elaborate laws, and established worship, and maxims of morality, attributed all this and their birth as a nation, so to speak, to one or more great men, all of whom, without exception, were supposed to have received their knowledge from some deity."

NEW TESTAMENT STORIES

The Miraculous Birth of Jesus Christ

As we read in the New International Version of the Bible, in the Gospel according to Saint Luke chapter 1, verses twenty six to thirty five

"God sent the angel Gabriel to Nazareth, a town in Galilee, to a virgin pledged to be married to a man named Joseph, a descendant of David. The virgin's name was Mary. The angel went to her and said, "Greetings, you who are highly favoured! The Lord is with you." Mary was greatly troubled at his words and wondered what kind of greeting this might be. But the angel said to her, "Do not be afraid, Mary; you have found favour with God. You will conceive and give birth to a son,

and you are to call him Jesus. He will be great and will be called the Son of the Most High. The Lord God will give him the throne of his father David, and he will reign over Jacob's descendants forever; his kingdom will never end." "How will this be," Mary asked the angel, "since I am a virgin?" The angel answered, "The Holy Spirit will come on you, and the power of the Most High will overshadow you. So the holy one to be born will be called the Son of God."

The following quotes, again from Doane, suggest that the story of a virgin birth is not unique to Christianity:

"We shall now see, in the words of Bishop Hawes 'that God should, in some extraordinary manner, visit and dwell with man, is an idea which, as we read the writings of the ancient Heathens, meets us in a thousand different forms.'"

"Immaculate conceptions and celestial descents were so currently received among the ancients, that whoever had greatly distinguished himself in the affairs of men was thought to be of supernatural lineage. Gods descended from heaven and were made incarnate in men, and men ascended from earth, and took their seat among the gods, so that these incarnations and apotheosises were fast filling Olympus with divinities."

As we look through the literature, we find references to a virginal or miraculous birth occur in connection with a mixture of people of history and of mythical characters from folk tales and legends, though quite often, there is uncertainty over whether a particular character was real or myth.[376] [377] Then in the case of Jesus Christ, there is the notion that he was both human and divine. This was a source of continuing tension beginning soon after Jesus's death. For example, Gnostics believed that Jesus's body only "seemed" to be physical i.e., his body was not real.[378] In the life of Zoroaster (also known as Zarathustra – possibly lived around 1500 to 600 BCE but other scholars think it could have been as early as 6480 BCE), the law–giver of the Persians, these ideas are also expressed:

> " He was born in innocence, of an immaculate conception, of a ray of the Divine Reason. As soon as he was born the glory from his body enlightened the whole room, and Plato informs us that Zoroaster was said to be "the son of Oromasdes, which was the name the Persians gave to the Supreme God " ; therefore he was the Son of God."

There are also stories in relation to Buddha, suggesting, again in Doane, that he was born of the Virgin Maya or Mary. However, other sources make it clear that although there was something unusual and divine associated with the birth of Buddha, his was not a virgin birth. Some even argue that Buddha may have even been an Avatar of Vishnu (see under Hindu mythology) or even a re–incarnation of Laozi – founder of the Chinese Taoist tradition (also known as Daoist – The Way).

Taoist myths state that Laozi (around 600 BCE) was conceived when his mother gazed upon a falling star. He supposedly remained in her womb for sixty two years before being born while his mother was leaning against a plum tree. Laozi was said to have emerged as a grown man with a full grey beard and long earlobes, both symbols of wisdom and long life. It was said that he had existed from all eternity and that he had descended on earth and was born of a virgin. His wisdom is distilled in the text of the Laozi, though its authorship remains uncertain. He was a senior contemporary of Confucius (551 to 479 BCE), who has his own stories referencing a miraculous birth :

"At his birth a prodigious quadruped, called the Ke–lin (Unicorn type creature) appeared and prophesied that the new–born infant " would *be a king; without throne or territory." Two dragons hovered about the couch of Yen–she (his mother), and five celestial sages, or angels, entered at the moment of the birth of the wondrous child ; heavenly strains were heard in the air, and harmonious chords followed each other, fast and full. Thus was Confucius ushered into the world. His disciples, who were to expound his precepts, were seventy– two in number, twelve of whom were his ordinary companions, the depositories of his thoughts, and the witnesses of all his actions. To them he minutely explained his doctrines, and charged them with their propagation after his death."*

Yu and Hau–ki, other Chinese sages and heroes have stories of virgin–births attached to them, as does Fo–Hi, another Chinese god.

"The ancient Babylonians also believed that their kings were gods upon earth. A passage from Menaut's translation of the great inscription of Nebuchadnezzar, reads thus :

" I am Nabu–kuder–usur. the first–born son of Nebu–pal–usur, King of Babylon. The god Bel himself created me, the god Marduk engendered me, and deposited himself the germ of my life in the womb of my mother." "

Stories also circulated about virgin and miraculous births in association with the origins of the Roman Emperors, particularly Julius Caesar, King

Cyrus, Alexander the Great and even Plato. In Greek and Roman mythology, it was commonplace to accord such characters as Hercules, Prometheus, Perseus and Bacchus with virgin births between a god as father and a human – mother. Virgin births also featured in the origins of both the Egyptian Pharaohs, giving a sense of being themselves divine, and in the genealogy of their gods. It was also customary with the heroes of the northern nations, such as Danes, Swedes, Norwegians and Icelanders), to see them as offspring of their supreme god – Odin.

In Hindu mythology, from India, we have the pantheon of Hindu Gods, with scripture references to Vishnu, the supreme, having numerous "Avatars" (different forms or manifestations). These are said to occur when the world is in need, when things are going wrong, in order to restore Dharma or the right way of living. Quite how this happens is not made clear and the character of these Avatars varies with the different Hindu scriptures consulted. In Doane, there is reference to Krishna as the eighth Avatar of Vishnu being born to a virgin – Devaki, though this is not made explicit in other versions and Krishna was in fact Devaki's eighth child. Theologians differ in whether such appearances should be regarded as "incarnation" – in the way it is applied to the birth of Jesus Christ. Though of course, Jesus is seen as a living person whereas the Avatars are not regarded as being of human flesh and blood.

When expeditions set out West from Europe and discovered the Americas, similar concepts were also found. This was often much to the surprise of the explorers, who thought they were the first people versed in the Christian story to have reached there. Again from Doane:

"Across South America there is the notion of a virgin–born god and worshipped as a saviour. In Mexico he is known as Quetsalcoatle, and

272

was regarded with the highest veneration. The Mayas of Yucatan had Zama, and was the only–begotten son of their supreme god, Kin– chahan. The inhabitants of Nicaragua called their principal god – Thornathoyo."

"There are similar concepts evident in Peru, Brazil and Guatemala. Whilst in North America, the native Americans had their notion of the Great Spirit. For the Iroquois this was a beneficent being, uniting in himself the character of a god and man called Tarengawagan. He imparted to them the knowledge of the laws of the Great Spirit, and established their form of government. Among the Algojiquins, and particularly among the Ojibways and other remnants of that stock of the North–west, this intermediate great teacher is fully recognised. He bears the name of Mlchabou, and is represented as the first–born son of a great celestial Manitou, or Spirit, by an earthly mother, and is esteemed the friend and protector of the human race."

The Crucifixion of Christ Jesus
The story of the crucifixion of Jesus is told in all four Gospels (Matthew, Mark, Luke and John) in the Bible, though there are differences in some of the details. (as there are differences in the recounting of many of the stories and events in the scriptures and writings of and about other religions).The central Christian message of the crucifixion relates to the notion of atonement; that is, Jesus died because of human sin and has somehow taken those sins upon himself so that humankind could be brought into salvation; that is, saved from permanent death and separation from God.[379]

This notion of atonement is present throughout the Old Testament in the practice of animal sacrifice – the sins of a person being transferred to the

273

animal and with the sacrifice of the animal, the sins are forgiven by God. A similar idea was taken up by St. Paul in his understanding of the crucifixion of Christ and subsequent resurrection, but in this case, the sins are transferred to Jesus, but as a once and for all event. Since then, there have been different models, theories and understandings of the concept of atonement, proposed by the early Christian fathers,[380] and which continue amongst modern day theologians.[381] [382] It is not necessary here to discuss these different propositions but to simply highlight the fact there are references to the concept of atonement, crucifixion, the crucifix (a person suspended on a cross) and resurrection, other than within Christianity.

Atonement

Quoting from Doane:

"In primitive ages, when men lived mostly on vegetables, they offered only grain, water, salt, fruit, and flowers to the gods, to propitiate them and thereby obtain temporal blessings. But when they began to eat meat and spices, and drink wine, they offered the same ; naturally supposing the deities would be pleased with whatever was useful or agreeable to themselves. In the course of time, it began to be imagined that the gods demanded something more sacred as offerings or atonements for sin. This led to the sacrifice of human beings, principally slaves and those taken in war, then, their own children, even their most beloved first–born."

"It came to be an idea that every sin must have its prescribed amount of punishment, and that the gods would accept the life of one person as atonement for the sins of others. This idea prevailed even in Greece

and Rome : but there it mainly took the form of heroic self – sacrifice for the public good. Cicero says : "

The force of religion was so great among our ancestors, that some of their commanders have, with their faces veiled, and with the strongest expressions of sincerity, sacrificed themselves to the immortal gods to save their country" "

"This idea of atonement finally resulted in the belief that the incarnate Christ, the Anointed, the God among us, was to save mankind from a curse by God imposed. Man had sinned, and God could not and did not forgive without a propitiatory sacrifice. The curse of God must be removed from the sinful, and the sinless must bear the load of that curse. It was asserted that divine justice required BLOOD."

"The belief of redemption from sin by the sufferings of a Divine Incarnation, whether by death on the cross or otherwise, was general and popular among the heathen, centuries before the time of Jesus of Nazareth."

"The idea of removing sin by the sacrifice of a god can be found among the Hindus even in Vedic times (1500 to 500 BCE).The Rigveda (a collection of Indian Vedic Sanskrit hymns dated around 1700 to 1100 BCE), represents the gods as sacrificing Purusha, the primeval male, supposed to be coeval ("contemporary with") with the Creator. "

This idea is even more remarkably developed in the Tandya– Brahmanas ("collection of ancient Indian texts"), thus:

"The lord of creatures (_prajā–pati_) _offered himself a sacrifice for the gods."

Quotes from Abbé Huc (in – Travels in Tartary and Thibet) says ;
"In the eyes of the Buddhists, this personage (Buddha) is sometimes a man and sometimes a god, or rather both one and the other a divine incarnation, a man–god who came into the world to enlighten men, to redeem them, and to indicate to them the way of safety. This idea of redemption by a divine incarnation is so general and popular among the Buddhists, that during our travels in Upper Asia we everywhere found it expressed in a neat formula. If we addressed to a Mongol or a Thibetan the question Who is Buddha? he would immediately reply: The Saviour of Men! "

"The Indians are no strangers to the doctrine of original sin. It is their invariable belief that man is a fallen being / admitted by them from time immemorial. And what we have seen concerning their beliefs in Crishna and Buddha unmistakably shows a belief in a divine Saviour, who redeems man, and takes upon himself the sins of the world "

"The idea of redemption through the sufferings and death of a Divine Saviour, is to be found even in the ancient religions of China. One of their five sacred volumes, called the Y–Kmg, says, in speaking of Tien y the " Holy One "

" The Holy One will unite in himself all the virtues of heaven and earth. By his justice the world will be re–established in the ways of righteousness. He will labor and suffer much. He must pass the great

torrent, whose waves shall enter into his soul; but he alone can offer up to the Lord a sacrifice worthy of him"

"Alexander Murray says : The Egyptian Saviour Osiris was gratefully regarded as the great exemplar of self–sacrifice, in giving his life for others." His being the Divine Goodness, and the abstract idea of good, his manifestation upon earth (like a Hindoo god), his death and resurrection, and his office as judge of the dead in a future state, look like the early revelation of a future manifestation of the deity converted into a mythological fable " Horus was also called " The Saviour." "

In Greek mythology, Hermes, Hercules, the son of Zeus, Æsculapius and Apollo were all given the epithet "The Saviour"

"The Persians believed that they were tainted with original sin, owing to the fall of their first parents who were tempted by the evil one in the form of a serpent. They considered their law–giver Zoroaster to be also a Divine Messenger, sent to redeem men from their evil ways, and they always worshiped his memory." In the life of Zoroaster the common mythos is apparent. He was born in innocence, of an immaculate conception, of a ray of the Divine Reason. As soon as he was born, the glory arising from his body enlightened the room, and he laughed at his mother. He was called a Splendid Light from the Tree of Knowledge"

Dr. Inman says :

"There are few words which strike more strongly upon the senses of an inquirer into the nature of ancient faiths, than Salvation and Saviour. Both were used long before the birth of Christ, and they are still common among those who never heard of Jesus, or of that which is known among us as the Gospels."

Crucifixion

"Quotes from J P Lundy [383]

The monk Georgius, in his _Tibetinum Alphabetum_ (p. 203), has given plates of a crucified god who was worshiped in Nepal. These crucifixes were to be seen at the corners of roads and on eminences. He calls it the god Indra." Huggins suggests this is inaccurate. [384]

"While speaking of "_a cross with a man on it_" as being carried by the Pagan Romans as a _standard_, we might mention the fact, related by Arrian the historian, that the troops of Porus, in their war with Alexander the Great, carried on their standards the figure of a man. Here is evidently the crucifix standard again.

"This must have been (says Mr. Higgins) a Staurobates or Salivahana, and looks very like the figure of a man carried on their standards by the Romans. This was similar to the dove carried on the standards of the Assyrians. This must have been the crucifix of Nepaul."

"Tertullian, a Christian Father of the second and third centuries, writing to the Pagans, says:"The origin of your gods is derived from figures moulded on a cross. All those rows of images on your

standards are the appendages of crosses; those hangings on your standards and banners are the robes of crosses."

"We have it then, on the authority of a Christian Father, as late as A.D. 211, that the Christians " neither adored crosses nor desired them," but that the Pagans "adored crosses," and not that alone, but "a cross with a man upon it." Jesus, in those days, nor for centuries after, was not represented as a man on a cross. He was represented as a lamb, and the adoration of the crucifix, by the Christians, was a later addition to their religion."

"In the " Celtic Druids" Mr. Higgins describes a crucifix, a lamb, and an elephant"

"If we turn to the New World, we shall find, strange though it may appear, that the ancient Mexicans and Peruvians worshiped a crucified Saviour. This was the virgin–born Quetzalcoatle whose crucifixion is represented in the paintings of the " Codex Borgianus" and the " Codex Vaticanus. EXPLORERS supRISED AND WHEN ASKING WHO IT WAS, RECEIVED THE FOLLOWING – Bacob (Quetzalcoatle), the Son of God, who was put to death by Eopuco. They said that he was placed on a beam of wood, with his arms stretched out, and that he died there."

"Crosses were also found in Yucatan, as well as Mexico, with a man upon them.

Resurrection and Ascension

The story of the resurrection of Christ Jesus is told in all four Gospels. After being crucified, his body was wrapped in a linen cloth and laid in a tomb, and a large stone rolled across the entrance and sealed for security.

On the first day of the week some of Jesus's followers came to see the tomb but found that the stone had been moved and the tomb was empty – Jesus had risen from the dead.

The story of his ascension as told in Mark's Gospel, says that Jesus was received up into heaven, and sat on the right hand of God. In Luke, it says Jesus was carried up into heaven and in the Acts of The Apostles (New Testament) it says Jesus was taken up (to heaven) and a cloud received him out of sight.

"Crishna, the crucified Hindoo Saviour, rose from the dead and ascended bodily into heaven. At that time a great light enveloped the earth and illuminated the whole expanse of heaven. Attended by celestial spirits, and luminous as on that night when he was born in the house of Vasudeva, Crishna pursued, by his own light, the journey between earth and heaven, to the bright paradise from whence he had descended."

"Samuel Johnson, in his " Oriental Religions," tells us that Rama, an incarnation of Vishnu after his manifestations on earth, "at last ascended to heaven" "resuming his divine essence."

Although Buddha is thought to have died by a tree, there are versions that suggest he :

"ascended bodily to the celestial regions when his mission on earth was fulfilled, and marks on the rocks of a high mountain are shown, and believed to be the last impression of his footsteps on this earth. By prayers in his name his followers expect to receive the rewards of paradise, and finally to become one with him, as he became one with the Source of Life."

"Zoroaster, the founder of the religion of the ancient Persians, who was considered " a divine messenger sent to redeem men from their evil ways," ascended to heaven at the end of his earthly career. To this day his followers mention him with the greatest reverence, calling him " The Immortal Zoroaster"

As recorded by Ovid:

"Aesculapius, the Son of God, the Saviour, after being put to death, rose from the dead."

There are numerous references to The Saviour Adonis or Tammuz, rising from the dead.

"Dr. Prichard, in his " Egyptian Mythology" tells us that the Syrians celebrated, in the early spring, this ceremony in honour of the resurrection of Adonis. After lamentations, his restoration was commemorated with joy and festivity."

And to the resurrection of Osiris:

"Alexander Murray says :

"The worship of Osiris was universal throughout Egypt, where he was gratefully regarded as the great exemplar of self–sacrifice in giving

281

his life for others as the Manifester of Good, as the Opener of Truth, and as being full of goodness and truth. After being dead, he was restored to life."

There are also references to Horus (son of Egyptian god Isis), Bacchus, Mithras (the Persian Saviour), Hercules and Atys (the Phrygian Saviour) all being resurrected.

"The ancient Scandinavians also worshiped a god called Frey, who was put to death, and rose again from the dead"

"The ancient Druids celebrated, in the British Isles, in heathen times, the rites of the resurrected Bacchus, and other ceremonies, similar to the Greeks and Romans. "

"Quetzalcoatle, the Mexican crucitied Saviour, after being put to death, rose from the dead. His resurrection was represented in Mexican hieroglyphics, and may be seen in the Codex Borgianus"

"In the words of Duubar T. Heath :

" We find men taught every where, from Southern Arabia to Greece, by hundreds of symbolisms, the birth, death, and resurrection of deities, and a resurrection too, apparently after the second day, i. e., on the third."

SUMMARY – SO FAR

The foregoing does not document an evolutionary process in terms of a linear progression from one simple idea to something more complex. Neither does it record the changes occurring to a single idea over time. But it does demonstrate that similar concepts have occurred in widespread parts of the world, at different times and in differing degrees of detail. Although the specifics may vary, cultures across the world have, over the millennia, developed similar ideas and similar approaches to questions of creation, good and evil, life after death, how to influence and control events and so on. This perhaps is not so surprising, although it is not known, certainly in ancient times, how much cross–fertilisation of ideas may have taken place.

Were the germs of these ideas present before the exodus out of Africa and then evolved along similar paths ? Did similar ideas arise spontaneously in different places and at different times ? Or perhaps did a "common brain" in the "common ancestor" function in similar ways and so develop similar questions and hence come to similar solutions ? The answer to these questions is unknown at present but perhaps there is a sense of timeliness for certain ideas. The notion of "memes" being transmitted through "the ether" (see chapter 3) seems a rather poetic description. As the ideas arrive and settle in different places, they are adopted and adapted in ways particular to that culture, time and place. Ideas evolving in incremental steps, from a common source or by a common process, building on what was known and present previously, but whose true origins are lost in the mists of time. So, what we become aware of now and throughout history are the "islands" of seemingly disconnected knowledge, whose "isthmuses" of connection have sunk and been covered beneath the "seas" and "waves" of evolution and development – progress !

The adoption and or adaption of ideas from other cultures and places is a natural part of human evolution. It may be one way of increasing respect and reducing tension between different cultures. Seeing the value and relevance of ideas and practices from other places and weaving them into the fabric of one's own culture (syncretism) seems to be a positive approach to explaining events and experiences occurring, at the time, in your neck of the woods. Thus, the meaning originally associated with the idea in its previous home can change, taking on a whole new significance in its new culture, whilst still appearing on the surface to be the same. Hence the importance of context and the need to be immersed in a culture if a true understanding of their beliefs and practices is to be gained. From this perspective, the inappropriateness of extrapolating a "myth" from one setting and assuming an equivalent significance in another, is clear.

Perhaps also there is a need for some sort of preparation. For people to accept the idea of a virgin birth, crucifixion, resurrection and ascension completely out of the blue, or to think these ideas up as explanations for the uniqueness of Christ would be difficult. But if the concepts are already in the "aether" and people are already acquainted with them, then adopting and adapting them to their own time and requirements is not such a big step. Such may be God's way of working and revealing himself at points in time when people may comprehend.

MAJOR RELIGIONS

Having looked at the idea of there being similar concepts, stories and practices between cultures and religions, we will now turn our attention to the particular history and development of some of the major identified religions. From the many major religions we might look at, I have chosen the top three, in terms of numbers of members and so are probably the more recognised – Christianity, Islam and Hinduism. I also include Judaism, because it too is probably readily recognised and also because of its significance in relation to Christianity. At first sight, it may appear that the origins of these traditions would be readily known, certainly for

Christianity and Islam. We know the date of Jesus's birth, at least to within a few years and similarly with the birth of the prophet Mohammad and hence the beginning of Islam. However, for Hinduism, although probably the oldest of the major religions and Judaism the position is not quite as clear cut.

We mentioned earlier, in chapter six, the difficulties and pitfalls in tracing origins, but in terms of the purpose of this book, it is necessary to identify some of the changes that have taken place within these faith movements, so their origins are of interest. Though it is more important to recognise that changes have taken place over time, rather than identifying the precise timing of events. Change is generally a process over time. Even if the precise timing of a decision can be identified, the consequences and implementation of any change can take years to occur and often does so imperceptibly. Even in the face of a cataclysmic event, the consequences may take years to work through.

In this section, we are seeking to discover how the major organised religions have emerged and developed. We use the sense of evolution, in this context, to indicate change rather than being an assumed progress from something primitive to something more sophisticated. Over time, particularly in isolated groups, their culture and ideas may have continued with little change, whilst others will have been touched by more outside influences. So for ethnic and folk religions, what is observed now, may or may not be how it has always been. However, it does seem that the major or organised religions do not represent a mature or sophisticated form of ethnic religion, but have emerged in their own right, though by incorporating some antecedent religious practices.

It is perhaps worth noting that one characteristic difference between folk and ethnic religions, and the major religions is that within the latter there is often a recognised set of philosophical or theological tenets that form the basis of belief. Though integral to the faith, it is described separately from the practicalities of operating or living the religion. This philosophical underpinning may undergo rigorous theological and academic scrutiny over the years, that is outside the mainstream experience of the members of the faith. Such ideas may be known to the leaders of the Faith, but are often not transmitted to the ordinary lay people, whose day to day lives may be guided by different understandings of the principals of their religion.[385] Thus, the level of understanding and practice of the ordinary lay person may actually represent another aspect of folk religion, in that it may, in part, be at variance with that of the hierarchy and theological elite and so represent a sub–set of the main beliefs as laid down by that Faith.

We will look at the four major faiths chosen, in order of age. Hinduism is believed to be the oldest, followed by Judaism, Christianity and Islam. It will become clear that none of them exist now in the same form as when they began. All have changed in various ways and it is these changes and additions that are the focus of this section, rather than the detail of the religion itself. So let us begin.

HINDUISM

As already mentioned, Hinduism by name didn't exist until the early nineteenth century, when the word was coined in the English language. Despite the fact that the rituals, practices and beliefs stemmed from traditions that had been in existence for several thousand years. The emphasis of Hinduism is on personal spirituality, though the story of Hinduism itself is closely linked to the social and political developments associated with the rise and fall of different kingdoms and empires. The early history of Hinduism is difficult to date, but Hindus themselves tend to be more concerned with the substance of their scriptures rather than specific dates. In general, they tend to see time as cyclical and eternal rather than linear. Nevertheless, various periods of Hindu history are identified and the following list represents one categorisation:

- Pre–history and Indus Valley Civilisation (until around 1750 BCE);
- Vedic period (around 1750 to 500 BCE);
- "Second Urbanisation" (around 600 to 200 BCE);
- Classical Period (around200 BCE to 1200 CE);
- Pre–classical period (around 200 BCE to 300 CE);
- "Golden Age" (Gupta Empire) (around 320 to 650 CE);
- Late–Classical period (around 650 to 1200 CE);
- Medieval Period (around 1200 to 1500 CE);
- Early Modern Period (around 1500 to 1850);
- Modern period (British Raj and independence) (from around 1850 CE).[386]

Western scholars regard Hinduism as a fusion or synthesis of various ancient Indian cultures and tribal traditions, originating from across India and the Indus Valley Civilisations, taking place during the Vedic period. After this, at the beginning of the Pre–classical period, there occurs something known as the "Hindu synthesis." At this time, all sorts of other influences and ideas are incorporated, including Buddhism, Jainism and various forms of asceticism in which the materialism of the world is renounced (Śramaṇic). Then, as we enter the Gupta period (fourth to

287

sixth centuries) we find a society ordered in accordance with Hindu beliefs, particularly based on a strict caste or class system. This created a peaceful and prosperous society that encouraged an upsurge in scholarship, thus fostering the emergence of the classical schools of Hindu philosophy, along with works, in classical Sanskrit, on medicine, veterinary science, mathematics, astrology, astronomy and astrophysics.

Following this period, in the eighth–century, power became decentralised in India, leading to regionalisation of religion and religious rivalry. Local cults and languages were enhanced, and we find the emergence of the Bhakti movement. This started in South India (now Tamil Nadu and Kerala), and spread northwards. It swept over east and north India from the fifteenth century onwards, reaching its greatest popularity between the fifteenth and seventeenth centuries CE.

The Bhakti movement developed locally around different gods and goddesses, such as Shiva and Vishnu, and was preached using local languages, ensuring the message reached the general population. It was inspired by many poet–saints, championing a wide range of philosophical positions and has traditionally been seen as an influential social reformation in Hinduism. It provided an individually focused alternative path to spirituality, regardless of caste of birth or gender. At this time, Buddhism lost its prominent position and began to disappear in India and the Hindu gods gained the ascendency.

From around the fifth to the thirteenth centuries, Hinduism flourished and expanded into South–East Asia where a powerful Indian colonial empire developed. This included a maritime empire centred on the island of Sumatra in Indonesia, and a program of temple building in Cambodia, Vietnam, Thailand and what is now Myanmar. Angkor Wat in Cambodia became one of the largest Hindu monuments in the world. Hinduism also developed links with Persia and Mesopotamia during the eras of the Sassanid and Abbasid caliphates. By the time of the British Raj (the rule of India by the British Crown), from the mid nineteenth century, the existing religious orthodoxies, particularly with respect to women, marriage, and the dowry and caste systems, were already being questioned.

Members of the elite began the Bengal Renaissance or Hindu Revival, espousing rationalism and atheism as well as developments in the arts, sciences and religion.[387] Many of the ideas of the movement were accepted by the government and society in general, even though as a movement it did not establish a great following. With the advent of Indology as a European academic discipline for studying Indian culture, came the notion of "Hinduism" as a unified body of religious practice and the popular picture of 'mystical India'. This was fertile ground for the promotion of the Ramakrishna Mission by Vivekananda, a devotee of Ramakrishna, a prominent figure in the Bengal renaissance. In the twentieth century, Hinduism has gained prominence as a political force and a source for national identity in India. It provides the basis for the Jana Sangha and Bharatiya Janata Party (BJP) in electoral politics.[388]

Scriptures

Understanding the organisation of the Hindu scriptures and writings is pretty complicated. Broadly speaking, the original scriptures are known as the Vedas and are classed as Śrutis, literally – "what is heard" literature. These are considered to be author–less, transmitted verbally across the generations and fixed. This is in contrast to other writings such as the sub–divisions of the Vedas and other supportive texts which are classed as Smriti – literally "that which is remembered," usually attributed to an author and traditionally written down, but constantly revised. This distinction is not precise with some of the Upanishads (see below) for example being classed as Śrutis and others as Smriti, though not all scholars agreeing the classifications. Each Smriti text exists in many versions, with many different readings. Smritis were considered fluid and freely rewritten by anyone in the ancient and medieval Hindu tradition.The smriti texts of the period between 200 BCE to 100 CE proclaim the authority of the Vedas. Acceptance of which became a central criterion for defining Hinduism, over and against the heterodoxies (see below) which rejected the Vedas.

Texts also refer to successive ages designated respectively as golden, silver, copper and iron. During the golden age people were pious and adhered to dharma (law, duty, truth) but its power diminished over time until it had to be reinvigorated through divine intervention. These interventions take the form of a god visiting the earth as an avatar of Vishnu (Avatars are manifestations in other forms – so there have been nine Avatars of Vishnu so far, the most familiar being Krishna, the eighth). With each successive age, good qualities diminish, until reaching the current iron or dark age, designated as one marked by cruelty, hypocrisy and materialism.[389] Most of the basic ideas and practices of classical Hinduism derive from the new smriti literature.

The basic classification divides the Vedas into four categories, with the Rigveda being the oldest work, probably from the period of 1700 to 1100 BCE. However, no actual evidence of this text exists prior to the thirteenth century CE:

- Rigveda (RV) – reciting hymns
- Yajurveda (YV) – performing sacrifices chanting songs

- Samaveda (SV) – chanting songs

- Atharvaveda (AV)

Each Veda is then sub–classified into four major text types – the Samhitas (mantras and benedictions), the Aranyakas (text on rituals, ceremonies such as newborn baby's rites of passage, coming of age, marriages, retirement and cremation, sacrifices and symbolic sacrifices), the Brahmanas (commentaries on rituals, ceremonies and sacrifices – the oldest being around 900 BCE), and the Upanishads (text discussing meditation, philosophy and spiritual knowledge). The Upasanas (short ritual worship–related sections) are considered by some scholars to be a fifth part. The Vaishnava Upanishads are minor Upanishads of Hinduism, related to Vishnu theology. There are fourteen Vaishnava Upanishads in the Muktika anthology of 108 Upanishads. It is unclear when these texts were composed, but were probably all written down by the first century BCE.

Other auxiliary texts, are accepted as commentaries to help understanding of the main scriptures. These include: the Vedangas (developed towards the end of the Vedic period, around or after the middle of the first millennium BCE). These are six auxiliary disciplines developed in ancient times and connected with the study of the Vedas, Parisistas and Upavedas, including the Āyurveda. The Ayurveda relates to the medicinal use of plants, trees and shrubs etcetera. The Puranas are a vast genre of encyclopaedic Indian literature about a wide range of topics particularly myths, legends and other traditional folklore. The first Puranas were written during the "Golden Age" and intended to establish a mainstream religious ideology to help pre–literate and tribal groups embrace the Hindu faith and culture. By transforming some of the other

291

Smriti writings and absorbing many of the other traditions, Puranic Hinduism became established as a significant strand of Hinduism.

In addition, there are the writings associated with the schools of Hindu philosophy (darśanas). These include six systems or ṣaḍdarśana:

- Samkhya, an atheistic theoretical exposition of consciousness and matter.

- Yoga, a school emphasising meditation, contemplation and liberation.

- Nyāya or logic, which explores sources of knowledge.

- Vaiśeṣika, an empiricist school of atomism.

- Mīmāṃsā, an anti–ascetic and anti–mysticist school of orthopraxy.

- Vedānta, the last segment of knowledge in the Vedas and by adopting from the other schools came to be the dominant current of Hinduism in the nineteenth century.

These are known as the Astika (orthodox) philosophical traditions which accept the Vedas as an authoritative, important source of knowledge. However, there are other sources of philosophy that share similar philosophical concepts but reject the Vedas, and these have been called nāstika (heterodox or non–orthodox) Indian philosophies, and include Buddhism, Jainism and others. The Prasthanatrayi is a collective term for the writings associated with the Vedanta school and include the Principal Upanishads, the Brahma Sutras and the Bhagavad Gita. (The Bhagavad Gita was composed in the fifth to second century BCE and introduces the three ways to spiritual freedom and release – bhakti marga (the path of faith/devotion), karma marga (the path of works) and jnana marga (the path of knowledge).

In the eighth century CE Adi Shankara is credited with unifying and establishing the main currents of thought in Hinduism by bringing all the Vedic communities together. He tried to remove the non–smriti aspects that had crept in, by arguing that any of the different Hindu gods could be worshipped. He established that worship of the various gods was compatible with Vedas as all were different manifestations of one god Brahman. By the twelfth century CE, further changes were unfolding and followers of the Bhakti movement moved away from the abstract concept of Brahman of Adi Shankara and promoted a more emotional and passionate devotion towards the more accessible Avatars of Vishnu, especially Krishna and Rama. From this point through to the sixteenth century CE attempts were being made to bring all the six systems of Hindu philosophy together.

Summary

It can be seen that Hinduism is neither one simple faith belief nor has it remained unchanged since its cosmopolitan beginnings. Its ethos is that it will continue to evolve as it responds and adapts to changes around it. But its core beliefs continue:

- that all creatures have souls and these are re–incarnated (in accordance with the degree of good "Karma" exhibited during the earthly life) until it reaches salvation and is one with the supreme being
- there are many paths that lead to salvation
- a belief in Dharma – which means moral, sacred, and obligatory duties that directly arise from God and which are

vital to the creation, preservation, and destruction of the worlds and beings, and for the order and regularity of the worlds

- the truth of the Vedas, Brahman – the supreme being and creator (of which the other gods are manifestations ✣)

- God is the creator and upholder of Dharma[390]

✣ [Brahman manifests as the Trimurti (triumvirate) – Vishnu, Brahma and Siva in their roles as the preserver, creator and destroyer respectively. He is partnered by Shakti the mother goddess. There are hundreds of other gods, with Garuda, the eagle god and Lord Ganesha, the elephant god attending Vishnu and Shiva respectively. Hindus also worship many saintly persons. Despite this pantheon of gods, which have accumulated over the centuries, often stemming from local communities or particular groups, Hinduism is thought of as a monotheistic faith as Brahman is regarded as the One Supreme Being]. There are many other aspects to Hinduism and its complexity means no one can know every minutia of it, so there is no single unified understanding of Hinduism beyond that of these general underlying values. Though even these are not understood in precisely the same way by all those calling themselves Hindus.

JUDAISM

As we begin to explore the history of Judaism, it is worth starting by looking at some of the terms associated with Judaism and being Jewish. However, it is difficult to disentangle the use of such terms as Judaism, Jews, Jewish, Israelite, and Hebrews and it is not at all clear when such terms came into existence, and in the case of Hebrews, ceased. Discovering the origin of Judaism is fraught with all sorts of difficulties.[391] The concept of Judaism does not easily fit into conventional Western categories of religion, ethnicity, or culture, as it encompasses all these things and has been shaped by a history of recurring conquest and exile.[392] Also, there are different understandings of what qualifies someone to be called a Jew. Rabbinic Judaism defines this as being born of a Jewish mother.

Other strands of Judaism such as Liberal Judaism are happy with either parent being Jewish, whereas Karaite Judaism emphasise the need for the father to be Jewish. Traditional Rabbinic Judaism accept that "once a Jew, always a Jew" but other strands argue that if one converts to another faith or rejects the religious aspects of Judaism, then one's Jewishness is forfeited. In addition, not all Jews identify themselves as Jewish, and some who identify as Jewish are not considered so by other Jews. The name Israel first appears in non–biblical sources around 1209 BCE, in an inscription relating to the Egyptian pharaoh Merneptah, on the Merneptah Stele (an erected stone tablet) and in the Hebrew Bible in Genesis Chapter 39 verse 29. It refers to the renaming of Jacob, who wrestled with an angel, who gave him a blessing and renamed him Israel because he had "striven with God and with men, and prevailed."

Other definitions suggest that: "Israelite" refers specifically to the direct descendants of any of the sons of Jacob. His descendants, as a people, were collectively called – "Israel." During the period of the divided monarchy (see below) "Israelites" was only used to refer to the inhabitants of the northern kingdom – The Kingdom of Israel. This group formed the ethnic stock from which Samaritans trace their ancestry. Its use then changes to cover the people of the southern kingdom – The Kingdom of Judah, after the Babylonian exile (586 BCE). Modern Jews are named after and also descended from the southern Israelite Kingdom of Judah. "Hebrews" on the other hand denotes the Israelites' immediate forebears who lived in the land of Canaan, the Israelites themselves, and their descendants, including Jews and Samaritans.

The name "Jews" refers to the descendants of the Israelites who coalesced when the Tribe of Judah absorbed the remnants of various other Israelite tribes. Thus, for instance, Abraham was a Hebrew but he was not technically an Israelite nor a Jew, Jacob was both a Hebrew and the first Israelite but not a Jew, while King David (as a member of the Tribe of Judah) was all three, a Hebrew, an Israelite, and a Judahite. A Samaritan, on the contrary, while being both a Hebrew and an Israelite, is not a Jew.[393]

These definitions are not universally agreed, but they give a flavour of the complexity that surrounds the Jewish heritage. This mixed picture of identities flows from the varied history and experience of a nation whose origins are equally difficult to disentangle. The biblical description of the origins of the Jewish people, whilst not necessarily providing a detailed totally historically accurate account, does help build a narrative of how things evolved:

Starting with Abraham and Sarah, (about 4000 years ago is the estimated dating of Abraham), they have a son Isaac who has two sons Jacob and Esau. Jacob was renamed Israel after a mysterious incident in which he wrestles all night with God or an angel and prevails. He then has twelve sons – Reuben, Simeon, Levi, Judah, Dan, Naphtali, Gad, Asher, Issachar, Zebulun, Joseph and Benjamin who become the twelve tribes of Israel. Mannasseh and Ephraim, the two sons of Joseph are also incorporated into this set up, though there are different accounts of how this may have played out. Jacob and his sons are forced by famine to go to Egypt, though Joseph was already there, and an indispensable servant to the Pharaoh. Over the years, the family and the Pharaoh die. But the new Pharaoh becomes worried about the vast increase in the number of Israelites and enslaves them. He also kills all new born baby boys. But baby Moses is hidden and grows up.

At the age of forty Moses kills an Egyptian, and escapes into the Sinai desert. He makes a new life there and acquires a family but at the age of 70, God, saying his name as Yahweh, sees him through a "burning bush" and persuades him to take the Israelites out of Egypt. Moses does so, crossing the Red Sea and into the desert – the Exodus. They camp near Mt. Sinai and Moses goes up to receive the Ten Commandments from Yahweh. He comes down and the people agree to follow God's laws. Because they then disobey God, they have to wander around the desert for 40 years. Moses dies and it is Joshua who leads them into the land of Canaan, in accordance with the promise made to Abraham by Yahweh. Land is allocated to the tribes by lottery. This narrative can be found in the Book of Genesis in the Bible and it records the entry into Canaan as a violent take over.

At this point, we will turn to the archeologically based version, which identifies a number of time periods, as follows:

- Bronze Age: Early and Intermediate: 3500 to 1600 BCE
- Bronze Age – Late: 1600 to 1200 BCE
- Iron Age I: 1200 to 1000 BCE
- Iron Age II: 1000 to 586 BCE
- Neo–Babylonian: 586 to 539 BCE
- Persian: 539 to 332 BCE
- Hellenistic: 333 to 53 BCE

Other academic terms often used are:

- First Temple period (around 1000 to 586 BCE)
- Second Temple period (around 516 BCE to 70 CE)

The main characteristic of Bronze Age civilisations relates to the production of bronze (an alloy of eighty to ninety percent copper and usually tin but other metals can be used) tools and weapons. This knowledge reached different societies at different times, presumably by spontaneous discovery arising in different places and also by groups interacting with each other. In addition to the knowledge of bronze working, many of these societies were also well advanced in other ways – in intensive year–round agriculture, use of writing systems and many had invented the wheel for both transport and pottery making. Some also had well organised systems of government, economic and civil administration and social structures and engaged in architectural projects as well as studying astronomy, mathematics and astrology. So in Mesopotamia we have the kingdoms of Sumer, Akkadians, Ur, Assyrians, Babyonian and Kassite kingdoms with evidence of the presence of Arameans and Chaldeans. In Anatolia (modern day Turkey) we have the Hittites and the Sea Peoples. In the Levant – including parts of Egypt, Iraq, Israel, Jordan, Lebanon, Palestine, Syria, and Turkey – there were Canaanites, Philistines, Pheonicians and Amorites. The Arameans were a semi–nomadic and pastoralist people originating in what is now, modern day Syria, during the Late Bronze Age and the Early Iron Age. They were entirely absorbed into the Neo–Assyrian Empire by the eighth century BCE. Other kingdoms in the area included Ammon, Edom and Moab, and itinerant groups such as Habiru and Šośu.

Some scholars believe that with the presence and intermingling of all these peoples in the area, it is likely that the Israelites emerged out of a

298

fusion of some of these different groups, many of which had their own kings and their own gods and religions. The emerging Israelites would thus initially have adopted the local polytheistic practices. But gradually, they developed a move towards Henotheism. This is the name used to describe religions that worship many gods, but actually have a particular one that is regarded as the highest and most important. This, it is argued, may be the process by which Yahweh comes to prominence as the god of the Israelites. It would be natural that many of the characteristics attached to these local gods, will be adopted and applied to describe Yahweh. Some scholars suggest the Canaanite god – El, may have been the "role model" but there were other local gods included Baal, Ishtar, that may have been the "prototype."

This, I suggest, would be part of an evolutionary process of trying to understand the nature of Yahweh, and using words and established practices to express this understanding, as far as they were able to at that particular time in history. It seems to me that we are part of that tradition – using our words, knowledge and experiences to express our understanding of the nature of God. At this stage, Yahweh was not yet exclusive, but during and following the Babylonian exile (see later), a monotheistic religious understanding emerges with Yahweh being regarded as the one true God.

Thus, scholars argue that the Israelite ethnic identity originated, not from the Exodus and a subsequent conquest, but from a transformation of the existing Canaanite–Philistine and other local cultures. The dates for this are uncertain, perhaps as early as 1500 BCE or more likely more towards the end of the Late Bronze Age. During this period there was a strong Egyptian influence imposed on a Canaanite city–state system with Jerusalem as a successful city–state, overlorded by the Egyptians.

This arrangement and Canaanite society began to break down during the Late Bronze Age and Canaanite culture was gradually absorbed into that of the Philistines, Phoenicians and Israelites.The process happened gradually with a strong Egyptian presence continuing into the twelfth century BCE.

We will now go back to the biblical version:

Eventually the Israelites ask God for a king, and Yahweh gives them Saul. David succeeds Saul, and manages to establish a united monarchy. Under David's son, Solomon they construct the Holy Temple in Jerusalem. On the death of Solomon and during the reign of his son, Rehoboam, the kingdom is divided into – the northern kingdom called Israel or Samaria, formed by the tribes of Reuben, Simeon, Levi, Dan, Naphtali, Gad, Asher, Issachar, Zebulun, and Joseph (as those of Mannasseh and Ephraim), and the southern kingdom or Judah (Judea) based around Jerusalem and formed by the tribes of Judah and Benjamin.

The biblical and archeological versions then fairly well correspond from this point, though the underpinning reasoning may be different. The biblical version emphasising the events that follow are a result of the Israelites disobeying God and so their subsequent experience is actually punishment.

The kings of the northern Kingdom of Samaria are uniformly bad, permitting the worship of other gods and failing to enforce the worship of Yahweh alone, and so Yahweh eventually allows them to be conquered and dispersed among the peoples of the earth; and strangers rule over their remnant in the northern land. In Judah some kings are good and enforce the worship of Yahweh alone, but many are bad and permit other gods, even in the Holy Temple itself, and at length Yahweh allows Judah to fall to her enemies. The people are taken into captivity in Babylon, the land left empty and desolate, and the Holy Temple itself destroyed. Yet despite these events Yahweh does not forget his people. He sends Cyrus, king of Persia to deliver them from bondage. The Israelites are allowed to return to Judah, the Holy Temple is rebuilt, the priestly orders restored, and the service of sacrifice resumed. Through the offices of the sage Ezra, Israel is constituted as a holy nation, bound by the Torah and holding itself apart from all other peoples.[394]

The archeological explanation just describes what happens as a matter of history, as a matter of fact, that disputes happen between different and rival groups and nations. So by the Iron Age, the Israelites had split into the kingdoms of Israel (or Samaria) in the north and Judah in the south, centred on Jerusalem. The two kingdoms shared Yahweh as their national god. By the late eighth century BCE, both Judah and Israel had become subjected to Neo Assyrian rule. But Israel rebelled and was destroyed in 722 BCE, and its people were either deported or they fled to Judah.

In the seventh century BCE, Jerusalem prospered and dominated its neighbours, by working with the Neo–Assyrian Empire. This was probably part of the occupiers policy to establish Judah as an Assyrian vassal state controlling the valuable olive industry. But in the last half of the seventh century BCE, Assyria suddenly collapsed, leaving a power vacuum, to be fought over by the Egyptian and Neo–Babylonian Empires. This led to the destruction of Judah in a series of campaigns between 597 and 582 BCE and Jerusalem was destroyed by the Babylonians in 586 BCE. The leaders of Judah were taken into exile in Babylon though most of the ordinary people stayed put in Jerusalem. For them, life after the fall of Jerusalem probably went on much as it had before and may even have improved by acquiring the land and property of the deportees. This division set the scene for tension and rivalry, once some of the exiles began to return after 539 BCE.

When Babylon fell, in its turn, to the Persian Cyrus the Great in 539 BCE, Judah became an administrative division within the Persian empire. It is likely that this Persian connexion, facilitated some cross fertilisation of Zoroastrianism and the Israelites worship of Yahweh. One of the first acts of Cyrus was to commission the exiles, led by Nehemiah and Ezra, to return to Jerusalem and rebuild the Temple. These were the descendants of the original exiles, and had never lived in Judah. Nehemiah was charged with rebuilding Jerusalem and Ezra provided the spiritual leadership. The Babylonian conquest had resulted not just in the destruction of Jerusalem and its temple, but damaged the religious infrastructure which had been built up under the Davidic kingship. This forced the leaders of the exile community – kings, priests, scribes and prophets – to reformulate the concepts of community, faith and politics. So on returning to Judah, they took charge of setting the spiritual direction of their "new nation." They did this by consolidating the monotheistic idea of Yahweh being the one true God and that the people were "Israel" and were to remain distinct and apart from other nations.

This "new Israel," comprised descendants of the inhabitants of the old Kingdom of Judah, returnees from the Babylonian exile community, Mesopotamians who had been exiled themselves to Samaria in earlier periods, Samaritans, and others. But not all Jews returned. Many remained as a community or "diaspora" in Samaria and also many migrated to Egypt.

Ezra and Nehemiah attempted to re–integrate the various factions into a united and ritually pure society, inspired by the prophecies of Ezekiel and his followers, but met with some resistance. However, they did lay the foundations for an Israelite nation and Judaic religion and the beginnings of a written scriptural canon. They consolidated the monotheistic idea and clarified the importance of adhering to the established religious laws. So by the mid–fifth century BCE, Judah had become, in practice, a theocracy, ruled by hereditary high priests, but with a Persian–appointed governor, charged with keeping order and ensuring the taxes were collected and paid.

Although the complex politics and detailed history of the region is not central to the book, the following is a brief synopsis of what happened in the area, after 404 BCE when the Persians lost control of Egypt. Alexander the Great overran Persia in 333/332 BCE. Thus, began the start of the Hellenistic period in the Levant. This was followed by Ptolemy I who, as the ruler of Egypt in 322 BCE took over Judah in 320 BCE. He was followed by the Selucids from Syria, in 198 BCE. Their ruler, Antiochus IV Epiphanes (174 to 163 BCE) tried to impose Hellenic cults on Judea, thus sparking the Maccabean Revolt and the eventual establishment of an independent Jewish kingdom under the Hasmonean dynasty. In 63 BCE the Roman general Pompey conquered Jerusalem and brought the Jewish kingdom into Roman jurisdiction. In 40 to 39 BCE, Herod the Great was appointed King of the Jews by the Roman Senate, and on his death in 6 CE the Roman Emperor Augustus, brought it back under Roman administration. More heavy taxes and insensitivity towards the Jewish religion followed, leading to further revolt (the First Jewish–Roman War, 66 to 73 CE), and in 70 CE the Roman general (and later emperor) Titus, captured Jerusalem. He destroyed the Temple and brought to an end the Second Temple period.

In 135 CE, the area became known as Syria Palaestina until about 390 CE. This was followed by the Byzantine period under Christian control until in 634 CE various Arab groups, with one usurping the previous one, were in control until the first Crusade in 1099 CE. During the next 200 years the Christians and Muslims waged war until 1291 CE when Muslims again gained control in the form of the Mamluks and then from 1517 CE to 1917 CE, the Ottomans held sway. All of this "activity" fuelled what is known as the Jewish Diaspora. The terminology used to describe the emigration of Jews out of Judah and Jerusalem to spread across the globe and set up Jewish communities all over the world. Although much of this migration would have been in response to fear and economics, there was also a sense in which they were responding to the Old Testament prophecy of Jews bringing "light to the nations of the world."[395] After the first World War, the area was ruled by the British as the Palestine Mandate and the State of Israel was formed in 1948. This encouraged many Jews in the Diaspora to move into the new Israel.

Scriptures

The evolution of the scriptures and other writings are an integral part of the development of Judaism and the various different groupings of Jews that emerged. During the Period of the Second Temple (538 BCE to 70 CE) Judaism was divided into theological factions, particularly the Pharisees, Sadducees and Zealots as well as numerous smaller sects including the Essenes and closely related traditions such as Samaritanism and Hellenistic Judaism. Also, towards the end of this period there were various messianic movements (sects looking for the coming of a messiah or saviour). Christianity was also in its infancy.

With the Roman destruction of the Second Temple in 70 CE, adjustments had to be made, as both the temple, (which had served as the centre of teaching and study), and the autonomy of Judah as a nation, had been destroyed. Many legal and religious rulings were required, to take account of these new circumstances. This period, from about the 3rd to 6th centuries CE marks the transformation of Pharisaic Judaism into Rabbinic Judaism.[396] Although the Rabbis traced their origins to the Pharisees, Rabbinic Judaism was required to make significant changes in order to safeguard Jewish religious life.

The experience of adapting religious observance to circumstances, that had been made by those Jews of the Diaspora when in Babylonian exile, provided a useful foundation on which to build. Such adjustments could only be accomplished by writing everything down. This also encouraged the commitment to paper of the original Oral Tradition of the Torah (Pentatuch). All these writings became known as the Mishnah and were complete by about 200 CE. These statements had corresponding explanations which were gathered together as the Gemarah. This is thought to have been completed by about 600 CE. These writings were brought together and became known as the Talmud. There were both Jerusalem and Babylonian versions but due in large part to the censoring and burning of manuscripts in medieval Europe, the oldest surviving complete copy of the Babylonian Talmud is dated to around 1342 CE.

These writings supplemented the books of the Hebrew Bible or Tanakh. It is thought that much of this – the written Torah (the books of Genesis, Exodus, Leviticus, Numbers, and Deuteronomy) and the other nineteen books (the material is arranged into twenty four books in the Hebrew Bible but thirty nine books in the later Protestant Bibles) had been assembled, revised and edited by the fifth century BCE. During the Hellenistic period, by about 200 BCE, the Hebrew Bible was translated into Greek and known as the Septuagint – because of the seventy scholars who had been involved in its compilation. The original Hebrew manuscripts would have been preserved through being copied by scribes, and the Septuagint is likely to have been based on these. However, it is thought that the "quality control" was not necessarily all it might have been.

So between the sixth and tenth centuries CE, a group of scholar scribes, known as the Masoretes embarked on a project to create a copy as close to the originals as they could. This would become the recognised version, and is now known as the Masoretic text. It was produced through a series of very skilful and "quality controlled" copying processes based on previous copies, older manuscripts and transcription of the oral tradition. But in addition, these scribes have also added pronunciation pointers and vowel markings (which original written Hebrew did not have), ordered the text into paragraphs and sections and also made comments in the margins. Once copied, the originals and those containing mistakes were destroyed.

Whilst this ensured only genuine copies survived, it does mean that todays scholars cannot be sure how faithful to the original these copies are. However, comparisons have been made with other versions, with the early Greek (Septuagint) and also with manuscripts found amongst what are known as the Dead Sea Scrolls. These are a collection of scrolls and scraps of manuscripts, found by accident in 1946/7 and dating from between about 150 BCE to 75 CE, in a cave at Qumran. This is situated in the current West Bank region, close to the Northwest bank of the Dead Sea. It is thought to be a place of dwelling for the Essenes, a Jewish sect. They are thought to have written some of the Jewish scriptures and collected others and preserved this collection. Over the following ten years other material has been found that dates from between the eighth century BCE to the eleventh century CE and that probably originated from other sites.

Analysis leads scholars to believe that for much of the material there is fairly close agreement between these scrolls and the Masoretic text. Many of the differences, it is thought, are due to the possibility of there being a possible common source of the Hebrew Bible that has given rise to various earlier versions, rather than being due to poor copying or bad translation, as may be the case with the Septuagint. There is some evidence to suggest that scribes perhaps felt free to choose according to their personal taste and discretion between different readings. This common source is known as an Urtext, though no such Urtext has ever been found. The oldest manuscripts of the Masoretic tradition come from the tenth and eleventh centuries CE. These are in the form of the Aleppo Codex from the later portions of the tenth century CE and the Leningrad Codex dated to 1008 to 1009 CE.[397] The current Masoretic Text was used as the basis for translations of the Old Testament in a variety of Christian Bible translations, including the King James Version and American Standard Version.

In addition, we find further writings produced between 600 and 1200 CE – the Midrash. This term refers to the activity of Judaic biblical interpretation of the Hebrew biblical texts, the method used in such interpretations and to the collection of writings produced by such interpretations. The Halekah (the way) literature refers to the legal stuff – Major Codes of Jewish Law and Customs – concerning what behaviour is sanctioned by the law. There are also procedures and principles set down as to how authoritative opinions and interpretations should be arrived at.

Summary

Jewish philosophy comprises the study of philosophy and Jewish theology and the interface between them. This provides fertile ground for a variety of opinions and interpretations of the material content of both subject areas. The Torah is considered by many Jews as a dynamic work, containing as it does, many words which are undefined and many procedures which have no detailed explanation or instructions. So this, with the story of how the current text of the Hebrew Bible has evolved, together with the variety of associated explanatory writings, leaves open the possibility of numerous different interpretations. It thus seems clear, there is no one single agreed and comprehensive version of what might define the Jewish belief system. Numerous attempts, for example Maimonides' thirteen principles of faith, developed in the twelfth century, have been made to formulate statements reflecting Judaism's core tenets. All of which have met with criticism and scholarly disagreements of such statements.398

Both differences of scriptural interpretation, as well as the story of the evolution of the Jewish people themselves, are partly responsible for the rise of the different movements within Judaism, such as : Orthodox Judaism (Haredi Judaism and Modern Orthodox Judaism), Conservative Judaism, and Reform Judaism (called Liberal or Progressive Judaism) Karaite Judaism, Reconstructionist Judaism and Humanistic Judaism. The major sources of difference between these groups are : their approaches to Jewish law, the authority of the Rabbinic tradition, the significance of the State of Israel and the relative importance accorded to the Oral Tradition of the Torah.

All this demonstrates the evolutionary nature of Judaism – its scriptures and its people.

CHRISTIANITY

Christianity emerged from within Judaism. Jesus Christ was Jewish. Largely preaching to the Jewish people, though indicating by his teaching and his actions that his ministry was to both Jew and gentile (none Jewish people). All his first twelve disciples were also Jewish. The first Christians therefore were ethnically Jewish or Jewish converts. They regarded "Christianity" as a continuation of traditional Judaism with the key initial differences being the Christian belief that Jesus was the resurrected Messiah,[399] and their noteworthy care for each other.[400] But this emerging Christian sect, came into being within a climate of paganism, polytheistic cults, monotheistic Judaism and the existence of multiple groupings both within and without Judaistic traditions. Therefore problems arose in terms of identity. There was high risk of corruption of their beliefs by contamination from other groups or the incorporation and weakening of their beliefs by other religions. So the early Christians faced a number of challenges.

These challenges were many and varied, but perhaps we can identify four in particular. The first being the need for Missionary activity to spread the word about Jesus. The second was fear of persecution. The third being a fear of contamination by other beliefs and the fourth being the need to agree some sort of shared orthodoxy ("correctness") in understanding what should be believed. All of these are still current today, to varying degrees. Such challenges would not have necessarily been seen as clearly as they are laid out here and not all would have been identified initially or addressed in a strategic and organised manor. But as we look with hindsight from a historical and academic perspective, it is possible to define them and describe the solutions adopted to address them. Although there would have been many more specific day to day issues the early Christians would have had to contend with, most can probably be placed under one of these four headings. So let us take each in turn, starting with the missionary focus of the Apostles work. Jerusalem was the initial centre of the Church under the leadership of Peter, James the brother of Jesus (known as James the Just) and John.

Mission – Establishment and Consolidation of the Church

The early Christians felt an urgent need to spread the Gospel ("Good News") about Jesus – his teachings and his resurrection, particularly as they were under the impression that the Kingdom of God – the Second Coming (when Jesus would return), was an imminent reality. The focus at this stage was on telling people about Jesus, not thinking of establishing a Church organisation or consolidating such an organisation. These pressures came later, as they realised that Jesus's second coming was not as imminent as they had first understood. In adjusting to this fact, their evolving understanding of the theology, led them to believe the Kingdom of God was indeed already here and within them, as they tried to follow Jesus' example. But being afraid, they met secretly in people's homes. They were afraid of many things, including – persecution, cynicism, ridicule, being contaminated by other beliefs and having their organisation infiltrated by non–believers.

Despite their initial fears, the apostles and then particularly Paul (who was not an Apostle but a pharisee charged with persecuting the Christians, until his conversion experience on the road to Damascus) with others, took the message to the world. The essence of the message they delivered was that the Grace of God made the Christian life possible, the love of God was the objective sought by Christians rather than their own desires and charity towards other human beings in need, should be their concern.[401] It is recorded in the Book of Acts, in the New Testament, that Paul undertook three or possibly four missionary journeys, including travelling through Asia Minor, Macedonia and northern Greece and visiting cities like Ephesus, Antioch, Philippi, Thessaloniki and Corinth and then Rome. All places that already had Jewish populations.

Alongside this missionary activity, and with more people becoming attracted to following the Christian life, there was a need for some sort of organisational structure. They developed what is called an episcopalian structure – Bishop (who oversaw the spiritual needs and gave leadership to a large area containing a number of churches), Presbyters (ordained clergy – theologically trained and qualified providing spiritual guidance to local churches) and Deacons – lay people i.e., non–ordained) who helped administer local churches. This system was already established by the second century CE and by at least the fourth Century CE, and possibly earlier, they started building publicly recognisable churches. In 326 CE Pope Sylvester I consecrated the original Basilica of St. Peter, in Rome, which was built by Constantine the Great over the tomb of the Apostle. The Sistine Chapel was built between 1473 and 1481 CE. The Hagia Sophia, in what is now Istanbul, was begun in 537 CE. The era of great cathedral construction, in Europe, began during the eleventh Century onwards.

As the Gospel spread, meetings of Bishops were called to help resolve questions of doctrine (beliefs about understanding the scriptures and the nature of God) and of Church authority. At the same time as such meetings were intending to provide some unity and agreed consistency in belief, we see different groups or movements emerging independently. One such were the Syriac Christians who became known as The Church in the East or Persian Church. It was centred on Edessa (now known as Urfa, in south east Turkey, close to the northern Syrian border). They used the Syriac language, which was a variety of Middle Aramaic closely related to the Jewish Palestinian Aramaic spoken by Jesus. They developed their own writings and their own form of the Bible in Syriac – the Pershitta Bible. Initially this excluded some of the New Testament books but by the fifth Century CE these had been included. This Syriac Church branch, situated within the Sasanian Empire, organised itself separately and in 424 CE declared its leader independent of other Christian leaders. It also became known as the Nestorian Church (see below).

Other emerging movements included monasticism. So around 270 CE we have Anthony (later designated as St. Anthony) starting as one of the early hermit monks. He was so revered that many people sought him out and he organised small groups of people to live together. Following his death, his followers built The Monastery of Saint Anthony as a Coptic Orthodox monastery (one of the oldest Christian monasteries in the world). This occurred around 280 to 300 CE in the Eastern Desert of Egypt, about 200 miles south east of Cairo. St. Pachomius, also began to built houses in which hermits could gather. This was the beginning of organising monks into a form of communal living which preceded the establishment of the first flush of monasteries around 318 CE. Saint Catherine's Monastery was founded between 527 CE and 565 CE in Egypt's Sinai Peninsula. Its importance relates to its priceless collection of early codices and manuscripts, second only to the Vatican Library. It contains Greek, Georgian, Arabic, Coptic, Hebrew, Armenian, Aramaic and Caucasian Albanian texts. This archive has played an important part in uncovering parts of the history of Christianity and its origins.

All the monastic movements, beginning in 529 CE, with the foundation of a number of monasteries by Saint Benedict of Nursia, including the

Abbey of Monte Cassino, aimed at renouncing material wealth and living more simply. Benedict wrote a set of rules of conduct – Rule of Saint Benedict – which eventually became the gold standard of all monasterial houses. They focussed on renouncing personal possessions, doing good works for the community, working in agriculture and other occupations and spending time in worship and prayer.

As the dominance of the Benedictine monastic way of life declined towards the end of the twelfth century, we see the establishment of the Franciscans (St. Frances of Assisi in 1205 CE), Carmelites, Dominicans and Cistercians. The Cistercians, as did other off–shoot groups, felt that monastic life had lost touch with its roots and so wanted to get back to the original ideals. So a group of Benedictine monks from the monastery of Molesme founded Cîteaux Abbey in 1098 CE. Later, Bernard of Clairvaux became Abbot and was influential in the movement and in the future of Christian thought. A similar motivation led to the formation of the Trappists in the seventh–century at La Trappe Abbey in France. St. Augustine's Monastery, founded in 1277 CE in Erfurt Germany, was once home to Martin Luther who lived as a monk from 1505 to 1511 CE.

In the English Reformation of the sixteenth century, when the Church of England broke away from the authority of the Pope and the Roman Catholic Church, all monasteries were dissolved and their lands confiscated by the Crown. Despite such widespread destruction, we now have monastic traditions within both the English Anglican and Protestant Churches. How times change ! Other faiths also developed their own monastic traditions – Hindus, Buddhists and Zoroastrians for example.

Africa – Christianity began in Egypt in the middle of the first century CE with Alexandria developing into an important centre. By the end of the second century it had reached the region around Carthage and gradually spread over the next two centuries into the Horn of Africa and Ethiopia. There were several important African theologians whose writings and opinions influenced the early theological and organisational development of Christianity, including Cyril, Clement and Origen – all of Alexandria, and Tertullian, Athanasius and Augustine of Hippo. Several centuries later, Islam became a dominating force, followed by a period of Christian re evangelisation in 1500 CE, with European colonisation. By 2002 CE, The Encyclopaedia Britannica (cited in Wikipedia) [402] estimated that Christians formed forty percent of the continent's population with Muslims forming forty five percent.

Thus, Africa had gone from having a majority of followers of indigenous, traditional religions, to being predominantly a continent of Christians and Muslims. The process was actually one of significant and sustained syncretism between the African traditional religious beliefs and the other major faiths. This resulted in modifications to particular theological understandings and practices within local and national communities.

India – According to tradition, Christianity is thought to have been brought to India by Thomas the Apostle, around 52 CE, together with the Apostle Bartholomew. The resulting communities developed strong ties with Nestorian Christianity (see later) – the Church of the East, when they arrived as settlers and missionaries from Persia around 295 to 300 CE. Bartholomew is also credited with introducing Christianity to Armenia, which is thought to have become one of the first countries to officially adopt Christianity as its state religion, in 301 CE. So by the sixth century CE, Christianity was fairly well established.

China and The Mongols – By the seventh Century, Nestorians were also influencing the Mongols in Central Asia before reaching China in about 635 CE, under a Persian cleric named Alopen. After thriving for approximately 200 years, it underwent persecution from Emperor Wuzong of Tang (reigned 840 to 846 CE) along with all foreign religions, including Buddhism and Zoroastrianism. The Church disappeared from China in the early tenth century, coinciding with the collapse of the Tang dynasty. It underwent a revival during the Mongol conquest of China in the thirteenth century, as a number of Mongol Leaders and their wives were Nestorian Christians. After another round of persecution, Jesuits (a Roman Catholic order of priests) once again initiated mission work inside China, introducing Western science, mathematics, astronomy, and cartography around 1582 CE. This led to controversy over the degree of syncretism between western Christian theological understanding and the incorporation of traditional Chinese Confucianism. There were further Christian missional incursions over the next three hundred and fifty years.

Japan – Roman Catholicism arrived in Japan in the middle of the sixteenth century, though some scholars believe Nestorian Christianity had arrived and then retreated centuries earlier. After it was outlawed, during the Edo period (1603 to 1868 CE), Christianity went underground but continued in secret and emerged again when Japan was opened up to western influence in 1853 CE.

Southeast Asia – comprising, present day countries of Myanmar, Thailand, Laos, Cambodia, Vietnam, the Philippines, Malaysia, Brunei, Singapore, and Indonesia. This region was heavily influenced by China, particularly in the north and east and by India in the west and south. Both Buddhism and Hinduism were also introduced to the mix of local religions. There is evidence of Christian influence from around the seventh Century. This influence seems to be from the Nestorian Christians as well as by Catholic travellers such as Marco Polo, in the Middle Ages. The Nestorian influence continued until serious volumes of trade and exploration came with the Spanish and Portuguese in the sixteenth century and with the Dutch in the form of the Dutch East India Company in the seventeenth Century. France and Britain followed and the Americans came in the nineteenth century.

Europe – Christianity was spreading into western Europe by the fourth Century, with tradition having it that Saint Ninian evangelised the Picts in Scotland in about 397 CE. St. Patrick began his mission to Ireland in 432 CE and by 550 CE St. David was converting Wales. In fairly quick succession we have Columba in Scotland in 563 CE establishing the monastery at Iona, St Augustine in Canterbury (sent by Pope Gregory) in 596 CE and St Aidan, founding Lindisfarne in 634 CE. Cuthbert (around 634 to 687 CE) later became Abbot of Lindisfarne Priory. By 664 CE the Synod of Whitby united the Celtic Christianity of the British Isles with Roman Catholicism and in 731 CE The Venerable Bede had written the history of the English Church. By the tenth Century, in Europe, Poland (966CE), Kievan Rus (988CE) and Hungary (1000 CE) became Christianised, and most of Europe had become Roman Catholic by the fifteenth century.

North and South America – Following the arrival of Christopher Columbus in 1492 CE, European colonisation, in the form of the Spanish and then the French, brought Roman Catholic traditions with them. The British introduced Protestant Puritanism when the "Pilgrim Fathers" aboard the "Mayflower" arrived on the American continent in 1652 CE. Over the centuries, countries often received further injections of

Christian influence, occurring through contact from Christian traders, explorers and the later emphasis on war and colonisation by European Christian nation states. There would often be missionaries tagging along with this type of contact. In addition, migrants who had been exposed to the Christian story in other places, would take this experience with them as they journeyed to other parts of the world.[403] Roxbrough [404] describes, in relation to the development of Christianity in South East Asia, that:

"those who changed religion did so for many· reasons. Such as : a quest for social or personal security and identity in the face of social change, a search for personal salvation and for a religion which appeared to better cope with the modern world to which they aspired, a faith which seemed to allow scope for traditional religious and material concerns and aspirations, which also addressed other needs. Those who kept the faith and witnessed to Christ in successive generations sometimes did so because they had come in to what was now a tradition; yet the ability of Christianity to regenerate itself across time and culture in Southeast Asia as elsewhere has a great deal to do with what it has always claimed to be about. Mixed motives are a commentary on the human condition, to be acknowledged as part of the story; yet they do not explain or explain away ultimate questions of truth and value which must also be considered."

Such a description might equally apply to other situations in which one faith movement seeks to influence those of another culture or country.

Persecution

The early Christians were initially ignored by the Roman authorities, whilst ever they were not perceived as a threat. Most Emperors were tolerant and it was only later that Christians became "scapegoats" and persecuted for the disasters that befell the Roman Empire. Decisions about persecution were taken by individual emperors and so in 64 CE, Nero blamed the Christians for the great fire of Rome and began his persecution. Caligula, Trajan, Domitian and Marcus Aurelius all had Christians killed for one reason or another, though seemingly on a small, individual scale. But in 250CE, Decius declared that everyone should make public sacrifices to the gods in the presence of a Roman magistrate and obtain a signed and witnessed certificate. Whilst not specifically targeting Christians, it did mean that Christians had to choose between their faith and their lives. A great many chose martyrdom.

Diocletian and Galerius were the last emperors to enforce Decius' declaration and the persecutions finished with the Edict of Milan in 313 CE, issued by the Emperors Constantine and Licinius. Constantine had seen a vision (as mentioned in the previous chapter) prior to the battle of Malvern Bridge and this led him to champion the cause of Christianity, though he probably did not get baptised until just before his death. This ordered the state's toleration of Christians, though it was not until 380CE, under the Edict of Thessalonica, signed by Emperor Theodosius I, that Christianity became the official state religion. This was a turning point for Christianity in that it now moved from being a marginalised secret sect to an open and officially acceptable mainstream religion.

However, this is not the end of the story of Christian persecution. Not only have Christians been persecuted by people of other faiths and none, there are examples of different groups of Christians murdering each other. For example, the St. Bartholomew's Day Massacre in 1572 CE in France when a group of Catholics murdered thousands of Huguenots (French Calvinist Protestants), during the French Wars of Religion (1560 to 1598 CE). In English history, both Catholics and Protestants have murdered each other, depending on what brand of the faith was on the throne. There have been many martyrs, both Catholic and Protestant who have died for their beliefs, though the context in which such martyrdom takes place is not usually simply one of religious intolerance but a cocktail of political intrigue and the exercise of power by vested interests.

Other examples of religious persecution include: the nine Crusades between 1095 CE and 1272 CE, during which the Roman Popes urged Christians to re–take Jerusalem from the Islamic Caliphate. All sorts of atrocities were perpetrated by both sides. Numerous Orders of Knights were created, and these later developed their own spiritualities. There have also been the inquisitions, in which people were questioned, tortured and executed on the basis of their beliefs or understanding of Church doctrine. If your belief did not conform to what, at that time, was considered Orthodox; that is, "correct", you might find yourself being tortured and killed. The Inquisitions began around 1184 CE, initially to root out "supposed heresies" of Roman Catholic doctrine. They were initiated to counter the spread of Cathar and Waldensian movements, so initially began in France. Later they examined the Hussites and the Beguines.

The most well known is the Spanish Inquisition but there were major movements in Portugal and these practices stretched to their colonies in Africa, Asia and the Americas. The Roman Inquisition began in 1542 CE. The last execution of the Spanish Inquisition was in 1834 CE. By the end of the Middle Ages, most European areas, with the exception of England and Castille had Inquisitorial movements. The Protestant Reformation proved to be prime hunting grounds for the Grand Inquisitors !

Christians are still persecuted in different parts of the world, as are members of other faiths.

Adulteration and False Teaching

Concerns about contamination of the faith led Paul, James and Peter and others in their letters to churches (53 to 57CE) such as Corinth, Colossae and Ephesus, to warn of false prophets and false teachers. They included pointers on how to recognise these and how to remain true to the faith. Of particular concern were the influence of the Pharisees, Sadducees, Marcionists, Zealots, Essenes and the Gnostics. The Pharisees stuck rigidly to the Jewish religious and theocratic way of Life. This was not totally compatible with the new Christianity, even though the new Christians accepted the Jewish scriptures to be authoritative and sacred and retained many of the Jewish practices such as baptism. The Sadducees did not accept the resurrection of the dead. While Marcionism rejected all Jewish influence on Christianity. The Zealots advocated armed struggle to overthrow the Romans and the Essenes had particular ideas relating to keeping the Sabbath day and the meaning of stones and plants, making predictions and the names of angels.[405]

Gnosticism was not a single movement, but a collection of different systems and beliefs that were eventually designated as a heresy by Irenaeus (from around 130 to about 202 CE). There were two particular areas of disagreement. Firstly, the Gnostics claim to some "special/ secret" knowledge passed down just to them either by the Apostles or the Holy Spirit. Evidence of this has been found in manuscripts found in Nag Hammadi in 1945 CE. [406][407] Secondly, they also held to the view that the material body did not matter, only the divine nature. Because of these sorts of challenges, it was felt necessary to establish an agreed orthodoxy – common understanding of the nature of Christianity.

This was done in a number of ways:

- by accepting the teachings of the apostles as true and if there was no evidence that they taught a particular thing then it was regarded as not true. (in this instance no evidence is evidence of it not happening !)

- setting up the hierarchical system of bishops, presbyters and deacons as the authentic voice of the apostles and transmitters of the true teachings. This established the understanding that truth was handed down in succession from Jesus to the Apostles and to the Bishops and then later through the Popes

- because of the oral tradition of passing on stories and information, as the apostles died it was necessary to write things down for fear of losing the narrative, and hence a New Testament canon of writings was compiled – gospels, apostolic and Pauline letters and so on

- it was also felt necessary to establish a set of core beliefs (creed or credo – see later)

It took several centuries to establish a definitive biblical canon and credo and to settle some of the theological disagreements. The process for doing so was by a series of meetings or Ecumenical Councils of Bishops. Such Councils often comprised a number of separate sessions conducted over a period of several years. For example, The Council of Trent met for twenty–five sessions between thirteenth of December 1545 CE and fourth of December 1563 CE.

Establishing Theological Orthodoxy and a Biblical Cannon

By the first century CE, Christian communities had been established. Besides Jerusalem, communities could be found across Asia Minor, Greece, Italy, North Africa and Egypt, at places such as: Antioch, Alexandria, Carthage, Ephesus, Constantinople and Rome. But a number of fundamental theological areas of disagreement surfaced between different groups and the centres of Christianity. Some of these questions, not in any particular order, related to :

- the Judaisation of Gentiles – discussed at the First Council of Jerusalem around 50 CE.
- the timing of Easter
- virgin birth
- nature of belief
- nature of Jesus – Arianism ,Nestorianism, other isms.
- nature of God
- resurrection
- the Trinity
- eucharist
- role of Mary
- grace and faith versus works
- Church authority

Because of differing views on many of these areas, in the early days, the Roman Emperors (probably given their experience of knowing that differences can lead to unrest and violence !), felt that for the good and stability of the Empire, it was necessary to attempt to unify aspects of belief. Their priority was unity rather than truth! Their solution was to call ecumenical councils of all the Bishops (from the East and West of the Empire) to discuss the issues. Many were held over the succeeding centuries.

By following the time–line of the early Church councils addressing the controversies of Arianism and Nestorianism – disagreements on the nature of Jesus and his relationship with and to God, we can see some of the changing nature of Christian belief and the processes by which agreements were reached. The arguments were often heated and personal and the frequent political and vested interests involved, made for lively affairs. So much so that any agreements were often at the expense of the formation of splinter groups or the excommunication or even death of those finding themselves in the minority. Some of the councils were called to resolve genuine differences of interpretation and understanding. Other disagreements arose over terminology, as words, especially in translation, meant different things to different people. Particularly so, when used to try and genuinely express and explain things that are inexplicable, such as the idea that Jesus was both divine and human !

There is evidence that this idea of Jesus being both human and divine was around possibly by the second century.[408] So it seems reasonable, to think that the Apostles would have had some sense of Jesus being both God and also human, as they experienced him on a day to day basis, even if they could not articulate what this actually meant or how it worked. Who can blame them ? We still cannot fully comprehend what this really means today, even though we use words that aim to convey this astonishing concept. But by the fourth century, arguments about what this meant had crystallised into what became known as the Arian and Nestorian Controversies.

Arianism

The Arian Controversy [409] is the name given to the arguments relating to differences of opinion between Arius, a presbyter in Alexandria and Athanasius, the assistant bishop of Alexandria. The disagreement focussed on the nature of Jesus, as the Son of God and his relationship with God as Father. Essentially, the argument was whether Jesus was the Son of God in the sense of being of God and being eternal and being present from the beginning and so being the same substance as God or whether he had been created as a man by God but then elevated to be called God. The latter was Arius' position. Arius argued that because the Son, had been created, he was therefore subordinate to the Father. This idea is summarised in the statement "there was a time when the Son was not."[410] Adoptionism was a variant of this which held that Jesus was an ordinary man, born of Joseph and Mary, who became the Christ and Son of God at his baptism.

To resolve these differences, Emperor Constantine called the First Council of Nicaea (now Isnik in Turkey) in 325 CE. After heated discussion and some violence apparently, the Arian view was rejected. A statement of belief, called the Nicene Creed, was formulated in which it was made clear that God and Jesus the Son of God were one and the same substance (that is; consubstantial). This is called the Homoousian Christology – "of the same substance," as opposed to the Homoiousian Christology – "of a similar substance." (this difference in terminology is said to have caused Edward Gibbon, [411] in *The History of the Decline and Fall of the Roman Empire*, to point out, that Christianity was nearly split by "an iota" – "i" – the smallest letter in the Greek alphabet !). This decision resulted in Arius and his supporters being excommunicated as heretics. However, the Syriac Church, as the Eastern Church, did not accept this decision and continued to support the Arian position. So the controversy continued bubbling away. This condemnation of Arius was then reversed by a further council in Jerusalem in 335 CE.

So another council was called in 381 CE – The First Council of Constantinople. This re–affirmed the anti–Arian stance of the council of Nicea and made amendments to the creed. The new creed – Niceno– Constantinopolitan Creed, (see Appendix 1) included an implication that the Holy Spirit was also of the same substance as God and as Jesus the Son of God and so encapsulated the idea of the Trinity. This concept had been around since the first century, as God the Father, God the Son and God the Holy Spirit. Thus, all were of the same substance or essence (ousia) and so one and the same and yet three persons (three separate beings – hypostases), all at the same time. (the "what" and the "who" aspects of the nature of God). Explaining the inexplicable ! This concept was consolidated in the Athenasian Creed, probably written sometime between the fifth and sixth centuries CE. This was about 100 years after Athenasius, though bearing his name as it is in line with his anti–Arian stance. It is also contrary to Sabellianism (which viewed the three parts of the Trinity as three different aspects of God rather than as three persons in one Godhead).

Nestorianism

Nestorius (around 386 to 450 CE) was Archbishop of Constantinople from 428 to August 431CE. His "crimes" were to reject the title of Theotokos, "Mother of God", for Mary, the mother of Jesus, referring to her as the "Mother of Christ." This implied he did not think that Christ was truly God. He also suggested that the divinity and the human aspects of Christ's nature represented two distinct hypostases (natures) that existed together. Nestorius's opponents, particularly Cyril of Alexandria, charged him with detaching Christ's divinity and humanity into two persons existing in one body, thereby denying the reality of the Incarnation; that is, the virgin birth.

These theological arguments were also flavoured by the rivalry of the two centres from which the arguments arose – Antioch (Nestorius) and Alexandria (Cyril). This rivalry had been increased by the First Council of Constantinople granting the 'See' (or diocese – the Bishops geographical area of responsibility) of Constantinople, primacy over the older 'Sees' of Alexandria and Antioch. It was fuelled further in 428 CE when Nestorius was made Archbishop of Constantinople. Thus, when the First Council of Ephesus was called in 431 CE to settle the matter, the odds were already stacked against Nestorius. Descriptions of what happened at the Council and how the matter was handled, bring into question the fairness of the process. Nevertheless, The council ruled against Nestorius, stating that the divinity and human aspects of Jesus resided in a single hypostasis; that is, Jesus was both divine and human at one and the same time, as the Incarnate Christ. This doctrine is known as the Hypostatic union. The consequences of this declaration from the Council resulted in Nestorius' s excommunication. Emperor Theodosius II confirmed this decision.

However, the Church of the East, (the Syriac Church – later becoming known as the Nestorian Church.) never accepted this condemnation, and so the ruling did not settle the matter. The Nestorian position was promulgated with expansion of the Syriac Church into South East Asia, as described earlier.

Other theological models devised to try and explain the concept of Jesus Christ being both divine and human include: miaphysitism, dyophysitism, monophysitism.[412] Docetism and Apollinarisism both emphasising the divinity of Christ at the expense or even denial of his humanity, with Docetism even suggesting that Christ only "appeared" to be human. If hypostasis is understood as meaning "person" or "the real existence," the following definitions in Wikipedia might (or might not !) help to differentiate them:

"The First Council of Ephesus debated miaphysitism (two natures united as one after the hypostatic union) versus dyophysitism (two coexisting natures after the hypostatic union) versus monophysitism (only one nature) versus Nestorianism (two hypostases i.e., Jesus being two people – one divine and one human)."[413]

The Council of Ephesus supported the miaphysite position but the later Council of Chalcedon, in 451 CE accepted the dyophysite position. Their statement being known as – the Chalcedon Definition. This is the accepted position of most of the Eastern Orthodox and Western Roman Catholic and Protestant Churches today.

The differences between some of these theological concepts are not easy to understand, even now, particularly between miaphysitism and dyophysitism. One has to wonder at the reasons why people became so heated by what after all might seem to be rather semantic differences. The situation is not helped when the same word is used in different contexts and so risks being misunderstood, particularly as words can also lose something of their nuance when translated. This of course can be crucial when so much rides on everyone understanding what is intended. At one level, one has to marvel at the intellectual gymnastics and depth of thought the theologians brought to these arguments, but also wonder at what impact any of this had on the lives of ordinary Christians.

If todays experience is anything to go by, it is likely that many would be unable to follow the arguments, even if they could read and had access to the documents, and probably most would not particularly care anyway, as they would find the struggles of day to day existence more pressing. However, where these arguments lead to pronouncements by the Church about what was orthodox and allowed and what was not, they would have an impact when it came to the era of Inquisitions. One's life could depend on it ! It is not clear what training was given to the presbyters and deacons and what processes were devised to pass this information down to the "people in the pews" ! Potentially, ordinary folk could be in a position of ignorance or possess out of date knowledge and so be at risk, when asked, of giving the wrong answer and facing punishment. Although, these extremes are generally a thing of the past, certainly in the west, there are still consequences for individuals.

However, this ruling of the Chalcedon Council caused schism between the Oriental Orthodox Church from the Eastern Orthodox and western – Roman Catholic Churches. The Oriental Orthodox rejected the Chalcedon Definition and accepted miaphysitism. Hence they propagated this theology as the Oriental Orthodox Church spread to Armenia, Egypt, Eritrea, Ethiopia, Sudan and parts of the Middle East and India.

This is not the only major split within the Church's chequered history. As a result of the fortunes of the Roman Empire and the changes in power between the Hellenistic east and the Latin west, the fortunes and influence of the eastern and western parts of the Church fluctuated too, as did the Church's relationship with the ruling Emperors. With the eventual fall of Rome in 476 CE, the establishment of the first Holy Roman Emperor, Charlemagne in 800 CE and the gradual formation of what became known as the Holy Roman Empire, the relationships between the Eastern Church and the west became more strained. So, with all this uncertainty and upheaval and the ongoing disputes over theological doctrine, Pope Leo the Ninth made a play to be Pope of all the Church – East and West, in 1054 CE. This did not go down well and resulted in schism between the Eastern Orthodox and Western Roman Catholic Church, which has persisted to this day. This of course did not put a stop to continuing unrest and claims to authority.

Tensions grew between the role of Emperor and that of the Pope. Excommunications of rulers were common and military campaigns often enjoyed papal support. Between 1309 and 1376 CE, the popes resided in Avignon. When they returned to Rome, a further schism occurred and in 1378 CE the pope went back to Avignon, by which time there were three popes vying for authority ! The schism ended in 1417 CE at the Council of Constance.

By the early 1500s CE, we have the beginning of the Protestant Reformation. Although this crystallised around perceptions of corruption and ostentation of the Roman Catholic Church and its drift away from the authority of scripture, in truth, many of the ideas taken up by Martin Luther, as he pinned his ninety five theses to the door of All Saints Church in Wittenberg, had been around for over 150 years. John Whycliffe (1320 to 1384 CE) had translated the Bible into English and inspired the protestant Lollard movement. So as the protestant movement gained further momentum, a further schism ensued as the Protestant Churches formed in separation from the Roman Catholic Church. They established their own interpretation of scripture and forms of worship which they held in more simple meeting places. These changes coincided with the formation of the Church of England, in 1534 CE, necessitated by Henry the Eighth's personal circumstances. Initially, the Church of England retained its Roman Catholic rites and traditions, but over time, moved to a more protestant reformed position.

Further splits among the protestant Churches occurred over differences in theology and so over the next 400 years a number of new religious movements began (dates approximate), including :

1525 Anabaptist movement begins

1534 Jesuit order founded by Ignatius of Loyola

c. 1556 Unitarian ideas promulgated and first used in English in 1673

1561 Menno Simons, founder of Mennonites

1572 John Knox founds Scottish Presbyterian Church

c. late sixteenth century – Congregationalism – independent/ autonomous Churches

1609 Baptist Church founded by John Smyth

1648 George Fox founds the Quaker movement

1693 Jacob Amman founds Amish sect (break from Anabaptists and Mennonites)

1738 Methodist movement, led by John Wesley and brother Charles

1747 Shakers formed – Ann Lee, an early founder moves to America 1774

1780 Robert Raikes begins Sunday schools for poor, uneducated English children

1784 American Methodists form Methodist Episcopal Church

1801 Cane Ridge Revival in Kentucky initiates the Stone Movement

1830 Joseph Smith starts Church of Jesus Christ of Latter Day Saints (Mormons)

c.1848 Christadelphian movement began based on teachings of John Thomas

1863 Seventh–day Adventist Church formed

1865 Methodist preacher William Booth founds the Salvation Army

1879 Church of Christ, Scientist founded in Boston by Mary Baker Eddy

1906–1909 Azusa Street Revival in Los Angeles begins Pentecostal movement

1907 The Church of God in Christ is formed as a Pentecostal body

1930 Rastafari movement founded

1931 Jehovah's Witnesses formally separate from the Bible Student movement

1948 World Council of Churches is founded

1954 Unification Church (Moonies) founded by Reverend Sun Myung Moon

1972 United Reformed Church (URC) formed as an attempt to begin a move to unifying some of these different factions. So Congregationalists, Presbyterians and Churches of Christ agreed to form the new Church. However, a number of individual Churches refused and so formed their own organisations.

Chart 2 shows some of the divisions that have occurred over the centuries, as Christianity and the church have developed.

Even today, differences of theology can produce heated debate. Though such disagreements don't usually lead to martyrdom, they can still result in splits within the Church. For example, the debate over women priests within the Anglican Church (even though the practice is well established in other

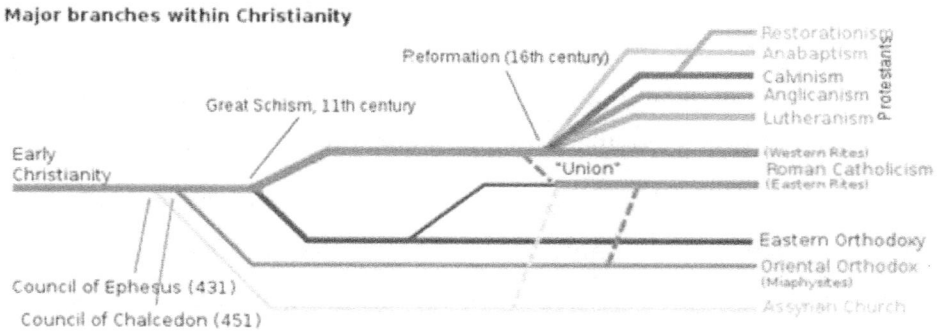

CHART 2 SPLITS DURING THE EVOLUTION OF CHRISTIANITY

branches of the protestant Churches) and concerns over homosexuality and same sex marriage. The fall out of these discussions continues to cause people hurt.

Just a few other things worthy of note, in reviewing the evolution of Christianity :

1. Many of the early Churches claimed origin and Apostolic succession, from visits by one of the Apostles, such as: Antioch by Peter and Paul, Alexandria by Mark,

Constantinople by Andrew, Cyprus by Barnabas, Ethiopia by Matthew, India by Thomas, Edessa in eastern Syria by Thaddeus, Armenia by Bartholomew, Georgia by Simon the Zealot and Rome by Peter."[414] However a consensus developed over recognising Antioch, Alexandria, Constantinople, Rome and Jerusalem as the main centres of early Christianity. Various statements in the Councils of Nicea and Constantinople specified an equality between the Bishops of these "Sees" (the area overseen by a Bishop), though Rome was tacitly accepted, by Rome, as being more equal than the others ! Although Rome eventually assumed its present standing and had its role within world wide Church respected, the rationale was not universally accepted.

2. The first recorded use of the title – Pope (meaning "Father") seems to be in connection with Heraclas the Patriarch of Alexandria (232 to 248 CE). By the third century, it was being applied generically to all bishops. In the Western Christian world, "Pope," as chiefly associated with the Bishops of Rome, began around the fifth or sixth century. In 1073 CE it was formally decided by Pope Gregory VII that no other bishop of the Catholic Church, save the Bishop of Rome, would hold the title

3. Around 500 CE Incense was introduced in church services. First plans for Vatican

4. 692 CE Orthodox Quinisext Council approved Clerical celibacy

5. 1543 CE the English Parliament bans Tyndale's translation of the Bible as a "crafty, false and untrue translation"

6. 1545 to 1563 CE The Council of Trent (or Trento, in northern Italy), was the nineteenth ecumenical council of the Catholic Church, convened as a counter to the Protestant Reformation. It issued clarifications of Church doctrine : including scripture, sacred tradition, original sin, justification, salvation, the sacraments, the Mass, and the veneration of saints. The Council made the Latin Vulgate Translation the official version of the Biblical canon. The council abolished some of the worst abuses of Church activities and introduced some restrictions on: the sale of indulgences, the morals of convents, the education of the clergy, the non–residence of bishops

7. 1551 CE The Council of Trent confirms the dogma of the term Transubstantiation (describes the change from bread and wine to the actual body and blood of Christ in the Eucharist). First used by Hildebert de Lavardin, Archbishop of Tours, in the eleventh century. By the end of the twelfth century the term was in widespread use. The Fourth Council of the Lateran in 1215 CE spoke of the bread and wine as "transubstantiated" into the body and blood of Christ: "His body and blood are truly contained in the sacrament of the altar under the forms of bread and wine, the bread and wine having been transubstantiated, by God's power, into his body and blood"

8. 1646 CE Westminster Standards produced by Reformed & Presbyterian Churches

9. 1854 CE the idea of the Immaculate Conception (i.e., that Mary the Mother of Jesus was born free of original sin) defined as Catholic dogma

334

10. 1869 to 1870 CE First Vatican Council– doctrine of papal primacy and papal infallibility, With the Enlightenment, religion lost its attachment to nationality, says Fitzgerald, but rather than becoming a universal social attitude, it now became a personal feeling or emotion.[415]

Scriptures

The biblical canon is the name given to the collection of writings that make up the content of the book, Christians know as The Bible. The basic format comprises thirty nine books of the Old Testament and twenty seven books of the New Testament. Some forms of the Bible also contain The Apocrypha. This comprises a number of different books, that over time, have either been judged appropriate to be included, or not. Various of these other books are included in the Roman Catholic and Orthodox Bibles, though not all the same, but are excluded from Bible versions that tend to be used by protestant Christians. However, these books are also published separately in their own right.

Herein lies one of the first problems when considering The Bible. Why were some writings included and others not and who decided and when ? And having made such decisions, are they set in stone for all eternity ? Part of the answer relates to how one sees the role of God and the Holy Spirit in the process. Some argue that both the writing and the process of choosing were directed by the Holy Spirit. Thus, the biblical content is the word of God and without error. Others suggest that the writing and choosing were inspired by the Holy Spirit, but because human beings were involved, the results may not be perfect and so are open to interpretation, which, itself requires inspiration by the Holy Spirit. These fundamental questions are over and above the consideration that needs to be given to questions of translation, copying and the accurate dating and assurance of the validity of the scrolls and documents used to compile The Bible.

Briefly and broadly speaking, the time–line of an accepted consensus narrative of the history of the Bible, is as per the following list. However, it is important to note that the proposed dates and authors all have their supporters, but are also disputed by one group of scholars or another. Given the section on origins in the previous chapter, this is understandable but not of particular importance to our purpose, as precision is not the name of the game:

- The Hebrew Bible (Tanakh) was compiled and settled by about the second century BCE and translated into Greek in the form of the Septuagint, completed by about 132 BCE.
- 60 to 64 CE Early letters by James and Peter written
- ? 65 CE "Q document" written. This is a hypothetical Greek document that is thought by some scholars to have existed, though now lost, and to have been the basis for the writing of the Gospels of Matthew and Luke
- c. 70 to 80 CE Gospel of Mark is written
- ? 70 CE Signs Gospel written. This is another hypothetical Greek text which was suggested by the twentieth century theologian Rudolph Bultman, that may have been used in writing the Gospel of John, to prove Jesus was the Messiah. This theory is contested, and Bultman had heresy proceedings commenced against him
- c.70 to 100 CE Additional Pauline Epistles (letters)
- c.80 CE Didache written. This anonymous text constitutes the oldest extant written document that could be accepted as

an early catechism, dealing with Christian ethics, rituals such as baptism and Eucharist, and Church organisation

- c. 80 to 100 CE Gospel of Matthew, based on Mark and Q, most popular in Early Christianity
- c.80 to 100 CE Gospel of Luke, based on Mark and Q, also Acts of the Apostles by same author
- c. 80 to 100 CE Pastoral Epistles written
- c. 95 to 120 CE Gospel of John and Epistles of John
- c. 95 to 105 CE Book of Revelation written, by John (son of Zebedee) and/or a disciple of his

It is thought that the main parts of the biblical canon might have been understood as early as the second or third century CE, though conversations at the councils continued over particular books. Sometimes ruling them in and then ruling them out again. But by the fourth century it was pretty much sorted. In his Easter letter of 367 CE, Athanasius, Bishop of Alexandria, includes a list of the books that would become the twenty seven books of the New Testament. So it has been suggested that it was during the 382 CE Catholic Council of Rome, under Pope Damasus Ist, that the full biblical canon – the twenty seven books of the New Testament and the thirty nine books of the Hebrew Bible as the Old Testament, was accepted and set.

This is not universally agreed and others suggest that the first council to accept the present canon may have been the Synod of Hippo Regius in North Africa (393 CE) with further confirmation by the Council of Carthage in 397 CE. Certainly by the time of Jerome's Latin translation – The Vulgate Bible in 400 CE, the matter would have been settled. Nonetheless, full dogmatic statements from the Church about the canon were not made until the Council of Trent in 1546 CE. This approved the list already accepted in 1442 CE by the Council of Florence for the Roman Catholic Church. For the Church of England, a definite statement about the biblical canon, along with approval of the Thirty–Nine Articles (statements of Church doctrine) was made in 1563 CE and for the Greek Orthodox Church, in 1672 CE at the Synod of Jerusalem.

Once the biblical canon was defined, the story is then one of versions, translations and its use in a country's common language. So the different Orthodox and Roman Catholic Christians each had slightly different versions in terms of what additional books they included. This changed again with the Protestant Reformation and the creation of new sects and movements. Between about 690 CE and 1000 CE we see translations of parts of the Bible into Old English – Bede translated John's Gospel for example. In 1378 CE John Wycliffe produced a middle English version of the New Testament and by 1382 CE he completed the Old Testament. However, this was banned and anyone, for example Wycliffe's followers, the Lollards, found reading it, were persecuted.

- 1455 CE Gutenberg Bible, first printed Bible version of Vulgate – Johann Gutenberg

- 1526 CE William Tyndale used the Greek and Hebrew texts and Jerome's Latin Vulgate translation to produce the first complete edition of his New Testament. He was the first translator to use the printing press – this enabled the distribution of several thousand copies of his New Testament translation throughout England.

- 1536 CE Tyndale put to death and ecclesiastical authorities ordered his Bible burned

- 1535 CE The first printed English translation of the whole Bible was produced by Miles Coverdale using Tyndale's work together with his own translations from the Latin Vulgate or Luther's German text.

- 1539 CE Coverdale authorised by King Henry VIII, working with Thomas Cromwell, to produce a version for public worship – the Great Bible.

- 1560 CE Geneva Bible (1560), notable for being the first Bible divided into verses

- 1610 CE First complete Roman Catholic Bible in English – the Douay–Rheims Bible,

- 1611 CE Authorised King James 1st Version

- 1644 CE Long Parliament directs that only Hebrew canon be read in the Church of England (effectively removing the Apocrypha)

- 1881 to 1894 CE Revised Version based on Septuagint & Hebrew Masoretic Text

- 1946 to 1952 CE Revised Standard Version

- 1973 CE New International Version of the Bible

- 1989 CE New Revised Standard Version

Most translations are made by committees of scholars in order to avoid bias or idiosyncrasy. However, translations are sometimes made by individuals, such as the translation by J.B. Phillips (1958 CE) and The Living Bible (1971 CE) by Kenneth N. Taylor.

There are numerous problems facing a Bible translator, including: which text to translate, which is the closest version to the original and how can one decide such a thing as there are no original texts of any biblical book. Many books may have circulated in more than one version as a result of copying errors, and errors in copying the copying errors or through personal additions made by the copier. The finding of the Dead Sea scrolls has helped to support the idea that the Masoretic text is as close to the original as can be done and that the subsequent copies are virtually error free. If other manuscripts are found, these may help shed light on what texts are the earliest known. Where the text of a verse is obviously corrupt, several plausible reconstructions of the verse may have been created, some quite ingenious.

So one of the jobs of a Bible translator is to choose the most plausible reconstruction, discriminating between what is merely ingenious and what is in fact likely. Then, what does the word mean ? This is not the same as deciding the best way to say it in English, but it is necessary to first establish what the original text is saying before deciding how to express that meaning in the "target" language. It is bad enough doing this for the New Testament, most of which was written within fifty to a hundred years after the events, with the earliest copies we have being written two or three hundred years after the events. Whereas the books of the Old Testament were composed and edited over a period of about a thousand years. Some of the words are now unknown and the syntax of the poetry is complex.

Other sources that can help are the writings of the early theologians, such as the early Bishops, some known as the Apostolic Fathers, living in the first and second centuries CE. Through their writings in support of Christianity and in developing their views on aspects of doctrine, aided by the fact that it is likely they knew or were influenced by the original Apostles, they can shed light on the scriptural manuscripts, helping the translation. They included: Clement of Rome (around 35 to 99 CE), Ignatius of Antioch (around 35 to about 110 CE) and Polycarp (around 69 to about 155 CE) Bishop of Smyrna (now İzmir in Turkey), all of whom were possible students of the Apostle John. Other early theologians included: Justin Martyr (around 100 to 165 CE), Irenaeus of Lyons (around 130 to 202 CE), Clement of Alexandria,(around 150 to 215 CE), Tertullian (around 155 to 240 CE), Origen (around 184 to 253 CE), Athanasius of Alexandria (around 296 to 373 CE), John Chrysostom (around 347 to 407 CE), Ambrose (around 340 to 97 CE), Jerome (around 347 to 420 CE), Cyril of Alexandria (around 376 to 444 CE), and later Augustine of Hippo (around 354 to 430 CE) and Pope Gregory I (around 540 to 604 CE). Much later, there is the scholastic interpretations of scripture and theology such as the *Summa Theologiae,* written by Thomas Aquinas, in 1274 CE. He was a theologian and philosopher and his writings later become official Catholic doctrine.

More recently, in 1947 CE, material now known as The Dead Sea Scrolls [416] [417] were found in the caves of Qumran. These caves are situated in the Judaean Desert, near Ein Feshkha on the northern shore of the Dead Sea. The writings discovered were ancient Jewish religious manuscripts, dating from between the last three centuries BCE and the first century CE. They are of particular significance as some of the scrolls are copies of texts from the Hebrew Scriptures, and possibly up to 1000 years older than other extant versions. Others are books that did not find their way into the biblical canon, including the Book of Enoch, the Book of Jubilees, the Book of Tobit, the Wisdom of Sirach and Ecclesiastes for example. Other documents describe rules and beliefs of the community living in that area. However, the nature of the scrolls and their relationship to the Christian narrative is not without controversy.[418]

341

Summary

The reasoning behind this rather extended section on Christianity, is to demonstrate that it was not "perfectly formed" at the time of Jesus or by the end of the first century. This may seem obvious, but it seems to me that much of the conversation that takes place within religious circles almost assumes that Christianity is a fixed entity that we question or change at our peril, by flying in the face of what God has determined. If we follow the narrative as just set out, even though it is only skimming the surface of the machinations of the historical narrative, we cannot fail to appreciate how Christianity has influenced and been influenced by the last 2000 years of world development. Christianity has evolved through a mire of argument, fighting, war, persecution, prayer, self–sacrifice and a genuine seeking of the will of the Holy Spirit. Given that human beings have been at the heart of all this change, and that all these factors are still around, it seems only sensible to anticipate that changes will continue to happen into the future. In 1984 CE, Pope John Paul II and the Syriac Patriarch were able to express the view:

"The confusions and schisms that occurred between their Churches in the later centuries, they realise today, in no way affect or touch the substance of their faith, since these arose only because of differences in terminology and culture and in the various formulae adopted by different theological schools to express the same matter. Accordingly, we find today no real basis for the sad divisions and schisms that subsequently arose between us concerning the doctrine of Incarnation."

If we can envisage future change in this light, there is perhaps hope that the next 2000 years will be less confrontational, dogmatic and less violent – more loving and respectful, that is, more Christian !

ISLAM

In chapter 5, I gave a brief summary of the spread of Islam, so here I will just fill in some of the gaps and concentrate on demonstrating the evolutionary aspects of the development of the faith.

According to tradition, the Islamic prophet Muhammad was born in Mecca around the year 570 CE. He was born into a wealthy Arab family at a time when, as he perceived it, the various tribes were losing some of their traditional nomadic caring concerns for the poor and concentrating more on making money through trade. At the time, there were many local religions with numerous gods, and the centre of Mecca, with the Ka'ba – the Black Stone – was a centre of religious activity. Jews and Christians were also present in this area and so their beliefs were part of the religious mix to which Muhammad may have been exposed.

When he was about forty years old, in 610 CE, he began receiving what are called the divine revelations, delivered through the angel Gabriel. These revelations were the verses of what, once written down, would become the Qur'an. He initially shared these early experiences within his family, with his wife and her cousin, a Christian. As they became convinced these revelations were of God, he started to preach and share them more widely. This did not begin until 612 CE. The verses (Āyah plural āyāt) were revealed gradually over the next twenty two years, so were not complete until 632 CE. The revelations led him to proclaim a strict monotheistic faith and to highlight the social injustices in society. Karen Armstrong, in her book – *Islam A Short History* [419] – explains that many of Muhammad's tribe were already aware that a supreme god – Allah, had created the world and would judge humanity at the end of time. Just as the Jews and Christians held a similar belief.

As already mentioned, Arab culture traditionally showed concerns for others, but at the time of Mohammad's birth, there was a shift towards being more concerned with creating wealth. Therefore, the focus of the verses on social justice, caring for the poor and doing good works were not without controversy (see below). This was the substance of the revelations of the Qur'an and thus, Islam, meaning surrender, was born. Muhammad acknowledged that previous prophets – Abraham, Moses and Jesus – had also preached the same message. But he was doing so directly to the Arabs. Tradition had it that Abraham had in fact re–erected the Ka'ba, that was thought to have originally been built by Adam but had fallen into disrepair. It was said that Abraham visited Mecca to see his son Ishmael, whom he had fathered with Hagar. She was the handmaid of his wife Sarah. At God's command, Abraham had sent her away, believing that God would take care of both her and Ishmael. But more than just take care, God would raise up a mighty people from them. She had eventually found herself in Mecca.

However, many of Muhammad's ideas didn't go down well with his fellow Meccans, as they were both a threat to business and critical of their current way of life. Their response was more than just antagonistic and he found his life to be in danger. So in 622 CE, with his family and a few followers, he migrated about 280 miles north to the town of Yathrib, which later became known as Medina. This trip is known as the hijra or Hegira – migration, and is the time which Muslims point to as the start of the Islamic calendar – year 1 AH (anno hegirae = "in the year of the Hijrah").

Muhammad continued to preach and his following increased and by 630 CE he had taken over Mecca, which surrendered when they saw the size of his army. He then established a peace accord with many of the Arab tribes. This was the first time any sort of common cause had been established between the tribes who generally operated independently. Mecca remained the religious centre but Medina became Muhammad's home and administrative centre. In 632 CE, Muhammad died, creating the need for a successor. This matter was resolved, but issues surrounding the process and whether Muhammad had specifically identified a particular successor or not, has been a bone of contention ever since. Abu Bakr was selected, but others thought Ali ibn Abi Talib (Mohammad's cousin and son in law) should succeed, given that he was a family descendant and on the assumption that good characteristics are passed down family generational lines. These followers of Ali would form the Shiite faction.

Abu Bakr was the first of four caliphs ("a successor to the prophet" – as leader of the community, not in terms of succeeding in prophethood) whose rule was known as the Rashidun Caliphate or caliphate of the "rightly guided." Ali did become the fourth caliph – leader, but was himself killed in 661 CE. Between them, the first four caliphs oversaw the early Muslim conquests, advancing through the areas of Persia, Anatolia, Afghanistan, the Levant, Egypt, Syria, Palestine, Lebanon, Iraq and north Africa.

Ali was murdered by a member of another group known as the Kharijites, who formed as a response to their unhappiness with the process of succession management. They broke into revolt against Ali's authority after he agreed to arbitration with his rival, Muawiyah I, to decide the succession to the Caliphate, following the Battle of Siffin (657 BCE). The Khawarij opposed arbitration as a means to choose a new ruler on the grounds that "judgement belongs to God alone". They thought that arbitration was a means by which human beings could make decisions, whereas a victor of a battle was determined by God and so the right person would be chosen by God. They believed that any Muslim could be the Imam, the leader of the community, if he was morally irreproachable. If the leader later sinned, it was the duty of Muslims to oppose and depose him.

The group continued, but remained a minority, as they saw the use of violence as the remedy for dealing with those with whom they disagreed. They spawned a number of off–shoots, many of which have become extinct, but over the centuries they have continued to rebel against what they perceived as authoritative injustice and wrong religious thinking.

Before continuing with the narrative, we will deal with some of the different religious groups that have emerged since the death of Muhammad. The reason for doing so, is that despite the unifying feature of the Qur'an, disagreements over the succession planning and understanding of how Islam should be lived, have contributed to how Muslim culture has spread across the world and led to the establishment of rival groups.[420] New groups often emerged at a time of succession, sparked by disagreements over who was thought to have the necessary credentials, which, particularly for the Shi'a, included, amongst other things, being in direct blood line to the Prophet. Other issues included, whether the caliph was to be elected, how things should operate and whether the primary concern was the consolidation of power or the spread of Islam and adherence to the "truth" of its precepts, as they saw it.

The largest group are the Sunni Muslims who make up around eighty five to ninety percent of the Muslim population. They originate from the group supporting Abu Bakr and the processes for choosing the first and subsequent caliphs to succeed Muhammad. Their name means – people of the tradition and the community of Muhammad – taken after the Sunnah.[421] Their identity becomes more consolidated during the Abassid Caliphate after 750 CE when the four schools of jurisprudence (Madinahs) are formed – Hanafi, Maliki, Shafi'i, Hanbali. The role of these schools is to give legal rulings and interpretations of how to live, in situations not specifically covered by the Qur'an.

The second largest group are the Shiites. They take their name from "Shiat Ali" or the "Party of Ali" as in their view, Ali was the rightful successor to the Prophet Muhammad as leader (Imam) of the Muslim community. They were very unhappy about the way things turned out initially and then subsequently after Ali's death. Three distinct groupings have subsequently emerged – Ithna Asharis (Twelvers), Zaydis (fivers) and the Ismailis (severners). The Ithna Asharis are the largest group and are in the majority in Iran and Iraq, with smaller Shia communities in Afghanistan, India, and Saudi Arabia. Some of these groups also have their own Madhhabs or schools of thought.

Both Sunni and Shi'a groupings believe that how they responded to the question of the immediate succession to Muhammad was commanded by God and Muhammad; it was not their choice.

Other movements include :
- The **Alawites**
- – a distinct religion that developed in the nine/tenth century

- The **Druze** – a distinct traditional religion that developed in the eleventh century as an offshoot of Ismailism,

- **The Salafi,** is a reform revivalist movement within Sunni Islam developing in the late nineteenth century. They look back to the early traditional and original form of Islam – as they see it. The movement is often divided into three categories: the largest group are the purists (or quietists), who avoid politics; the second largest group are the activists, who get involved in politics; the third group are the jihadists, who form a minority and advocate armed struggle to restore the early Islamic movement.

- The **Wahhabi movement** originated in the Najd region of modern–day Saudi Arabia. It also advocates return to the early form of Islam but is regarded as more austere and extreme than Salafi. It was initiated by Muhammad ibn Abd al–Wahhab in 1792 CE who formed a pact with Muhammad bin Saud and hence it is more prevalent in Saudi Arabia.

- **Sufism** – means "wearing wool" as an indication of the desire to return to a simple life. They tend towards mysticism with veneration of pious forefathers. There are many different groups which often begin as followers of one particular holy man. Pilgrimage to the tomb of a saint has been an important part of Sufi devotion. So has the use, by ascetic Sufis (or dervishes) of repetitive phrases and actions, conducive to mystical experience. A well–known but extreme example is the whirling of the so–called dancing dervishes, a group founded in the thirteenth century. Sufism is open to any Muslim, without membership of a particular group.

So, returning to our narrative of Muslim expansion. This was not an orderly progression from one generation to the next. As the area of Muslim influence enlarged, different factions emerged in different parts and ruled, expanded, declined and were superseded. Similar activity could be taking place in different parts at the same time. That is, more than one dynasty was ruling over different parts of the Muslim world at one and the same time. Hence, there is often overlap between the dates of the various dynasties.

Umayyad Dynasty (661 to 750 CE)
After the death of Ali, Mu'awiya, who was the governor of Syria, took over as Ali's successor, in the face of opposition from the Shiites. They supported Hassan, the eldest son of Ali, whom Ali had designated. This confrontation established the basis for the lasting religious disagreements between Shia Islam and the Sunni Islam of the caliphs. Mu'awiya's power base was Syria and so Damascus now becomes the capital of the first Muslim dynasty and the centre of the new Arab empire. The Umayyad caliphate establishes a new principle – that the role of caliph shall be hereditary rather than elected. Thus it takes on the form of a dynasty rather than continuing the caliphate tradition. Further expansion occurs with the acquisition of: Rhodes, Crete, Kabul, Bukhara, and Samarkand. In 664 CE, Arab armies conquered Kabul, and in 665 CE pushed into the Maghreb – North Africa. They also went North and reached the Caucasus by 724 CE.

During the first heady days of Muslim expansion, it was taken for granted that Muslims would be of Arab descent. But many of the conquered populations – Persians, Berbers in North Africa and so on converted to Islam. So the armies are Arab and the non–Muslim inhabitants of invaded lands are taxed. However, by the early eighth century, as the empire expands, there are insufficient Arabs to provide the armies of occupation and war. So people of the occupied lands are recruited and converted to Islam, despite this having a negative impact on the treasury ! Resentment rises and with the difficulty of running and expanding the empire, tensions spill over and rebellion breaks out. This occurs particularly from within Persia, and the Umayyads are displaced by a new dynasty – The Abbasids. Male members of the Umayyad family are hunted down and killed, though one Umayyad leader escaped and established the Umayyad Caliphate of Spain as a new dynasty in Cordoba.

Abbasid Dynasty (750 – 1258CE)

Abul Abbas is proclaimed the first caliph of a new line and they establish a new centre in Baghdad in 762 CE. There is further conquest of the Mediterranean Islands, including the Balearics and Sicily in 827 CE. By incorporating elements of ancient Persian culture, a distinctive Arab civilisation begins to take shape. The particular features being Arabic as a language, Islam as a religion and a thriving cultural expansion powered by a creative centre for learning in Baghdad – The House of Wisdom. It was this concept of identifying with an Arabic civilisation, rather than a dependance on Arab ancestry, that the Muslims now took with them on their conquests. Muslim no longer equalled Arab.

This is the "Golden Age" of Islam and great interest and advancements are made in: astronomy, prose, poetry, the arts, philosophy, science, alchemy and mathematics. Ancient manuscripts were taken from the Byzantine Empire and Greece – Aristotle and Plato among them, and translated into Arabic. These eventually becoming key drivers for the emergence of the European Renaissance. Al–Kindi, whom we have met earlier in the book, served at the House of Wisdom studying and expanding knowledge of geometry and algebra. Commerce and industry also flourished under the caliphs, including – Harun al–Rashid (ruled 786 to 809 CE) of "A Thousand and One Nights" fame and al–Ma'mun (ruled 809 to 813 CE) and their immediate successors.

However, the political unity of Islam began to disintegrate. We see the Shiites forming political factions and by 765 CE we have the Ismaili group followed in 873 CE by the Twelvers. In addition, there are independent dynasties appearing which then ruled over various parts of the Muslim empire, for varying lengths of time. Thus, the Abbasid influence over the Islamic empire was gradually reduced to a ceremonial religious function, though they retained control over much of the Mesopotamian area. These new dynasties include: the Tahirid dynasty 821 to 873 CE, the Saffarids 861 to 1003 CE, the Samanids 819 to 999 CE, Turkic Ghaznavids (977 to 1186 CE) and the Great Seljuq Empire by 1037 to 1194 CE. These ruled over various parts of what was Persia and the neighbouring areas. The Fatimid dynasty then emerged and by around 960 CE had conquered Abbasid Egypt, building a capital there in 973 CE now known as Cairo.

Ayyubid Dynasty (1174 to 1250 CE) and Mamluk Dynasty (1250 to 1517 CE)

The Ayyubid dynasty was founded by Saladin in 1174 CE and centred in Egypt. Saladin proclaimed himself Sultan and ruled much of the Middle East during the twelfth and thirteenth centuries. But in 1250 CE, his dynasty itself was overthrown and the Mamluk Sultanate was born. The Mamluks immediately had to contend with the Mongols invading west and sacking Baghdad in 1258 CE and Damascus and Aleppo in 1260 CE. This marked the end of the Islamic Golden Age and effectively ended the powerful influence of the Abbasids. This power vacuum left the way open to the Mamluks, provided they could resist the Mongols. This they did and so brought some stability to Syria and Egypt until being overthrown themselves by the Ottomans in 1517 CE.

Mughal Dynasty (1526 to 1857 CE)

Expansion was still taking place with early forays into sub–Saharan West Africa and India around 1000 CE. The Delhi Sultanate was established and flourished from about 1206 to 1526 CE. Five dynasties ruled over the Delhi Sultanate until it was taken over by the Mughals. The Mughal Empire was founded by the Timurid prince Babur in 1526 CE. They marched into India and destroyed the Delhi sultanate and created their capital in Agra. One of the Mughal leaders was Shah Jahan. He commenced the construction of the Taj Mahal in 1632 CE as a tomb and memorial for his wife. The Mughal Empire subsequently covered vast areas including modern day Afghanistan, Pakistan, India and Bangladesh. Its decline in the early eighth century allowed India to be divided into smaller kingdoms and states. The Mughal dynasty was dissolved by the British Empire after the Indian Mutiny of 1857 CE.

Ottoman Empire (1299 to 1922 CE)

With the gradual demise of the Seljuk empire around 1300 CE, Anatolia was divided into a mosaic of independent states. One of which was led by Osman I. In 1299 CE he set about expanding his influence and the seeds of the Ottoman empire were sown. By 1453 CE they had taken Constantinople. The Ottomans had the advantage because of their use of gunpowder in the form of muskets and large cannons. They made Constantinople the capital of the Empire, which by the sixteenth century stretched from the Persian Gulf in the east to Hungary in the northwest; and from Egypt in the south to the Caucasus in the north. From this point, the empire gradually declined. With the opening up of new sea routes it lost its trading monopolies and the cost of war and maintaining its empire became too prohibitive to fund. It was officially wound up in 1922 CE by the order of the Government of the Turkish Grand National Assembly in Ankara, following the Turkish War of Independence (1919 to 1923 CE). The Caliphate was abolished in 1924 CE.

Safavid Empire (1501 to 1722 CE)
The Shiite Safavid dynasty rose to power in about 1501 CE and in the early sixteenth century assumed control in Persia under the leadership of Shah Ismail I. Despite defeat by The Ottomans, who were Sunni, at the Battle of Chaldiran in 1514 CE they maintained their power until 1722 CE.

So, during the fifteenth and sixteenth centuries there were three major Muslim empires existing at the same time: the Ottoman Empire in the Middle East, the Balkans and northern Africa; the Safavid Empire in Greater Iran; and the Mughal Empire in south Asia.

In modern times, some Muslim countries, such as Turkey and Egypt have sought to separate Islam from the secular government. In other cases, governments such as Saudi Arabia (founded only in 1932 CE) and Iran, following the Iranian Revolution in 1979 CE, seek to include religion in all aspects of political and social life.

※∩ ※∩ ※∩※

Scriptures
The key book for Muslims, the source of God's (Allah) revealed knowledge of Islam, is the Qur'an (meaning "recitation" – initially

transmitted orally through public readings, as most people were unable to read). The content, Muslims believe, came to Muhammad through the Angel Gabriel, beginning in 610 CE. The verses (āyāt) were revealed gradually over about twenty years, so was not completed until about 632 CE. Initially it was transmitted orally and memorised, but was later written down to ensure that after Muhammad's death, it would not be forgotten or mis – represented : – (the Arabic language is written without vowels, so speakers of different Arabic dialects and other languages reciting the Qur'an may apply phonetic variations that could alter the meaning of the text). The final and definitive text of the Qur'an was established under the third caliph, Uthman. At a meeting in 647 CE the dialect and form of the Qur'an were decided and one definitive copy was made, with all other previous copies being destroyed. This copy, became the template for all future copies. Copies were made and sent out to the different centres of the expanding Islamic empire. There are therefore no inconsistencies between any of the copies of the Qur'an across the world. The Qur'an is the only scripture.

The succinct core elements or practices of the Islamic faith, are the five pillars of Islam – Declaration of Faith, Obligatory prayer, Almsgiving, Pilgrimage to Mecca and Fasting for the month of Ramadan. However, some groups interpret these slightly differently and the Shia Twelvers and Ismailis have additional and different pillars. Other writings are also available and used to help clarify how to live a good life and provide guidance in matters not dealt with definitively by the Qur'an.

The first of these in importance is the Sunnah – "a path, a way, a manner of life". These writings contain what are thought to be the sayings, traditions and practices of the Prophet, as his understanding of the way to live out the Qur'an on a day to day basis. They thus constitute the elements required for people to emulate as they follow the Prophet as a role model. The content may be supported by a Hadith, but it need not be.

The second source is The Hadith. As explained by Prof. Shahul Hameed,[422] this refers to the record of reports of sayings, actions or the silent approval of the Prophet, as observed or witnessed by friends and family. Most of the sifting, classification and compilation of the canonical collections, took place in the eighth and ninth centuries.[423] These Hadith may be either authenticated and accepted or still under dispute but are seen by many as crucial to understanding the meaning of the Qur'an. However, not everyone accepts the Hadith at all. Some Sunni and Shia groups have both commonly accepted Hadith as well as ones that are different. Because some Hadith include room for debate and interpretation, the authentication of Hadith has became a major field of study in Islam.

The other important set of writings concern the Shariah. This is the collection of the divine Islamic law, as identified from the Qur'an and the Sunnah and compiled from interpretation through the practice of fiqh (Islamic jurisprudence) within the particular Madhhab (schools of thought). Their rulings are called fatwa.

For Sunni Islam, the imam (person who stands or walks "in front") is the person who leads the course of prayer and worship in the mosque and will be a leader of the local Muslim community.

For Shia Muslims, the Imam has a different connotation. Although he may fulfil the roles just mentioned, he is regarded more significantly as the Divinely Appointed One, being the true caliph or successor to the Prophet. He thus possess divine knowledge and authority, by which they can provide commentary and interpretation of the Qur'an. As such, they must be obeyed. They are designated as infallible by God and chosen infallibly by the preceding Imam.

Summary

Whilst the foregoing is a very superficial and selective account of the spread of Islam and the globalisation of Muslim culture, the evidence of variations in theological understanding amongst a variety of different groups mirrors similar situations within the other major faiths we have considered in this chapter. There is no "pure" version of Islam that all Muslims accept and the intolerance of the different factions is just as evident as it is in Christianity and Judaism.

CONCLUSION

Thus, on the one hand we have religion referring to the various traditional practices and rites as exercised by different groups, tribes and cultures and on the other, religion, as often considered in modern terms, referring to an institution – the organisational structure that administers and oversees the religious rites and practices of its members.

We have seen that in the former, most of the population would generally take part and comply with the religious practices, which are just accepted as an integral part of life as a whole. There is no sense in which such practices are seen as being separate from how people behaved on a day to day basis. They were just a natural part of how life was lived. There was no need therefore to have a name that distinguished them from other aspects of life. In the latter, there is more of a sense of personal choice – whether to join or to stay in (for those born into and brought up in that particular tradition). Even so, the tenets of the corresponding faith will generally emphasise that the practice of the religion should be seen as an integral part of life as a whole. But it is clear, that although the origins of these beliefs cannot be precisely defined, each of the major faiths have changed in their organisational development as well as in some of their stated beliefs. What was of paramount importance in their early years, has become less so over the years, under the influence of changing knowledge and changing cultural and environmental experiences i.e., these faiths have evolved. They are now different from what they were centuries ago.

Thus, it seems to me that such a history of evolution establishes a pattern that is likely to be ongoing. Namely, that such organisations and beliefs are likely to continue changing and evolving into the future. If this is the case, then surely it is inappropriate to think that how we understand scripture and the nature of God now, is so set in stone, that we can think any one view is "the correct" and only one – that one is right and so everything else is wrong. Other views and understandings emerge as humankind tries to make sense of things that are beyond our current level of knowledge, and may always be so, but we have a duty to respectively consider other positions and see them as part of our collective struggle, even if they don't seem to make sense or chime in with where we happen to be at this moment in time.

To be still arguing about doctrine and scripture in ways that try to claim one is right and another is therefore wrong suggests that we have totally failed to understand the lessons of history. Not only is this happening between faiths, but even under the same umbrella within each of the major faiths. No one KNOWS, we can only respect and take on trust that others are grappling with the same problems and trying to understand and articulate their own understanding. This may help some but do nothing for others. So should we not accept there are different roads to travel but that we can travel in parallel, arm in arm, accepting and respecting others' views. God will ultimately decide the end game !

As quoted earlier, from Pope John Paul II and the Syriac Patriarch:

"The confusions and schisms that occurred between their Churches in the later centuries, they realise today, in no way affect or touch the substance of their faith, since these arose only because of differences in terminology and culture and in the various formulae adopted by different theological schools to express the same matter. Accordingly, we find today no real basis for the sad divisions and schisms that subsequently arose between us concerning the doctrine of Incarnation."

And Karen Armstrong,[424] quoting from the Qur'an 29:46

"Do not argue with the followers of earlier revelation otherwise than in a most kindly manner – unless it be such of them as are bent on evil–doing – and say 'We believe in that which has been bestowed from on high upon us, as well as that which has been bestowed upon you, for our God and your God is one and the same, and it is unto Him that we (all) surrender ourselves.' "

Surely, the same needs to be said about any potential future schisms before they happen; that is, what ever differences have not been yet resolved or whatever differences arise in the future, we should be able to live with them and struggle together, rather than split apart and go our separate ways.

CHAPTER 8

FAITH AND GOD

Following the same format as previous chapters, we will start with definitions and origins before searching for the evolutionary aspects of faith and God.

FAITH

Now faith is a tricky concept. This short section seeks not to defend a faith position or to justify the rationality of having faith at all, but to acknowledge that the concept of "faith" exists[425] and that it has been extensively considered in both philosophical and theological circles. It is not possible to say whether faith itself is something that changes, certainly at a global level. But there is a sense within the major religions that an individual's faith increases/deepens, as they immerse themselves in their religion. This in turn, is said to bring them closer to God. However, it is probably reasonable to suggest that what has changed over time, are our understandings of the objects of faith, the arguments both in support for and against having faith and the philosophical perspectives on the nature of faith.

Definitions of Faith

Definitions vary, but essentially, when discussed within a religious context, faith can be taken to mean – "a state of confidence or trust in a particular system of religious belief, which specifically may include a 'supernatural being or god' of some sort". Whilst this definition largely covers what "faith" might generally be understood to be, there are a wide range of views as to how the idea of faith is perceived. At one end of the spectrum there are the new atheists, like Professor Richard Dawkins who regard faith as "blind and a belief in things that have no evidence and may even fly in the face of contradicting evidence." Such views are supported by many philosophers and scientists.

At the other end of the spectrum, faith is seen as something more than just a belief in a religious system or a doctrine but rather as a way of life and a willingness to subjugate oneself to the will of God. Paul Tillich, a twentieth century theologian, expresses this view in his book *The Dynamics of Faith.* [426] and there are scientists and philosophers who are also on this end of the spectrum. They would argue that faith is a rational and reasoned response and "warranted" according to the evidence.[427] Faith, they would suggest, is not something blind or irrational.[428]

It is not really possible to say when faith began, or identify the point at which it emerged or became necessary. Colin Wells [429] proposes that Faith evolved out of and as a consequence of the development of Reason. In his article "How Did God Get Started?" he traces a thread of development from the birth of Reason, the necessity for the emergence of faith, the transition from polytheism to monotheism and the eventual propagation of Christianity. "Reason" is the term used to describe the new way of thinking and discovering things about the world, thought to have started with Thales of Miletus and the other pre–Socratic philosophers of Ancient Greece, we met in chapter four. Uncovering explanations for how the world was, from within the natural world itself, meant that the role of "the gods" in human existence became less central. However, it also created a separation between the seen and unseen or the natural and supernatural worlds, which had not been identified before.[430] This separation involved questioning whether any sort of supernatural or divine dimension existed at all. Answering this question in the positive, involved the emergence of faith.

Attempts to define faith and to consider the nature of faith, comes under the umbrella of the discipline of Philosophy of Religion. A field of study examining matters concerning religiously related ideas. A number of theories have been advanced to provide explanations and justification for including faith as a valid means of seeking and declaring truth. However, it is not always possible to identify specifically separate arguments in relation to faith, religion or God. Thus, there is some cross over and blurring of the edges, with some arguments applying to all or failing to discriminate between them. These include:

- **Evidentialism** – states that one is justified in believing something if and only if there is evidence supporting that belief

- **Foundationalism** – initiated by the French philosopher, René Descartes, states that one is justified in believing something, and that belief is a justified belief, if it is based on a secure foundation of certainty of some sort. Defining this "certainty" is the crux of the argument. Descartes' certainty was in the fact of his own existence whereas John Locke's was based on experience. Others have postulated other foundations, such as empiricists who emphasise experience and rationalists who emphasise reason. Foundationalism holds that there are "basic beliefs" which may be accepted as self–evident or infallible, and on which other beliefs can be based. These may include aspects of Externalism and Internalism.[431]

- **Externalism** – states that one is justified in believing something despite not knowing or being aware of the nature of the evidence

- **Internalism** – states that one is justified in believing something provided that one knows or is aware of the nature of the evidence.

- **Reformed epistemology** – is the proposition that belief in God may be taken as a "basic belief". It represents a continuation of the sixteenth–century Reformed Theology of John Calvin, who postulated an "innate divine awareness of God's presence". Alvin Plantinga argues that a "basic belief in God" is warranted when produced by a sound mind, in an environment supportive of proper thought, and in accord with a design plan successfully aimed at truth.[432]

- **Reliabilism** – states that one is justified in believing something if one has arrived at the belief through some reliable process.

- **Philosophical Skepticism** – questions the possibility of certainty in knowledge. either by denying the possibility of all knowledge or suspending judgement due to the inadequacy of evidence

- **Fideism** – states that one is justified in believing something purely by faith, without recourse to reason. It questions the ability of reason to arrive at all truth. Fideism is supported, in particular, by Blaise Pascal, Søren Kierkegaard, William James, and Ludwig Wittgenstein. Fideism has been rejected by the Roman Catholic Church, which has adopted the view of St. Thomas Aquinas (around 1225 to 1274 CE), building on the views of St. Anselm of Canterbury, that truth is known through reason and faith together. The use of reason in enquiries into the basis of faith and the nature of God is called Natural Theology.

A similar idea is expressed by William James in his lecture The Will to Believe. James argued that some religious questions can only be answered by believing in them.

"One cannot know if religious doctrines are true without seeing if they work, but they cannot be said to work unless one believes them in the first place. Faith is found through experience of the spiritual, and so understanding of belief is only gained through the practice of it. Thus, reason has no place in this endeavour."

- Zwemer also articulates this idea in his book – *Origin of Religion – Evolution or Revelation,*[433] where he states that:
  - *"primitive religion always included two beliefs: that the Supreme Being is (both) Creator and morally good. This belief could not have been acquired by purely natural reason and research, as in their own experience they could not find a being who would have been in every respect, morally good. And this conclusion is derived from both the evidence of anthropology and that of the scriptures; the origin of religion is not by evolution but by revelation."*

A different approach is provided by Dominic Johnson who applies Error Management Theory (EMT) to religion.[434] EMT examines the costs of false positive and false negative decision–making errors. In applying this to religion, he postulates that :

> *"if there has been an asymmetry between false positive and false negative errors, over human evolutionary history, then natural selection would favour a bias towards the least costly error over time (in order to avoid whichever was the worse error). Applied to religious beliefs and behaviours, I derive the hypothesis from EMT that humans may gain a fitness advantage from a bias in which they tend to assume that their every move (and thought) is being watched, judged, and potentially punished by supernatural agents. Although such a belief would be costly because it constrains freedom of action and self–interested behaviours, it may nevertheless be favoured by natural selection if it helps to avoid an error that is even worse: committing selfish actions or violations of social norms when there is a high probability of real–world detection and punishment by victims or other group members. Simply put, supernatural beliefs may have been an effective mind–guard against excessively selfish behaviour – behaviour that became especially risky and costly as our social world became increasingly transparent due to the evolution of language and theory of mind. If belief in God is an error, it may at least be an adaptive one."*

This is not a million miles from Blaise Pascal's (a seventeenth century French Philosopher) classic wager [435] in which he indicates that reason cannot prove or disprove God's existence. However, reason would suggest it makes more sense to believe in God, whether he exists or not, as there is everything to gain if God does exist and nothing to loose if he does not.

Tillich also says this about faith:[436]

- *"Faith is the state of being ultimately concerned."* (page 1)

- *"Reason is the precondition of faith; faith is the act in which reason reaches ecstatically beyond itself."* (page 87)

- *"Faith stands upon itself and justifies itself against those who attack it, because they can attack it only in the name of another faith. It is the triumph of the dynamics of faith that any denial of faith is itself an expression of faith, of an ultimate concern."* (page 147)

The Bible of course has plenty to say about faith. One particularly pertinent quote from the Letter to the Hebrews is: *"Now faith is the substance of things hoped for, the evidence of things not seen."* [437]

Summary

All of these propositions have been criticised. It is a paradox and the nature of faith, that faith or its necessity, cannot be proven and must be taken on faith / trust, or not. Whether faith is justified, in the end, is a matter of personal faith or judgement. All the explanations and justifications in the end cannot get around the fact that we are left with a mystery that we either embrace or reject. We will deal with this at the end of the next section.

GOD

God either exists or he does not. Pretty much the whole of human existence has been influenced by how certain people have responded to that statement. Some of the more negative consequences that have followed an acceptance that God does exist, are the reasons for this book. How we behave and manage our lives depends on our response to this statement and not only how we respond, but also how others, who have the power to exert wider influences on our lives, respond or have responded.

The choice of responses open to us are represented in table 1.

	GOD ACTUALLY DOES EXIST	GOD ACTUALLY DOES NOT EXIST
BELIEVE / HAVE FAITH IN GOD	A	B
DON'T BELIEVE / HAVE NO FAITH IN GOD	C	D

TABLE 1

Whilst it is possible for us to choose a row, it is not possible to choose a box, as doing so is contingent on the truth of the two columns. This we cannot know. We will only know which box is correct, perhaps, depending on what happens after death or on the Day of Judgment (if there is one).

Definitions of God

No one definition can convey the wide range of concepts and contexts, in which the idea of "god" may feature. These include: the gods of early tribal religions, the physical gods made by humans out of natural materials, we have the anthropomorphic (human – like) creations as in the Greek and Roman gods with all their flaws, capriciousness and particular interests and the panoply of gods within Hinduism. There is also a supernatural concept of god, being "something that is other than and outside / beyond ourselves and the perception of our senses." Such a god may be understood as one that sits on the sidelines and watches the world play out, or one that acts, interacts and communicates with humankind.

This latter concept features particularly in the three Abrahamic regions – Judaism, Christianity and Islam. Such a god is seen as the creator of the universe, is worshipped and represents the highest perfection, incorporating concepts such as omnipotence, omniscience, omni– benevolence, omnipresence and other "metaphysical attributes" such as —simplicity, timelessness (free from the constraints of our space–time concept), immutability (unchanging), and impassibility (unaffected by the things he has created). The attribution of these terms provides fertile ground for endless discussions as to their precise meaning and their implications for doctrine, particularly in relation to compassion, suffering and evil.[438] Superimposed on any of these ideas can be a variety of anthropomorphic characteristics as seems to befit the understanding by the peoples, of the nature of their god. Each person may share some common concepts underpinning their understanding of God, but these may be influenced to a greater or lesser extent by individual experience and knowledge.

On a website about best definitions of God, there was a comment to the effect that by trying to define god, one is implying that god is like the definition, which by definition, is man made and therefore god must be man made ! One can understand the sentiment behind this statement, though the superficial logic of it does not stand scrutiny as an argument against the existence of God. But it does highlight the dilemma of trying to define and explain something that is conceived of as being beyond human kind's ability to fully comprehend – the central communication problem which has been the prompt for this book. I questioned earlier whether the created could fully comprehend the creator. Clearly for people of no belief such a question has no particular relevance, other than from perhaps a curiosity point of view. But in any discussions of these matters, it is important to establish some common understanding of the terminology in order to at least try to avoid unnecessary mis–understandings. In the introduction, I included the following as my understanding of the definition of God:

When referring to God, I refer to the concept of a "something that is other than and outside / beyond ourselves and the perception of our senses i.e., that God is supernatural". In this sense, God "exists" as an independent and separate 'being' that can act, interact and communicate with humankind, as it sees fit. This concept of a "supernatural" God, is opposed to the idea of "god," arising as a human creation, construct or idea that is referred to in the psychological, sociological and cultural literature relating to the study of religion and human behaviour. Clearly, this "god" has no means of independent thought or action.

So we perhaps come back to the quote from Dave Allen I used at the end of the introduction – "may your God go with you." Each of us will have our own picture and understanding of God. This may be adopted from the faith or religions we follow, or to which we have been exposed. We then try to integrate this picture into our position on these matters, whether as belief or not. For Christians, this individual experience can be expressed in relation to Jesus Christ, an idea captured nicely at the end of Albert Schweitzers book – "*The Quest Of The Historical Jesus*":[439]

"He comes as One unknown, without a name, as of old, by the lake–side. He came to those men who knew Him not. He speaks to us the same word: 'Follow thou me!' and sets us to the task which He has to fulfil for our time. He commands. And to those who obey Him, whether they be wise or simple, He will reveal Himself in the toils, the

conflicts, the sufferings which they shall pass through in His fellowship, and as an ineffable mystery, they shall learn in their own experience Who He is."

In addition to the complexity involved in trying to define God, there are also the ".....isms" of religious terminology to consider, including – theism, monotheism, deism, polytheism, pantheism, panentheism and atheism.

- **Theism** is the notion of a Supreme Being called God with Monotheism being the belief that there is only one God. The monotheistic religions include Christianity, Judaism, Islam, Baha'i Faith and some forms of Hinduism. Zoroastrianism is thought to be one of the first monotheistic religions, and so may have influenced the other major faiths. This concept usually includes the idea that the God is not only the creator but has a moral dimension and may also intervene in the ways of the world and in peoples' lives. This is in contrast to :

- **Deism** in which the understanding is God as creator but who then allows things to take their course. It rejects the idea of supernatural interventions like prophecies, miracles, and divine revelations. Deism holds that religious beliefs are founded on human reason and observed features of the natural world, and that in these the existence of a supreme being is revealed.

- **Polytheism** is the belief that there are multiple gods that are usually distinct.

- **Pantheism** is the belief that the physical universe is equivalent to god, and there is no division between a Creator and the substance of its created.

- **Panentheism** is an extension of pantheism in recognising that the physical universe is joined to a god or gods and that the divine pervades every part of the universe, extending beyond time and space.

- **Atheism** rejects the idea of a Supernatural Being. The actual idea of God or a god is an anathema and is rejected as being a figment of the human mind.

There are even more sub–categories identified with more nuanced distinctions !

※∩ ※∩ ※∩※

ORIGINS AND THE EVOLUTION OF THE IDEA OF GOD

If we thought seeking the origins of religion was problematic, attempting to discover the origins of the idea of god is even more so. In fact, we may as well face it from the start, it is impossible to answer the question – "where or how did the notion of God – "a being that is something other than ourselves and the universe" – arise or come from ? What theories or comments there are about this question, tend to be crafted in terms that are aligned with either the belief system or the professional stand point of the proposer. This is understandable and inevitable to a degree. Given that it is only possible to consider the question retrospectively, and from a position in which one may have an extensive knowledge and experience of a wide variety of religions and spiritual beliefs. It isn't possible to look prospectively from a position where the concept of god or spirituality did not yet exist.

This potentially colours any proposed theory of origin, with ideas and knowledge that have developed later. For example, it has been proposed that the idea of god(s) was developed to either explain happenings in the natural world or as a source of comfort to counter the fears and anxieties associated with suffering, imminent dangers, death and an uncertain future. These ideas may very well be part of the story, but there are other theories highlighting additional origins. For the purposes of the argument of this book, it is only necessary to touch on some of these, to get a flavour of the various proposals and a sense there has been an evolutionary development of these ideas. Some we have met previously and so will only be briefly re–capped here.

Animism

Animism, as advocated by Tylor and Frazer,[440] is the notion that all living and non–living things in nature contain a spirit which is capable of communicating with other sprits. In addition, there is also the notion of one spirit that pervades the world. This may be seen as a creator. These ideas are said to stem from the consideration of the nature of death, sleep, consciousness and from noting the figures/personages found in dreams. The next step considers that these spirits can live separately and independently and hence take on the role of gods. Thus a polytheistic situation arises, from which, one god may be vested with more importance and so a monotheistic state emerges.

Ancestor Worship

This is not a dissimilar idea to animism, in that it is said that each person contains a double of themselves. This double, spirit, soul or ghost, is what can leave the body in death or in sleep. The ghost idea is one put forward by Herbert Spencer.[441] So in this, the focus is on the worship of dead ancestors and tribal chieftains whose achievements in life were revered. Such reverence gradually elevates them to the status of a god. So initially, as gods, they are associated with a specific personage, but as time goes on, these original associations are forgotten and so the god becomes independent and an entity in its own right

Thus, as more people die, a collection of gods is formed – polytheism. As one of the gods assumes greater importance for whatever reason, a monotheistic situation arises. Grant Allen in his work – *The Evolution of the Idea of God An Inquiry into the Origins of Religions*[442] – develops these ideas but concentrates on the practical aspects of the religious rites and rituals that ensue and deliberately excludes the mythological aspects. Ignoring this dimension and hence any discussion of god, is criticised by Staniland Wake[443] and George Coe.[444]

This idea was expressed centuries before by Euhemerus (around 330 to 264 BCE), who wrote that gods were excellent historical persons who eventually became worshiped.[445]

Henning [446] expresses a similar idea of reverence being shown to certain ancestors and members of the tribe or group. However, he suggests that this was predicated on the economic and material success these persons had brought to the community, rather than on some pre–religious ideas. Furthermore, the practice of worship of these persons only occurred once society had some settled and sound economic basis of sustainability.

Ego Explosion

Steven Taylor [447] uses the spirit theory of animism as the starting point of his narrative towards monotheism, which, in brief, goes like this. During the pre–historic period when the notion of animism was widespread, integral to this understanding of nature was that humans regarded themselves as being part of both their community and the environment and that their community and the environment were part of them. A sense of everything being part of an integrated whole. The concept of "I/me" was not a significant feature of this world view. The idea of ownership or personal possessions was not part of the landscape, things were held in common and society operated on that basis – egalitarian, rather than hierarchical and everyone was part of the decision making processes.

However, this changed around 4000 BCE when the area covering north Africa, the near East, and central Asia, described by DeMeo in his research study, Saharasia,[448] suffered an environmental catastrophe. The area was changed from a wet, fertile grassland, home to a variety of animals, into a dry and desert landscape. The survival techniques needed to face the extreme hardship this imposed on the peoples living in those areas, promoted a sense of individualism and competition : a sense of us and them, rather than one of a common community. That is, the notion of the "Ego Explosion" – the increased importance of "I/me". Thus, they were transformed from a relatively peaceful matriarchal dominated society to one dominated by men. This resulted in repressive social institutions, destructive aggression and warfare. At the same time, there was also a loss of the sense of connectedness to the environment and each other and loss of the roll of spirits in their culture.

This resulted in a feeling of aloneness and becoming separate. Thus, theism evolved, as a psychological response to compensate for such feelings of loneliness. The idea of gods "out there" and apart from the world of human beings, that were observing and controlling its events from a higher realm. Presiding over different aspects of life such as war, love, travel, agriculture and so on. Thus, the matriarchal 'natural religions,' centred around an awareness of animating and spiritual forces, gave way to patriarchal 'high God religions.' These were characterised by a polytheism of dominating male gods with anthropomorphic characteristics, separated from nature and who demanded obedience and certain forms of moral behaviour. So these peoples had to leave their homeland and search for new, more hospitable environments. Their ideas of god and their changed views of self and their relationship to the world went with them.

The first archaeological evidence of theistic religion is found from around the fourth millennium BCE, among some of the peoples of the Middle East and Central Asia, such as the Ancient Sumerians, Egyptians, the Indo–Europeans and the Semites. Initially, their religions retained the polytheistic character until the idea of one god assuming greater importance – Monotheism – emerged. This is probably first seen in Egypt with Pharaoh Akhenaton in the fourteenth century BCE. He proclaimed that the only God was Aton, the sun God, and that all the old gods were obsolete. However, this idea didn't really catch on and it declined after Akhenaton's death.

Monotheism

In *The Invention of God*, Thomas Römer [449] traces the development of YHWH, as the god of the Jews, from a polytheistic environment to that of the one true God. The transition taking place between the Bronze Age and the end of the Old Testament period in the third century BCE, in the Hellenistic period. He locates the origins of Yhwh somewhere in Edom or in the northwest of the Arabian peninsula as a god of the wilderness and of storms and war. He became the sole god of Israel and Jerusalem in fits and starts, as other gods in the area were gradually sidelined. But he argues that it was not really until the destruction of Jerusalem and Judah in 587 BCE, that the Israelites, whilst in Babylonian exile, came to fully worship Yhwh as the one god of all, creator of heaven and earth.

Wells describes a similar sequence of events from animism, through polytheism to monotheism, but begins with the emergence of Reason as its stimulus. As mentioned in the faith section, Wells is of the belief that Reason was the reason for the development of faith and the subsequent evolution of the concept of god(s). We see that at the same time as the pre–socratic philosophers were looking to the natural world for explanations of the way things were, they were also postulating some sort of unifying force or single "concept" that underpinned the whole system and held the natural world together. For Thales, this was water. For Anaximander this unifying substance was apeiron and for Anaximenes it was air. Then Aristotle postulates his "un–moved mover" and Plato offered the "demiurge".

As more natural explanations were identified, the role of the polytheistic gods in the natural world was gradually marginalised. However, a vacancy for the role of a single unifying power, opened up in the supernatural or unseen realm. Such an idea was gradually incorporated into Jewish thought, as they subsumed the numerous "gods" of Canaan and surrounding areas into a monotheistic belief. Christianity capitalised on this by saying that not only do we believe in one god, but he is superior to all others and in fact all other ideas are wrong. In Islam, "there is no God but Allah."

Error Management Theory (EMT)

Johnson [450] argues that a belief in God may have evolutionary advantages in that believers may have been more cooperative. This resulted from the fact that with the advent of language and theory of mind [451] the concept of God could develop. The idea of being watched by a supernatural being that could meet out punishment, would make people less selfish as they would worry about being found out. If people knew, they would gossip and one's reputation would go down hill. Thus, even if God did not exist, the fact that the thought was present, might still confer an evolutionary advantage as co–operation, caring and sharing will generally have better outcomes. So these advantages may outweigh the disadvantages of having some constraints on one's personal freedom. Thus, a belief in God may be adaptive because it helps people avoid the social costs of selfish behaviour, which have, according to Johnson, been underestimated in the evolution of cooperation literature. However, Error Management Theory cannot explain why there is belief in supernatural agents, or why their punishment is feared.

Other Arguments

From McGiffert,[452] in his analysis of Spinoza, Hegel, Schleiermacher and Herder we get the idea of God as Absolute and an emphasis on God being love. For Immanuel Kant, it is not about the possibility of proving rationally God's existence but is about stating, that moral life is possible only if God exists.[453]

Cockerill also argues that God is both Creator and the Ethical perfect being, though for him, the creator element is the more important.[454]

Paul Carus [455] argues that the existence of God is scientifically demonstrable.

"Not as an anthropomorphic manifestation of a supernatural being, but beyond the world of matter and energy in the uniformities of nature, and in the intrinsic reality of those eternal laws of form, formulated in the logic of arithmetic, geometry, algebra, pure mechanics, and pure natural science."[456]

For Carus, God is omniscient but as superhuman rather than supernatural.[457]

In addition to the theories just outlined, theologians and philosophers have offered other arguments as proof that God exists. This is covered by the field of study known as Christian Apologetics.[458][459][460] This seeks to provide a range of reasoned, intellectual, evidence based arguments for the existence of God, without relying on ideas of mystery, faith, the supernatural and so on to justify belief. Each argument has both its proponents and its critics. I offer the headings, with a brief note on the four main ones for completeness, with web addresses for the enthusiastic to get into the detail.

From : http://www.existence–of–god.com/index.html [461] – we have :

The Ontological Argument

The ontological argument seeks to prove the existence of God from the laws of logic. It dates back to St Anselm, an eleventh century philosopher–theologian. It argues that once we mentally grasp the concept of God we can see that God's non–existence is impossible.

The First Cause Argument or Cosmological Argument

The first cause argument seeks to prove the existence of God from the fact that the universe exists. The universe came into existence at a point in the distant past. Nothing can come into existence, though, unless there is something to bring it into existence; nothing comes from nothing. There must therefore be some being outside of the universe that caused the universe to exist.

The Argument from Design or Teleological Argument

The argument from design seeks to prove the existence of God from the fact that the universe is ordered and so God must have created it this way. The atheist puts this down to chance. This can fall into the trap of seeing God as "god of the gaps."[462] That is, God is the source for all things science cannot yet explain. The risk is that God gets smaller as more explanations are forthcoming.

The Moral Argument

The moral argument seeks to prove the existence of God from the fact that there are moral laws. Moral laws have the form of commands; they tell us what to do. The existence of moral laws, demonstrates the existence of a being greater than us that rules over creation.

Lord, Liar, Or Lunatic?

Evidence for the Resurrection

Arguments for Atheism

Is Faith Just Wishful Thinking?

Is religious belief just an emotional crutch for those unable to cope with the reality of life without God?

The Problem of Evil

The Paradox of the Stone

Can God create a stone so heavy that he cannot lift it?

Religious Pluralism and Tolerance

Must we abandon the absolute truth–claims of Christianity in order to be tolerant of those of other faiths?

Can Evolution Explain Our Origins?

From the "Strange Notions" website https://strangenotions.com/god–exists/#1[463] – we get:

1. The Argument from Change

2. The Argument from Efficient Causality

3. The Argument from Time and Contingency

4. The Argument from Degrees of Perfection

5. The Kalam Argument

6. The Argument from Contingency

7. The Argument from the World as an Interacting Whole

8. The Argument from Miracles

9. The Argument from Consciousness

SUMMARY

One source of information about the nature of faith and the nature of God that we have not mentioned are the scriptures – whether these be the Bible, Qur'an, The Vedas or others. Depending on one's stand point, these either represent the revealed, inerrant Word of God, or the recollections and attempts by believers, inspired and guided (as opposed to dictated) by the Divine, to explain their experiences of God. Whether one accepts this or not, it is surely at best unscientific, to dismiss such testimony out of hand. Whilst coming to a personal judgement of the information and evidence of the likelihood of God's existence or not, in a situation where there is no definitive answer either way, the scientist will accept the possibility, however small, even if they live their personal lives on the basis that God does not exist.

The theories outlined in this chapter, offer much intellectual sense to the narrative of how ideas concerning religion, faith and God may have arisen. However, they all make assumptions about how earlier, ancient peoples might have thought and what might have been going on in their minds and their lives. I find it difficult to see how the idea of a supernatural creator or force or being, might arise simply out of one's experience of the environment through one's senses. It is now difficult or maybe impossible to imagine being part of a world within which there is yet no concept of a God.

Whether we believe in them or not, and if we do believe in something beyond/outside our normal experience but are unclear or have different understanding of what all this is about, everyone is aware of the idea of god, gods, spirits, supernatural forces, demons, devils and so on. We assume these ideas exist and either reject them, or think they might possibly be true but are unsure of their nature or we accept some of these things and create other ideas and concepts around them, in order to give them more meaning. But we cannot behave as though we have never heard of them and so come at some of these questions with a blank piece of paper. How did curiosity arise in order to begin formulating questions about why the sun gets up in a morning and disappear at night ? How did the notion of there being some unseen but substantive power behind it emerge?

Further consideration of these questions is for another time, but, as demonstrated in preceding chapters, evolutionary changes have occurred in the development of the understanding and ideas associated with God and religion. Thus, even if one comes down on the side of believing there is a God, it must surely be concluded that there cannot be only one single, unchanging definitive belief system that captures and defines all what God is and what he is about. Therefore, the approach of any particular belief system should be one that accepts the legitimacy of other belief systems, if people of faith are to fulfil their calling of love, caring and respect for their fellow human beings.(see reference) [464]

CHAPTER 9

A STUDY :

RELATIONSHIP BETWEEN EXPOSURE TO FAITH MATERIAL AND DECISION TO BELIEVE IN GOD

This chapter summarises the information obtained from a small survey carried out in 2018 CE. The study was prompted by one of the questions identified in the Preface. Namely: why is it that some people adopt a Christian faith and others don't, particularly when they have possibly been exposed to similar information and experiences of Christianity ? The Null Hypothesis[465] was therefore: "There is no difference in exposure to Christian influence between those who adopt a Christian faith and those who don't"

INTRODUCTION

Whilst this study is primarily driven from the perspective of Christianity and the development of a Christian faith, the question is also pertinent to other major faiths. This is particularly so in terms of Judaism and Islam, in that children are viewed as being born into the faith of their parents and are purposefully schooled in that religion from an early age. This tends to be a more intense and prescriptive engagement in a way that is not usually the case within Christianity. However, it is well documented that there are both Jews and Muslims who do not practice their faith and even disavow their traditions. Some studies have investigated factors that may influence an act of conversion (changing from no faith or from one faith to another) or apostasy (the rejection of the faith to become either just non–practising, an atheist or changing to another faith).[466] Such "apostasy" may come at a price. People may be ostracised by their family and community or may even be killed. So it is an intriguing question as to what factors influence the adoption or the rejection of religious beliefs in these circumstances.

A number of studies have looked at various factors that may influence both the adoption of a religious faith and what may maintain it. Such factors include – individual personality, religious socialisation, relationships with family, friends and within communities and demographic characteristics such as age, race and gender.

Religious socialisation refers to the process whereby religious institutions, their members and their families inform and educate others into the ways of that religious community. Religious socialisation is found to be important because it both provides the individual with a world view, as well as channeling individuals into personal communities that sustain that particular world view, through into adulthood. It taps into the demographic characteristics as well as individual occupational and political networks, with their associated ethics and world views. It also takes account of personal psychological variables, such as: perceived values of success, failure, self–image, power and responsibilities to others.[467 468 469 470 471 472]

There have been conflicting findings in identifying any relationship between faith and personality – the dimensions of introversion and extroversion, as measured by the Eysenk's Personality Inventory.[473]

Whilst the picture of a "model Christian" – someone who is demonstrative and enthusiastic about their faith and is eager to share their views with others and happy to take leadership roles within the church – may fit with what may be regarded as an extrovert personality, there is no consistency in studies to show that extroverts are more likely to be religious than introverts. Though introverts may experience some difficulties within a church context. The three main factors that seem to be important in a decision to adopt a religious faith are the influence of the local church, family and peers.[474]

METHODOLOGY

To examine the question from the Christian perspective I devised a questionnaire which can be found in the Appendices at the end of the book (Appendix 2). It was piloted with a few Christian and non–Christian friends and some questions refined before settling on the final version. The questions on categorisation of faith – Christian, atheist, agnostic etc and that relating to personality – introversion /extroversion – were both intended to be for self designation.

Not having the resources to do a random postal population based survey, I chose to ask Christians from churches both local to me and ones attending representative meetings from a wider geographical area. I felt it would be straightforward to obtain a Christian group, but non–Christians posed more of a challenge. I contacted the national atheist and humanist societies of the UK, Canada, Australia, New Zealand, Ireland and the United States and also the atheist/humanist societies of most of the Universities in the UK. The intention was to ask the secretary or other officials of these groups, identified via the internet, if they would be kind enough to circulate the questionnaire amongst their members. I did not have any replies from the Universities, but did receive a few responses from the countries contacted.

Some of them agreed, some declined and some questioned the validity of the study. The respondents were therefore all self selecting. The Christians volunteered themselves when a general request was put out at a number of meetings and the non– Christian (Not C) respondents decided whether to participate or not, if asked by their particular organisation. As I only received 21 questionnaires from non–Christians, I decided to recruit only a similar number of Christian respondents, 19 eventually responding.

RESULTS

CATEGORY	CHRISTIAN	ATHEIST	AGNOSTIC	HUMANIST	OTHER	TOTAL NON–CHRISTIAN (NOT C)
NUMBER	19	16	0	2	3	21
SEX: M	12	10	0	2	0	12
F	7	6			3	9
AGE GR						
Age 30–50	1	1				1
Age 51–60	4	5		1	1	7
Age 61–70	10	6		1	3	10
Age > 71	4	2			1	3

TABLE 1: PARTICIPANTS BY FAITH, SEX AND AGE GROUP

There were 40 respondents in the study. Where the figures do not add up to the total participants, it is because not everyone answered all the questions. The following tables summarise the data collection under the various headings of the questions. They are ordered in terms of demographics and the potential influencing factors used in the study questions. Out of the number identifying themselves as non–Christian, 16 said they were atheists, 2 were humanists and three were others. No–one designated themselves as agnostic. In view of the small numbers, in the following tables I therefore aggregate these groups to identify 21 as "NOT C". There were 19 identifying themselves as Christian. The age and sex breakdown is similar for both groups, but with more men than women in the study. This is perhaps surprising amongst the Christian group, as perhaps the impression is that more women than men attend church.

CATEGORY	CHRISTIAN	NOT C
introvert	6. (31 %)	14 (67 %)
thoughtful	12 (76 %)	16 (63 %)
extrovert	9 (47 %)	5 (24 %)
impulsive	1	1
scientist	8 (42 %)	15 (71 %)
artist	3 (16 %)	8 (38 %)
education attainment	degree / pos grad 10 (52 %)	degree / post grad 16 (76 %)

TABLE 2: FAITH BY INTROVERSION V EXTROVERSION, THOUGHTFUL V IMPULSIVE AND EDUCATIONAL ATTAINMENT

More of the NOT C were introvert – 67 percent v 31 percent and more of them were scientists – 71 percent v 42 percent. But also, more NOT C described themselves as being artistic – 38 percent v 16 percent. More Christians described themselves as having a more considered or thoughtful approach to problems 76 percent, as opposed to acting impulsively, compared to 63 percent of NOT C. A few people regarded themselves as being both scientific and artistic. More Christians described themselves as extrovert – 47 percent v 24 percent. More of the Not–C had studied to degree level or beyond – 76 percent v 52 percent.

CATEGORY	Mother Christian	Mother to church Regularly	MOTHER ATTENDS Regular most weeks	MOTHER ATTENDS Rarely: special occasion / family occasion
CHRISTIAN	16	10	8	5
NOT C	18	10	7	8
	FATHER Christian	FATHER to church Regularly	FATHER ATTENDS Regular most weeks	FATHER ATTENDS Rarely: special occasion / family occasion
CHRISTIAN	13	10	8	4
NOT C	11	8	5	5
	SIBLINGS	TOTAL NUMBER OF SIBLINGS	HOW MANY SHARE	HOW MANY DON'T SHARE
CHRISTIAN	17	32	10 (31 %)	19. (59 %)
NOT C	19	48	18 (37 %)	21 (43 %)

TABLE 3 FAITH BY PARENTAL FAITH AND CHURCH ATTENDANCE AND WHETHER SIBLINGS SHARED THE SAME BELIEF AS THE RESPONDENT

For both Christian and non–Christian participants, the number of mothers who were Christian and attended church regularly are similar. The number of fathers who were Christian and attend church regularly are a little less in non–Christian group. In the Christian group, fewer siblings share the faith of the respondent – 31 percent v 37 percent and also more siblings don't share the faith of the respondent – 59 percent v 43 percent than in Not–C group. faith of the respondent – 31 percent v 37 percent and also more siblings don't share the faith of the respondent – 59 percent v 43 percent than in Not–C group.

CATEGORY	ATTEND SUNDAY SCHOOL AS CHILD	FOR HOW LONG (average yrs)	ATTEND SUNDAY SCHOOL AS TEEN-AGER	FOR HOW LONG (average yrs)	GONE TO CHURCH RELATED YOUTH CLUB	FOR HOW LONG (average yrs)
CHRISTIAN	15 79 %	11	17	10.5	14	4.5
NOT C	11 52 %	4	9	3	4	3

	GO TO CHRISTIAN SCHOOL	TAUGHT CHRIS-TIANITY			GO TO go to CHURCH NOW. REGULARLY	GO TO CHURCH NOW RARELY/ SPECIAL OCCASION
CHRISTIAN	6 (31 %)	18			17	1
NOT C	15 (71 %)	21			2	13

TABLE 4: FAITH BY SUNDAY SCHOOL AND FAITH BASED YOUTH GROUP ATTENDANCE AND LENGTH OF ATTENDANCE AND WHETHER AS A CHILD THEY ATTENDED A CHURCH BASED SCHOOL AND WHETHER THEY ATTEND CHURCH NOW

79 percent of Christians went to Sunday school for longer (11yrs v 4yrs) compared to non Christians – 52 percent both as children and as teenagers. More Christians went to Church based youth clubs. Thus, current Christians had more exposure to Sunday school and Church based youth clubs than non– Christians. Practically everyone was taught Christianity at school but more of the non–Christians attended Church related schools (71 percent v 31percent). Most Christians go to church regularly (17 out of 19) whilst non–Christians, understandably, are more likely to go only rarely or on special occasions.

CATEGORY	VALUE IN TV / RADIO	VALUE IN NEWSPAPERS	VALUE IN IN-TERNET	VALUE IN SO-CIAL NET-WORKING	FUNNY SIGNS
CHRISTIAN	13	12	9	4	10
NOT C	5	12	11	3	5

TABLE 5. FAITH BY WHETHER ANY VALUE IS SEEN IN THE RELIGIOUS CONTENT OF VARIOUS TYPES OF MEDIA

Christians and non–Christians similarly see value in the religious content of newspapers, the internet and social networking media, but Christians also value much more the role of radio and television. Only 50 percent of Christians see value in the "funny billboards" often seen displayed outside churches but non–Christians expressed an antipathy towards them.

CATEGORY	CLOSE FRIENDS CHRISITAN	KNOW CHRISTIANS	DISCUSS MATTERS OF FAITH		
CHRISTIAN	18	18	17		
NOT C	18	18	14		
	FRIENDS EVER ASKED YOU TO CHURCH	IF YES – HAVE YOU GONE:	HAVE FRIENDS EVER OFFERED TO GO TO CHURCH WITH YOU	IF YES – HAVE THEY GONE:	HAVE FRIENDS INFLU-ENCED YOUR BELIEF
CHRISTIAN	9	8	7	8	9 (50%)
NOT C	3	2	2	1	4

TABLE 6. FAITH BY THE INFLUENCE OF FRIENDS

Both categories know Christians and have Christian friends and generally do discuss matters of faith with them. Whereas about half of Christians have been asked by friends to go to church or their friends have offered to go with them, few Not C have been asked. For Not C,

389

few feel their friends have influenced their belief but 50 percent of Christians feel they have been influenced.

CATEGORY	BECOME CHRISTIAN AFTER THOUGHT	TIME WHEN DECIDED	GROWN INTO THE FAITH	"ACQUIRED" – IT JUST SEEM TO HAPPEN	
CHRISTIAN	12 (63%)	10	7 (37%)	7 (37%)	
EXPRESS FAITH REGULARLY BY:	ATTENDING CHURCH	PRAYING	READING THE BIBLE	HELPING OTHERS	SPEAKING ABOUT MY FAITH TO OTHERS
CHRISTIAN	17	15 (80%)	14	18	14

TABLE 7: HOW PEOPLE CAME TO CHRISTIANITY AND HOW THEIR FAITH IS EXPRESSED

63 percent said they had adopted a faith after careful thought and 52% were aware of a time when they made that decision. Equal numbers (37 percent) felt they had either grown into faith or it had just happened without their necessarily having made a conscious decision. These were people who had grown up in the church and had just carried on as they grew up. 95% express their faith by helping others, 89% through regular church attendance 79% percent through prayer, and 74%, through reading the Bible and talking about their faith to others..

From Table 8, on the following page, we can see that all respondents had read some of the Bible, generally for study at school or personally. 50 percent of Christians had read it for pleasure. Less than half of respondents had read the whole Bible. Of those that had read parts of the Bible, similar numbers had read OT, NT and the gospels but with more Christians having read more than ½ and most non–Christians having read less than ¼. More Christians had read other religious books, but similar numbers had read more than 10.

※∩ ※∩ ※∩※

390

CATEGORY	READ ANY OF BIBLE	FOR PLEASURE	STUDY AT SCHOOL	PERSONAL STUDY
CHRISTIAN	18	9	9	12
NOT C	19	4	12	13
	ALL OF THE BIBLE			
CHRISTIAN	7			
NOT C	6			
	OT	OT MORE THAN ½	OT ½ TO ¼	OT LESS THAN ¼
CHRISTIAN	18	10	5	2
NOT C	11	1	2	7
	NT	NT MORE THAN ½	OT ½ TO ¼	NT LESS THAN ¼
CHRISTIAN	18	12	3	1
NOT C	12	4	0	7
	GOSPELS	GOSPELS MORE THAN ½	GOSPELS ½ TO ¼	GOSPELS LESS THAN ¼
CHRISTIAN	18	15	2	0
NOT C	13	4	2	6
	READ OTHER RELIGIOUS BOOKS	READ MORE THAN 10 OTHER RELIGIOUS BOOKS	TO FIND OUT	CURIOUS
CHRISTIAN	19	10	13	4
NOT C	13	10	9	4

TABLE 8 : FAITH BY READING THE BIBLE (OT = OLD TESTAMENT, NT = NEW TESTAMENT, GOSPELS – BOOKS OF MATTHEW, MARK, LUKE AND JOHN) AND ANY OTHER RELIGIOUS BOOKS, WITH REASONS

CONCLUSIONS

Although the groups were comparable in terms of age, sex and numbers it is not possible to generalise or extrapolate the findings because of the various potential biases within the study. Firstly, the numbers were small, the participants were all self selecting and also self designating in terms of their faith position and their personality and declaration of being scientific or artistic. Exposure to Christian influence is taken to be the role of parents, attendance at Sunday school, Christian youth club, church based school, friends and media. The actual degree of exposure has not been measured or assessed other than in terms of the stated number of years attending Sunday school, Christian youth club and Church based school. Likewise, quality is not addressed.

Nevertheless, the findings do seem to be in keeping with the literature. In that the Christians had more exposure to influences from Christian parents and their Christian church – Sunday school and youth clubs. Members of both groups had been taught Christianity at school, though more non–Christians attended Christian based schools. One curious finding is that fewer of their siblings shared their faith and more did not share their faith compared to non–Christian respondents. We obviously don't know anything about them, at what point they left the faith or the reasons for their current position on faith matters. There might also be many other factors that may have reduced the siblings' exposure. However, assuming their exposures would have been similar, the effect of religious socialisation seems to have been less influential for them.

More Christians felt their peers had influenced their faith and more of them had read more of the Bible or other religious books. 63 percent said they had adopted a faith after careful thought and half were aware of a time when they made that decision. Equal numbers felt they had either grown into faith or it had just happened without there necessarily being a conscious decision. These were people who had grown up in the church and had just carried on as they grew up. Most express their faith through regular church attendance and particularly by helping people, with nearly

80 percent expressing it through prayer, reading the Bible and talking about their faith to others.

Purely on the self–reported data reported here, it would seem that those in the Christian group have had a greater exposure to Christian influences than their counterparts in the Not–C group. This lends support for rejecting the null hypothesis.

There is some evidence therefore, for Churches to encourage the nurturing of families and children in the ways of faith, but they will need to take account of the impact of modernisation and globalisation, as suggested by Longo.[475] These cultural changes may increasingly present alternative attractions to those of organised religion and what are referred to as "religious compensators" – rewards in an after–life – may be insufficient to counter this.

We will discuss the ways in which God communicates in more detail in the final chapter, but this study suggests that one way people connect with God is through exposure to religious based activities and scripture. But additionally, for God to use all the culturally sensitive and appropriate means to continue his communications with humankind, people of faith will need to martial social media and all the technological innovations at their disposal.

CHAPTER 10

GOD'S COMMUNICATION

My purpose in this book, has been to consider whether God may have a communication problem, given the facts that:

- some people believe in God and others don't
- there is a variety of religions and differences in understanding of faith and God
- these differences occur both within specific faith groups and between the major faiths
- harms are caused by the ways humans have responded to these differences
- harms are caused by some faith groups asserting that their position and understanding is the "correct" one – and by definition therefore, other understandings are wrong
- and whether God does have a communication problem or not, to ask the questions "does it matter" ? and if so, "what should be our response" ?

I believe the consequences of these facts matter, as all of us may be affected to a greater or lesser extent, directly or indirectly. Thus, I think the answer to the question, but more importantly our response to the answer, is of concern to everyone. Particularly for people of faith, all faiths, who surely have a responsibility, on behalf of all people, for addressing those harms that result from how we deal with our different understandings of faith, God and religion.

I acknowledged from the outset that I was approaching this book from a Christian faith point of view. However, you will perhaps have gathered by now, that this will inevitably be MY OWN Christian perspective. But of course, this is "true" only at the time of writing. In keeping with the theme of the book my understanding continues to evolve and so my views as expressed here may very well change over time. As I have said before, there is a wide spectrum of understandings of God and faith, even within the Christian tradition, and so it is generally not possible to give THE CHRISTIAN view, that all Christians would unequivocally assent to. The things on which one might get universal Christian agreement, may be subjects such as child abuse or the principles of war crimes, genocide, discrimination and so on. But these are also subjects on which all right thinking people, whether they subscribe to a faith or not, are likely to agree. I use the word 'principles' here, as the decision on what specific activities may be included within these activities, may itself be a further source of different understandings.

But up to this point, I have tried to present the general content of the book largely from a neutral and factual perspective. However, this chapter, by its very nature, has to be approached from the stand point of faith. So from now on, "God," can no longer be the human construct beloved of academics and researchers, incapable of communication or independent action. God is "He" that exists and independently interacts with the world and human beings. Though in what ways and to what extent is, not unsurprisingly, a further area of debate and open to different understandings. In particular, is the divine present in the world or is the divine seen to be outside the material world ? These are known as the doctrines of immanence and transience, respectively. [476]

This chapter also involves some personal insights and so they inevitably come from my own Christian perspective. However, in keeping with the ethos of the book, they are offered in a spirit of sharing, not with any sense of claiming that my Christian perspective is THE way to understanding God. Hopefully by now, this should go without saying ! Please understand the following ideas and thoughts as they apply or not, to your own faith position.

So let us look at the nature and existence of this "being" we are calling God, who can involve himself with human beings and the universe he created. In doing so, I will draw on a range of scriptures – the Qur'an, Hindu texts and the Bible, though the main stand point will be my take on the Christian message, as that is my experience and knowledge. But all insights are offered in the spirit of respect and recognised mutual legitimacy, that are the underpinning threads of the book, and which I will come back to.

What is the nature of God – What is God like ?
In reference five of the introduction, I drew a distinction between a description of something, in that case an apple, and a definition. A description referring to the outward appearance, of which in the case of God, we have no idea. An old man with white hair and beard sitting on a cloud somehow doesn't seem to cut it ! In contrast, a definition refers to the nature or essence of something. In terms of God, although we don't fully know what his nature or essence is, we do have some pointers.

We have already identified some complex ideas that people use in connection with God. Words such as omniscient (having infinite knowledge), omnipotent (having unlimited power), omni–present (being present everywhere at the same time), omni–benevolent (showing perfect goodness) and being eternal (everlasting). These same concepts are found in the Hindu texts and in the Qur'an as well as both the Old (the Jewish scriptures) and New Testament sections of the Bible (see below). But such words are applied in order to try and explain something that is inexplicable. They are based on our assumptions of what we imagine a creator God must be like. It is rather like the researchers in the seventeenth century, referred to in chapter 6, who determined what characteristics define religion and then used them to determine what rituals and practices of earlier cultures were religious or not. Such methods were later criticised as leading to a distorted and incomplete picture of the religious aspects of those cultures. We find ourselves in the same position in trying to define God.

We are limited by our human language and capacity to comprehend something incomprehensible and so have to devise words and concepts that try to define something that by its very nature is indefinable. We do not have a lot of choice of course if we are going to try and gain some insights into God's nature that we can then share and discuss with others. But we should be wary of having once applied these characteristics to God, of then using them as yardsticks for measuring God's performance and even his existence ! Such as – a God who is omniscient, omnipotent and omni–benevolent could not allow suffering but since suffering does occur, then God cannot exist ! Or, if he is omniscient, then it must be God's will that such and such an event has occurred, whether it is for good or ill.

As we do not and cannot know the mind of God, we cannot use our definitions to determine what we think God can and cannot do or how he should behave, in any given situation. We can only accept that all events are just part of the way things are. That is not to say, that God does not or cannot influence in some ways the way we deal with any particular situation, but I do not believe it is "his will" that bad stuff happens. Believing that it is, does not seem consistent with what Christians perhaps, would now understand as other aspects of "His" nature. Though of course, I could be completely wrong !

What are these other aspects ?

People may say that we can see God in creation, in beauty and through other similar experiences. But such understandings and perspectives are only available through the eyes of faith. One first needs to have faith in God, if such interpretations of experience are to be meaningful. So, I think we can only know something of these other aspects of what God is like, either by some sort of clear, divine, personal revelation or indirectly from scripture.

Well known examples of dramatic personal revelations include: Paul's experience on the road to Damascus (The Book of the Acts of the Apostles: chapter 9 verses 3 and 4 – The Bible) and the Prophet Mohammad's vision of the Angel Gabriel that preceded God's revelations of what became the Qur'an. But many "ordinary" people have received a specific, unique personal experience that they can only account for in terms of it being of God. However, for those of us who have not had such a dramatic experience, it is probably a matter of reliance on the scriptures.

398

THE NATURE OF GOD:

HINDUISM

In Hinduism, in terms of the nature of the scriptures, *"the knowledge in the Vedas is believed to be eternal, uncreated, neither authored by human nor by divine source, but seen, heard and transmitted by sages."* [477]

For Hindus, the nature of God is embodied in the idea that Hinduism is a 'pluralistic' religion, with God being understood in terms of being both a "formless principle" and having "form" or personality. When formless, God is referred to as Brahman – the One Supreme God with uncountable divine powers. This supreme God energises and pervades the entire universe, being both in the world and beyond it ("immanent and transcendent" – in the jargon). Thus, the essence of God is found in all human beings and is inside each and every person, as the nature of our true selves – called 'Atman.' In this sense, God can be experienced. Achieving this experience is the ultimate goal of one's soul.[478] The essence of God pervades plants, animals and includes mountains, rivers, trees, and other planets. But Brahman can become manifest in many different forms, with each deity being understood to exhibit omnipotence, omnipresence and omniscience. They are also seen as exhibiting personal qualities, such as: truth, goodness, compassion, power, knowledge, wisdom, comfort and freedom to those who seek it.

It is the love of the devotee that "turns" the formless God into the form of his or her desire. Hence the many different images of God. From the notion of the Divine essence being in all things, animals are often seen in representations of Brahman. For example, Ganesha is presented as having an elephant head, while Hanuman has features of a monkey, and rivers such as the Ganges are given Divine status. The three main forms of Brahman are: Brahma the creator, Vishnu the sustainer and Shiva the destroyer.

These different manifestations of God are often given representation in the form of "murtis." These are "statues" which act as visual aids for contemplating the nature of, as well as, communicating with God. They are imbued with the presence of God, so the worship and prayers offered, focus on the being represented by the images. The murtis are not God and God is not limited to the murti. These deities offer a different aspect of the divine to focus on, such as Ganesh as the Remover of Obstacles, and Hanuman as the Embodiment of Strength and Perfect Devotion.

Hinduism is also Henotheistic. This means the worship of one God without denying the existence of other Gods. And so Hinduism recognises and promotes respect for other religions and acknowledges the truths in their teachings.[479] [480] This aspect, together with its pluralistic nature, means that Hinduism accepts there is a variety of concepts of God with many pathways to know him. It emphasises that as we are all different and so the way we think of and approach God will necessarily be different. Hinduism gives the freedom to approach God in one's own way, encouraging a multiplicity of paths, not asking for conformity to just one.

Hinduism teaches that it is the heartfelt love for God that is more important than any conformity to specific ritualistic practice.

JUDAISM

In Judaism, the nature of God is understood through reading the Torah, and characteristics attributed to God, include:[481]

- God as the one true God" as per the Shema a Jewish prayer, said twice daily:

- "Hear, Israel: The L–rd is our G–d, The L–rd is one."

- God as the creator of the universe

- God as Omnipresent – being everywhere all at the same time
- God as Omnipotent – being able to do everything and anything

- God as Omniscient – knowing all things

- God as Eternal

- God as Holy and Perfect

- God has no form and no gender – all references to God's hands, hearing, seeing, walking and so on are figures of speech to make G–d's actions more comprehensible. Thus, as in Islam (see below), it is not permissible to create images of God.

- God as King and Father – in contrast with Islam, Jewish understanding is that man is created in God's image. There being three people involved in procreation – mother and father to provide the physical body and God providing the soul, personality and intelligence.

- God is Just and Merciful – he holds a perfect balance between Justice and Mercy.

ISLAM

For Islam, the nature of God is best conveyed by the following quotes from the Qur'an [482]: The key belief, which forms one of the 5 pillars of Islam, is that Allah is unique and the one and only God.

- "Say He is Allah the One'"(Qur'an:112:1)

- "Say, Allah is the Creator of all things, and He is The One, the Irresistible." (Qur'an from13:16)

- "This is a proclamation for mankind, that they may be warned thereby, and know that He is One God, and that people of understanding may remember." (Qur'an: 14:52)

Allah is understood to have 99 "names," or rather attributes,[483] [484] [485] indicating his characteristics and actions :

- "To Allah belong the Most Beautiful Names, so call Him by them, and disregard those who blaspheme His names. They will be repaid for what they used to do." (Qur'an 7:180).

- "Allah, there is no god but He, His are the Most Beautiful Names." (Qur'an 20:8).

- " He is Allah; besides Whom there is no god; the Sovereign, the Holy, the Peace–Giver, the Faith–Giver, the Overseer, the Almighty, the Omnipotent, the Overwhelming. Glory be to Allah, beyond what they associate."(Qur'an 59:23)

- "He is Allah; the Creator, the Maker, the Designer. His are the Most Beautiful Names. Whatever is in the heavens and the earth glorifies Him. He is the Majestic, the Wise." (Qur'an 59:24)

- These attributes include : The Most Gracious, The Most Merciful, The Most Holy, The All Subduer, The Creator and so on. By understanding these attributes, one has some insight into the nature of God from a Muslim perspective, though it is understood that the human mind is limited and so cannot comprehend fully.[486] Some of these are described in a little more detail by Oman Khalid in an article – "10 Facts about Allah – its time for you to know your Creator a bit more !"[487] One corollary of thinking of Allah as 'The All Knowing, The Omniscient", is that Muslims believe in a destiny, set by Allah:

- "Say, "Nothing will happen to us except what Allah has ordained for us; He is our Protector." In Allah let the faithful put their trust." (Qur'an: 9:51)

Another point worthy of note, is that associated with images of God. In contrast to the Hindu approach, Muslims do not create images of Allah. This is seen as imposing limits on Allah, from our own limited abilities, imagination and comprehension. This position also relates to the Christian view that man is created in God's image, one that Muslims do not share.

CHRISTIANITY

The current Christian understanding of God would be as creator of the world, universe(s) and everything, and that "we are all created in God's own image." (Genesis 1:26–28). Though this concept is not confined to Christianity.[488] Such an understanding of the nature of God arises from the writings of the Old Testament. However, our understanding of other aspects of God's nature have evolved from Old Testament times to those conveyed by Jesus, described in the New Testament. From the Old Testament, the writers define him in ways that try to explain how they saw God's relationship with his people, Israel, and his role in history. Given the wars between Israel and her neighbours – the Philistines, the Babylonians, the Assyrians, the Egyptians, the tribes living in Canaan and so on – it is understandable that what emerges is a picture of a war–like and vengeful God that kills and destroys whole nations.

However, this is not a picture that most Christians would now recognise, nor would they interpret God's involvement in history in quite this way. Our understanding of God nowadays is more likely to conform to the characteristics we associate with the scriptural words of Jesus in the Gospels (books of Matthew, Mark, Luke and John in the New Testament part of the Bible). For example, in the Gospel of John, chapter 14 v 9, Jesus says to Philip, who has asked Jesus to show them the father, that "he who has seen me, has seen the father." So we can gain some insights into the nature of God by considering the nature of Jesus – loving, caring, self–sacrificing, understanding, forgiving, angry at injustice, supporting and being able to have a personal relationship with each of us as individual people. This is a clear example of how Christian understanding of God has changed over time.

These characteristics paint at least a partial picture of what God may be like. As Paul says in his first letter to the Corinthians:

"For now we see through a glass, darkly; but then face to face: now I know in part; but then shall I know, even as also I am known" (1 Corinthians 13:12 King James Version {KJV}).

SUMMARY SO FAR 1
Having given a brief résumé, of how different faith groups understand the nature of God, it is important to realise that not all members or all the Church hierarchies understand these concepts in exactly the same way. Nor do they necessarily see the implications and practical expressions of these ideas in exactly the same way. Hence the need to find ways of managing different understandings in ways that are respectful and non–confrontational.

We can now look at what other factors may influence a decision to believe in God or not.

Believing in God
As human beings, we can decide to believe in God or not. Many factors may influence such a decision. From the study data reported in chapter 9, exposure to Christian influence from parents, peers, church related activities and reading the Bible seemed to influence the acquisition of faith. Such factors may influence the thinking of members of other faiths,

in the way they educate their children in the faith. Personal encounters with other religious experiences including pilgrimage, miracles, prayer,[489] visions and God, may directly also play a part. Such exposure may take place over a period of time, even many years, as if on a journey, or it can happen suddenly – "a conversion experience" as it is often called. The epithet "being born again" is applied by some Christians, to this acquisition of faith and belief in God. An additional question here of course is – which comes first – faith and a belief in God or an experience of some sort which then leads on to faith? There is no "right or universal" way. It can happen either way round, but for each person, the journey and process will be unique.

These types of experience are not confined to or occur only within Christianity, with people converting or changing from their current status of faith or non–faith to any of the major and other well known religious traditions. As mentioned in chapter 1, such conversion can be a stressful and distressing experience, depending on how one's circle of friends or faith community respond.

However, if one does adopt a religious faith, one tends to interpret certain experiences as being of God. So for example, many people have told me that when they have been in particular short term financial difficulties they have received an unexpected tax rebate or a cheque from a friend or even a stranger, for the amount they needed. Churches needing to raise mega–amounts of money for a particular project have found that the amount has come, not without effort, but much more readily than they might have imagined. Others have found that problems, on which they had exhausted all their own solutions, be it with a relationship, work related or whatever, when they offer it to God in prayer, have found a means of resolution which they could not have foreseen.

But other people may rationalise these explanations and put them down to coincidence, or the way things happen or just part of the natural cycle of life – particularly as such occurrences are not outside the norm – people do get tax rebates, institutions do raise money and life related problems do get resolved one way or another. Therefore, they see no reason to invoke God as an explanation. These things can and do happen to anyone. It thus comes down to a matter of belief. Whether for you, God is that 'supernatural, independent being' that is involved and acts within the world, or not.

The examples given are clearly all positive ones. People tend to see God's hand in experiences where it seems that a difficult situation has been resolved positively, particularly where prayer may have featured strongly. God is seen as being part of the solution, rather than negatively as author of the problem. In general, we tend to see the origins of problems either as just part of the way things are, or as having been created by ourselves or others. But in some circumstance, there may be people who understand the problems and difficulties, particularly if they are "bad things" that might be called evil, as the work of the devil. This is a whole other area of theological debate, which has its own range of understandings, including questions of whether the devil is a separate independent entity or whether evil is just seen as the absence of good. But this is not a topic for this book.

But within the western, Christian influenced cultures, there is one area, where God may be thought to be involved negatively, and that is, suffering. That either, somehow God is responsible for the suffering or at the very least, responsible for not stopping or preventing it. That is, to all intents and purposes, God allows it to happen.

This question of suffering may be a major factor in influencing whether someone may or may not believe in God, or even turns their back on long held beliefs. Suffering is often the starting point for discussions concerned with the existence of God. The argument goes, that if there is a compassionate God, why does he allow suffering ? This may be extended to suggest that even they, the person offering the argument, as a human being, would not allow suffering to occur – if they were God (implied). Though of course, some human beings have and do, cause all manner of suffering. The focus of the question is then often directed specifically to examples of young children who are born with severe handicaps or acquire, early in life, some form of terminal illness.

So if there is a loving God, why does he let or will these things to happen? Some of the "answers" offered include: God is testing us, healing will occur if we have enough faith or if enough people pray or pray more often, they must have done something wrong, it is an opportunity for people to show care and love or for ill people to show faith or fortitude or be an example of how God is helping them "fight it" and a whole myriad of other "weasel words" or platitudes. None of these responses seem to convincingly answer the question. The bottom line is that we do not know.

The other major faiths have their own perspectives:

In Islam, the belief is that suffering can be due to sin, and that Allah wants this suffering to erase these sins and teach the need to adhere to Allah's natural and moral laws. It can also be punishment or a testing of faith.[490] The assumption is that Allah is most merciful and compassionate and knows best and so whatever he does, is with a legitimate purpose, although his actions cannot be fully comprehended by our incompetent minds. He cannot be questioned for His acts.[491]

In Hinduism suffering stems from changes or modifications between the mind and body. Such modifications manifest as pain and suffering but also include: attraction and aversion, union and separation, desires, passions, emotions, ageing, sickness, death, rebirth amongst others. Suffering is seen as an inescapable and an integral part of life. The purpose of religious practice is seen by various schools of Hinduism, to resolve human suffering. So the Vedic religion did not focus upon suffering initially, but upon securing peace and happiness in the mortal world with the help of gods in heaven and patrons upon earth through rituals and sacrifices. In contrast, the Upanishadic approach is to focus on the hidden causes of suffering and try to resolve them internally by cultivating purity, fortitude, equanimity, stability, balance, detachment and indifference through austerities, restraint and renunciation. The Bhagavadgita identifies the instability of mind as the chief cause of suffering. The true solution to suffering therefore lies in achieving freedom through self–restraint, mental stability, detachment, renunciation and absence of desires.

Hinduism also identifies desires and demonic nature as the root cause of human suffering and the resultant bondage to the cycle of births and deaths, as the ultimate suffering. Demonic nature means selfish actions done for the sole purpose of selfish enjoyment. The battle has to be fought in the mind and body as the mind is the seat of all desires and intentions and hence for a human being it is the battlefield.[492]

In Judaism, punishment and purification are suggested as reasons for suffering but acknowledge these arguments are inadequate, in situations of innocent suffering. There is therefore room within Judaism for protest to be levelled at God when suffering is thought to be undeserved. Among those who reproached God for inflicting suffering unjustly were Abraham and Job. One way of coping with the moral imbalance in the world is to formulate a doctrine of reward and punishment in the afterlife. But there is emphasis on acting compassionately to relieve other people's suffering and that with the coming of the Messiah, illness, poverty, and even death will be abolished.[493]

Of course it is human to ask, why suffering, but our response needs to be formulated in the abstract, that is, when we are not directly involved. Questioning our faith only at the time when the suffering is affecting someone we know and love, but not having done so when that same suffering affects other people day in and day out, whom we are not aware of, is the wrong time. If our experience and analysis has lead us to a position of faith, then that faith will need to include the uncertainty and discomfort associated with all the difficult and unanswerable questions. For me, the analysis which can lead to such a faith, from a Christian perspective, starts not with the question of "whether I believe in God" or the problem of "suffering" but with the person of Jesus.

There is a lot of information concerning the historical Jesus both in the Bible and other writings. Not withstanding the debates around the authorship, timing and accuracy of the actual texts of the Gospels touched on in chapters seven and eight, the nature of Jesus, the tone and direction of his teaching and his assertion that he was the "son of God" are clear. In addition, there was his willingness to pursue, what he understood to be God's will for him, by putting his own desires to one side and committing himself totally to doing what he believed God wanted him to do. Namely, to show that living the way of love is possible, even under the direst of conditions of occupation of his country by the Romans, that such love is available to everyone and that sins (the bad and selfish things we say and do) are forgiven. Truly living and proclaiming this message was so radical (as it is and would be to–day if we could do it – love/care/respect ALL those we dislike and distrust !) that it challenged the religious hierarchy and the status quo.

They saw their only way of stopping Jesus preaching his message was to engineer his execution. Even though there was nothing obviously in it for him and he could have fled Jersualem, to remain true to his message Jesus was willing to accept the inevitable consequences of challenging the system. However, amazingly his crucifixion was not the final act. There are multiple examples of different people seeing Jesus – the risen Jesus – again. But how ever we understand this, and there are various suggestions as to whether Jesus was the same as he was before the crucifixion or was in some way different, particularly as he appeared to be able to pass through locked doors, there was no doubt amongst his followers that something remarkable had taken place.

So, if Jesus, as a person, is convincing enough for you to follow, then a belief in God is a natural consequence, although not everyone takes that step. Even some people who regard themselves as Christian, and follow Christ as his disciples, do not necessarily see God as a supernatural independent entity. However, for the majority that do make that step, then part of the package is the need to accept a number of uncertainties – the way God has set things up, the question of suffering, the fact that some believe and others don't and for those that do, the variety of understandings within the faiths and between them.

Other faiths will approach the question of belief in God from their own perspectives. They will have other starting points for trying to get to grips with a concept that is beyond our complete understanding. For all faiths, it is important to use our God given intellect to discover as much as possible about our universe and the way it works and how it came into being. But all will need to recognise, that ultimately, the purpose behind the universe is likely to remain unclear. That the complete nature of the "origin" of that purpose; that is, God, is likely to remain beyond our capacity to understand and our ability to know or find out. We don't know why God has set things up in the way he has. We don't know the answer to why people suffer. We don't know the nature or timing of the "end–game" – if there is one. We can offer all sorts of suggestions – and we have touched on some – God is testing us, it is a means of creating faith and so on, but in the end, these are just "weasel words". The bottom line is we don't know. Speculation and questioning help us develop our understanding and insights of God, but they are not final answers.

For Christians, it is here that faith begins. Paul Tillich, a twentieth century German theologian in his book *The Dynamics of Faith*,[494] states that faith is uncertain so far as the infinite is concerned and that this uncertainty cannot be removed, and so must be accepted. But to accept this, risks failure; that is, being wrong. And so to have faith, takes courage, because if one is wrong, then the whole meaning of one's life falls apart. Thus, doubt and uncertainty are integral and necessary aspects of faith.

Such faith and courage can, by extension, be applied to our discernment of God's communication, with its uncertainties, doubts and different understandings.

Since God has chosen to set things up the way he has and why that is, is something we cannot fully know or understand, the three fundamental issues seem to be: firstly, to whom does God communicate ? Secondly, how does God communicate his purposes, in as meaningful a way as possible, to beings with a limited capacity and language for comprehending and articulating such matters? Thirdly, how do we recognise God's communication to us and how can we respond ?

With Whom Does God Communicate ?

With Everyone !

From the discussions throughout the book, it seems reasonable to think that God has been communicating with all of human kind ever since human beings began to evolve. This contrasts with the impression gained from the Bible, that God has only communicated with a few specially selected individuals or a specific group at certain moment(s) in history. Thus, it is possible to think that many of the rituals and beliefs of other faiths and ethnic religions also represent the struggles of those people and peoples to understand the nature of God. Any understanding that does emerge, will be limited by the language, knowledge and experience available in a particular culture, in a particular place at a particular time. Such understandings then evolve over time, to varying degrees, as we have seen.

In chapters 7 and 8, we followed some of the evolutionary developments of the concepts of religions and God and the transformation from polytheism to the notion of a monotheistic position. This principally happened within the Jewish community, who cottoned on to the idea of there being One God, who had a direct relationship with them. However,

413

the way they understood this was often wrong, distorted and rejecting (see the Old Testament stories of the prophets like Jeremiah, Amos, Elijah and so on, to whom God gave a message to deliver to the Israelites to get them back on track!). Ultimately, these messengers couldn't bring about a sustained change in behaviour and so perhaps, God felt the time had come when there might be sufficient people around who could understand a new way of doing things. Thus, he risked revealing himself directly, in the person of Jesus Christ.

This act did not solve everyone's relationship with and understanding of, God. As we know. Jesus was crucified, and although as Christians we believe he rose again from the dead, (however, we might understand the nature of that) we still have problems with the variety of understandings of God's communications. These still cause some of the harmful consequences mentioned at the beginning of the book. But this is not the end of the story. God continues to communicate. After Jesus, God didn't cease communicating with anyone other than the Jews or those that called themselves Christians. As indicated in the previous section, he uses all sorts of people, other faiths, doctrines, situations and methodologies to provide insights into himself to us personally, as individuals. The prime example being his communication with Mohammad in 610 CE !

Looking back, we have the benefit of being able to see how the Christian faith has evolved since the time of Christ. We have new insights into the ways we think God is still communicating with peoples whose religious beliefs and rituals continue to develop, within their own traditional religious frameworks of understanding the nature of God.

※∩ ※∩ ※∩※

How Does God Communicate And In What Ways ?

From a Christian point of view, God communicated his nature primarily by showing himself through the character of Jesus. Our understanding of this amazing act, is supplemented by our own personal experiences of relationship with him. These experiences are built up in a number of ways, some of them applicable to other faith systems too:

- Through prayer
- Through reading the scriptures and studying their background and interpretation
- Through an appreciation of wonder at God's creation
- Through people and situations – God works through all people by their good actions, writings, words etcetera, whether they are religious or not and in all the situations in which we find ourselves, including, modern media portals – facebook, YouTube and so on and at local, national and international levels. Through relationships and political dialogue. The questions we should ask are, 'What, if any, is God's message in this situation ? Can we identify any words or actions, that are humane, caring, respectful and that follow the "general drift" ?.[495] The people offering such sentiments or orchestrating such situations may very well be called prophets, as they are transmitting God's message, even though many may not realise it. We can recognise them by their "fruits' – see below
- Through one's own personal experience and sense of encounter with the divine
- Through the appearances of visions or Apparitions – for example the story of the appearance of the Blessed Virgin

Mary to Bernadette in Lourdes and the associated 70 authenticated "miraculous" healings. (Why some people seem to be healed and others not, is discussed in the next section).

- Through miracles – a sense of wonder at how often normal and ordinary "coincidences" can happen when they are needed.

- Through the miracles of Jesus, which have all been understood or explained in a variety of ways

- Through a conversion experience

- Through the workings of the "Holy Spirit" [496]

- Through the rainbow, as God's promise not to flood the whole world again

From the Bible, we find many references that indicate the presence of God in the world, which by inference are part of his communication. For example:

From Psalm 85 we learn that Righteousness goes before God and prepares the way for his steps. Righteousness means and includes characteristics like goodness, virtue, virtuousness, uprightness, decency, integrity, worthiness, rectitude, probity, justice, honesty, and honour. So where and when we come across these qualities, we can be sure that God is in that situation.

From Ephesians chapter 1 verse 9 we learn that God spoke through Jesus Christ, and so by following Jesus as his disciples we will receive and recognise God's communication.

In the book of Amos chapter 7 verse 14, we learn that God speaks through the words and actions of prophets but more importantly we learn that these prophets are very often ordinary people like us, who have responded to God's call. Amos says he was not a prophet or a prophet's son but a shepherd and keeper of sycamore trees.

We can recognise God's prophets as they bear good fruit i.e., they do good things Matthew chapter 7: verses 15 to 20. This good "fruit" includes: love, joy, peace, forbearance, kindness, goodness, faithfulness, gentleness and self–control – Paul's letter to the Galatians chapter 5: verses 22 to 23. So if their words and actions follow the 'general drift' of Christ's teachings and love, then we should take note of what they say.

This list is not exhaustive, but provides a picture of the spectrum of the day to day personal encounters and situations through which we can experience God communicating with us. But what is the mechanism by which we actually feel or sense that God is communicating with us specifically and at a particular moment ?

How do we know or discern what God is communicating ?

Herein lies the nub of the whole faith conundrum. The crucial question is: how do we recognise and understand God's communication with us specifically and at a particular moment, clearly enough to justify acting upon it ? We need to answer this, if we are to avoid the harmful consequences of any faith community or individual, acting just on what they "think" God might be saying to them, and so justify their actions, whatever they are. Part of the assessment will be to bear in mind that what God asks of us, is likely to be for the Common Good,[497] and so will "follow the general drift."

It is both difficult to know and be sure of a definitive answer. When we look at the Old Testament, it is written as if God actually talks to people like a mate. Moses appears to have heard the voice of God, just as if it was his pal standing next to him, and so there was no mistaking what God wanted him to do. Abraham was clear that God was asking him to sacrifice his son Isaac. The Virgin Mary spoke to Bernadette in Lourdes and told her what to do and say, and so she had no doubt. Despite being a girl of fourteen, uneducated and poor, she was able to stand firm in the face of fierce questioning, and unwaveringly tell the authorities what she had seen and heard. For me, it doesn't happen like this, with such clarity, and I suspect, from talking to friends, it doesn't happen like this for most of us. So how does it happen ?

In my experience, it seems to be through my being conscious of a specific thought or idea that comes into my mind. It may be in connection and as a result of any of the experiences outlined earlier or it may seemingly come out of the blue. Its outstanding characteristic is that it is **PERSISTENT**. No matter what I try to do to get rid of it, it keeps coming back. It may be associated with an identifiable need (external to myself) or related to some other aspect of life – work, relationships, volunteering, purchases, where to live and so on.

It may relate to something straightforward. Driving up a hill, seeing a lady carrying some shopping – the thought occurs – should I give her a lift ? Is this of God ? Who knows ! What about all the other times you went up the hill and saw people you might have stopped for, but the thought never occurred and the many times after this occasion when the thought doesn't happen? After all, you are not a taxi service for everyone walking up the hill and there is a bus service ! So you can stop or you can pass them by – whether rationalising your decision by saying you are too busy, the dog is in the back, there is too much traffic to just stop there and of course you don't need a reason, just ignore it and give it no more thought ! This time–limited thought will go. The opportunity passes and did it matter and was it of God ? Who knows ! However, you do decide to stop. The lady is so grateful, she has just missed the bus and there is not one for an hour and she is worried she will be late to meet her children from school, or she just bursts into tears as she is lonely and you are the first person in a long time to show her any kindness, or it is totally uneventful and she just gets out where she needs to and says thank you. Is that of God ? Who knows ! There may be no particular additional pay offs from responding to this situation. There may have been no additional deficits from not acting – other than she would have been a little later and a little more tired by the time she did get home. If she had been late for her children and they had been upset, or had run off or something else had happened to them – you are unlikely to ever know.

But of course, we cannot run our lives by thinking of all the possible dreadful scenarios that might ensue if we do or not do a certain thing. What if by stopping to pick her up, a cat runs in front of your car and you hit it, but you wouldn't have been there at that time if you hadn't stopped ! Or your neighbour has fallen and has been stranded in the garden all afternoon and you could have helped if you had got back earlier ! If we start to think like this, we become paralysed into inaction.

If we do recognise such promptings or nudges as of God – how should we then respond ? Act on them – don't analyse too much. Sometimes it will work out, other times it won't. Sometimes our help will be accepted and at other times it will be declined, but that is OK, we cannot be right all the time! This is the stuff of "mission on the front line" – a notion described and promoted by The London Institute for Contemporary

Christianity (LICC).[498] The mission front–line being anywhere and anytime you meet anybody or come across a need. This provides the opportunity to respond, when the situation coincides with the prompt or nudge from God. However, it is also important to realise that we cannot respond to every situation we feel sympathy for. So some sort of filtering is required. Not easy, particularly if the situation is an immediate one.

Sometimes God will do the filtering – you may only recognise the need in a situation afterwards and may wonder why there was "no prompt or nudge" and the thought did not occur to you at the time. Even though in previous similar situations it did ! I remember situations with patients where I felt it appropriate to offer prayer – sometimes they gratefully accepted and other times respectively declined. But sometimes, though the circumstances were similar – someone receiving bad news or dying perhaps, the thought didn't occur to me at all. Afterwards I would wonder why. So it isn't always clear cut. All we can do is the best we can. We are not responsible for the outcomes, only for having responded as best we can. We can only let go of ourselves, place ourselves at God's disposal and step out in trust.

Thoughts may also arise in connection with any of the other areas of our lives in which we may need to make decisions. But God does not micro–manage our lives. Whether we choose a plain carpet or a patterned one is our decision. Though there may be ethical considerations in choosing the carpet. Does one cost so much more than the other, that the difference will affect your ability to give to charity or visit your family in America ? Is one made with child labour ? Is one made abroad and so has an increased carbon footprint to get to you ? Are such questions of God ? Who knows?

But such questions may exercise anyone making such a purchase. These thoughts are not the prerogative of Christians. For people of faith however, such thoughts – that still small voice of conscience, prompt or 'nudge' may be interpreted as of God. But for Christians, these thoughts do not occur in isolation but on a foundation of faith. This faith, which grounds our motivation, responsibility and moral and ethical values, both influences the occurrence of the thoughts and our likely response to them. Many folk, whether people of other faiths or none, are concerned about justice, equality, the environment and fairness etcetera and respond accordingly, even though they would not necessarily recognise God as the source of their thought or decision. But where Love is, there is God ! In the words of the Gregorian Chant, written more than a thousand years ago:

Ubi caritas et amor, Deus ibi est.

Where charity and love are, God is there.

But all thoughts don't necessarily have these sort of potential overtones. It is OK to just decide to go to the cinema or a football match or take your daughter fishing or take your partner for a meal. Not all thoughts are of God and not all things that are not of God are bad. Discerning the difference may depend on the context and any other ideas that come into your mind at the same time !

We also need to be wary of getting trapped into feeling guilty about the decisions we make. Often we are choosing between several competing issues, none of which is necessarily clearly better or worse than any other, and there are always mixed motives for doing things. Pleasing ourselves or getting pleasure from doing something for ourselves or for other people is not necessarily a reason for rejecting it ! In such

421

circumstances, we can only focus on the prime reason motivating us to choose any particular one. If we do feel guilty, it doesn't necessarily mean the decision is bad or against God ! It could just be our personality or our upbringing or a result of the distorted teaching, of our faith organisation, which we mentioned in chapter 1. So although a sense of guilt can be a useful pointer towards the right decision, it may just as easily not be anything of the sort and we may just have to live with it ! We can only do the best we can to try and discern what is of God – see later.

This presents us with something of a paradox (an apparent contradiction). How can God be intimately concerned with each of our lives, in the ways suggested above, without micro–managing our lives and directing our every move ? On the one hand we read in the Gospel of Luke, chapter 12 verse 7: "Indeed, the very hairs of your head are all numbered. Don't be afraid; you are worth more than many sparrows" and on the other, God does not seem to micro–manage our lives. How is God able to influence our lives, whilst at the same time endowing us with "free–will" – the freedom to choose?

The issue of "free–will" is itself a source of major philosophical debate. There is a tension between what we define as an all powerful and all knowing God and the existence of free will.[499] [500] Arguments centre on whether the concept of determinism (all actions being pre–determined because either all actions follow on from a previous action or, looking at it from a theological perspective, God knows everything and so all actions are inevitable) is compatible or not with free will. These ideas are encapsulated in the theories of compatabilism and incompatabilism, but with various "partial" theories in between the extremes of Yes they are compatible or No they are not. That is, academics, theologians and philosophers are not agreed. Understandably, given that the tension results from the way God seems to have set things up, and we do not know the mind of God. But we do have to make enough sense of it in order to live with it.

Thus, explaining this 'balance' between God's influence and our freedom to choose, is again pretty much impossible but I offer the following, as one way of looking at it. This has to be purely from my own Christian understanding. I would be happy to receive responses from other faith perspectives.

God loves us and is concerned for each of us. By adopting a faith position, we wish to align our desires with those of God. However, there are three problems – firstly, we have difficulty in being clear what God's purposes for us are and so may get it wrong, even with the best of intentions (though we may not always know), secondly, we have difficulty in relinquishing control of our lives and offering ourselves completely to God and thirdly, we may not relish the prospect of doing what we feel God is asking of us and so deliberately go against what we otherwise feel God desires of us. Thus, we may find ourselves not fulfilling God's ultimate purpose for us, and not always recognising it. So, God communicates with us, through the Holy Spirit, as a thought that comes into our mind, that persists despite our best efforts to ignore it.

So finding ourselves with this persistent thought, how do we clarify its provenance – discern whether it is from God ? The thought may be self evidently good. Giving the lady with the shopping a lift is self evidently a good deed. Action is required there and then otherwise the opportunity is lost. Over analysing all the potential negative consequences of offering the lift or not, as outlined above, will result in inaction and the opportunity passes by. By not acting, and then experiencing a sense of guilt and later thinking about this experience, may influence our decision making in the future and make it easier to act instantly. By reflecting on our decisions, we learn about ourselves, so influencing our future actions. In the real world, such instances require an instant response. We can ask ourselves if the action is loving and caring and the instant answer is yes it is. There is not really much room for error, other than for whatever reason, ignoring this "nudge."

For more complex issues, depending on the nature of the thought and the circumstances, our response may not need to be quite so immediate. We may have more time to consider it. There may be time to consult other people, talk with trusted family, friends and colleagues. Particularly, there will be time to pray, not just in the heat of the moment but at some particular opportunity we may set aside deliberately, to consult with God. Time during which we can listen to further promptings and receive other thoughts that may help clarify our understanding of the nature of the original thought. It may be necessary to actually take steps towards acting on the thought, as part of the checking out whether it is from God.

So we may need to actually apply for a job, for example, to test whether God is asking us to go for a new post. If we don't get the job, we may then feel it was not the right thing, though this may not lessen the disappointment or the bruising of our ego. But there may be other spin offs from following the process – preparing a curriculum vitae (CV) which will be ready for the next application, testing whether we are prepared and willing to let go to trust and "risk" answering God's call, practising interview techniques, learning to cope with disappointment and so on. There may be no obvious benefits, though I do believe no experience is ever wasted.

Another example. The thought may relate to "healing" a relationship by risking making the first move in speaking to someone or saying sorry for a break down in that relationship. It may be appropriate to ask "what would Jesus do" in that situation or consider whether any potential outcomes "follow the general drift". None of these strategies may lead to absolute certainty, but may go far enough to help in finding out. It may be that one only feels sure whether the prompt was of God, with hindsight.

We may choose to ignore the thought altogether. If the thought was of God, however large or small, the fact that we have rejected it will not stop God using us in the future. It may mean it will be for a different purpose, but of course we will never know !

So God influences our lives through, the Holy Spirit, implanting these thoughts. The Holy Spirit may also engineer the circumstances and the environment, through the decisions of other people, that encourages our decision making to be in line with his, but ultimately it will be our choice. Thus, the future is uncertain, as it depends on each and every choice that each and everyone of us makes. So the situation we each find ourselves in – personally or collectively – has not been engineered directly by God, but rather unfurled and evolved over time as a consequence of the decisions we and others have made. With all the characteristics that are part of being human, from our altruism to our manipulative self–interest. So whilst we are responsible for our responses to these thoughts, the options may be influenced by the situations in which we find ourselves and the decisions made by others, both directly and indirectly, locally or remotely.

This may not sit comfortably with our human idea of an omnipotent God, the idea that God is not in complete control. However, it seems to me that God is in control. In the sense that he has chosen to set things up in this way and so hand over his control to the decision making of human beings. This is the ultimate sacrifice. The blue print for human beings to give our lives over to his control, as Jesus did. It follows the same pattern as when he chose to come to earth as a man and freely placed himself at the mercy of human beings. This, as we know, ended in Christ being crucified. Ultimately however, God took back control in the resurrection. Perhaps the pattern is that God will follow our human decision making until such time as he takes back control at the time of the Second Coming. Who knows ? These matters are all part of the uncertainty that is faith.

The idea that this sort of mechanism can apply to the billions of people on the planet, all at the same time, is mind blowing. But presumably possible for God, even though we cannot get our own heads round it nor understand the reasons it seems to be this way !

Of course, in circumstances in which only ourselves are affected, then we can afford to risk getting things wrong. But for decisions that potentially affect others, for example Prime Ministers wrestling with decisions about whether to take the country to war or not, the process of discerning whether the thoughts are of God or not is much more crucial. There is no guarantee even then, the decision will be right. There may be no right decision. Only the seemingly best, that one can make at the time, in all sincerity and honesty.

It may not happen quite like this for everyone, and most of us are not making Prime Ministerial level decisions. But somehow, we do become conscious that a thought comes into our mind, associated with the question – is this of God ? Our ability to recognise them, our percentage "hit rate" can be improved with practice. That is, by regular prayer and Bible study, and taking appropriate advice. In this way, we can learn more of God and develop a closer relationship with him. As in all matters, practice will help, developing our skills of discernment. In these ways our faith is deepened and we become closer to God.

This proposal seems to be a reasonable theory for how we can become conscious of God's personal communication to us. However, for others, this conscious awareness may take a different form, only they will be able to say. Some people say that if God sent a miracle then it would help them believe. But God did send his son Jesus, who did do miracles. However, people rejected him even at the time. People still reject him now and even many Christians play down the miracles and think of "rational" explanations to escape any hint of the "supernatural" as explanation! How one interprets such experiences, depends on where one stands in relation to faith and belief in God. One can perhaps either see miracles at every turn or not recognise one if it hits one in the face. There are many examples of people having seen Apparitions of the Blessed Virgin Mary.

Amongst these are appearances to three shepherd children in Fatima, Spain in 1917 CE, to Bernadette in Lourdes, France in 1858, and in Medjugorje in Bosnia Herzegovina in 1981 CE, where they purportedly are still occurring. The Church has not always acknowledged these claims as authentic and not all Christians are fully convinced by them. If Jesus came again, would we recognise him as such ? Would we welcome or shun him ? So, although such experiences may be useful to some people, there are no universal guarantees that "supernatural" occurrences will be recognised for what they are. These are further examples of the different understandings that abound in connection with our relationship to God, faith and religion.

The notion of God involving himself in the world, in this manner, and communicating with us directly and individually raises other questions. I was once challenged, by a fellow Christian, in response to my saying that I felt God had "called " me to be a doctor. She wanted to know, how could I account for the fact, that if I believed I was called by God to be a doctor, what was special about me ? The implication seemed to be that I was in a materially comfortable situation, which I attributed to God looking favourably on me, whilst others perhaps were in "dead–end" jobs or unemployed and were otherwise materially deprived or living in abject poverty, in war zones and so on. So, what sort of a God would, seemingly arbitrarily, decide who was doing what and where ? Wasn't it rather arrogant to think God had particularly chosen me ?

At the time, I didn't have an answer to these questions. I am not sure I do now, but it is something that has continued to exercise my mind and conscience. But thinking about it now, I think it is a similar issue to that of suffering or of healing. That is, if some people have to suffer, or a few selected people are miraculously healed or some people are more materially comfortable off whilst many people struggle, what are our response choices ? Is it either that this is just the way life is and there is no God, or that God is just capricious and makes arbitrary, unfair decisions ? Implying, that if the latter, then I do not want anything to do with such a God. Or is there something else ?

If our analysis from prayerful consideration of Jesus, has lead us to a place of faith and hence a belief in God, then what are we to make of this, yet another paradox ? Of the seeming unfairness, injustice and inequity in the worldly experiences of people and the notion that God is involved in the world and in the lives of human beings ?

Firstly, capriciousness and unfairness do not seem to be characteristics of God, as discussed earlier. Therefore, the unfairness and inequalities in the world cannot arise from God's arbitrary decision making. Secondly, we have indicated that the situations in which we each find ourselves, follow from a combination of the way God has set things up and by his giving us free–will. The consequences result from the decisions human beings are making now and have made over the centuries. In one sense, all new born babies are victims of the decision making of their parents, of their grandparents, of previous generations and of other people all over the world, whose decisions have influenced the environmental, cultural and material nature of the situations into which they are born. Into this melée comes God. So although we don't know why he has set things up the way he has, we have seen how he communicates with us as individuals. God will communicate with everyone, no matter what their situation. But an individual's response, will depend on their experience of God, faith, or religion and their cultural and environmental experiences, wherever and whatever they are.

So when we respond to the thoughts/promptings we believe are of God, then we are responding to what we believe God is wanting of us at that particular time, however great or small. If it relates to a job, then we are in order in using the epithet : we have been "called" by God, if that helps us communicate why we believe we are where we are. We do not know the reasons why particular thoughts and prompts come into our minds, just as we do not know the thoughts or prompts that others receive. All we can do, is accept the thoughts we have, discern as best we can whether we perceive them as being of God and make a choice of responding or not. So it isn't about being feeling singled out or special, it is purely about trusting and accepting the pattern of the way God seems to communicate and having the faith to risk a response, in whatever circumstances we happen to find ourselves.

All we can know, is that if this is our faith position, then we have the choice of responding to the "thoughts" we have, or not. These thoughts are not "one off"/"once in a lifetime" situations and we can be sure other thoughts will follow, which too will require a process of discernment and response. Some, no doubt, being a consequence of the earlier choices we have made. We can only hold on to our faith and do our best. In the words of the old children's hymn : "Jesus bids us shine, you in your small corner and I in mine."[501] One can only respond to the thoughts in one's own mind, in one's own circumstances.

So my becoming a doctor is not the end of the matter. It requires me to exercise that role to the best of my ability and for the furtherance of God's purposes, by continuing to respond to future opportunities and situations arising as a consequence. Both as a doctor and the use to which I put any material benefits.

This is part of the courage Tillich was referring to. That in the face of all the uncertainty surrounding these questions, faith, if lived out daily in accordance with God's desires, is not a crutch or veneer that allows one to ignore or forget the harsh realities of life, but is the strength that keeps one true to the discipleship of Christ. This is not to say that people of no faith don't do good stuff, of course they do. Good stuff is of God, whether the doer recognises it or not. But they risk missing out on knowing that sense of love, forgiveness and support that comes from a personal relationship with God.

So, does God have a communication problem?

We have seen through the earlier chapters of the book that knowledge is only as good as its pragmatic usefulness and the latest update. That it is always likely to change as research reveals new insights. This theme applies, not only to our knowledge of science, philosophy and the cosmos, but also to our understanding of religion, faith and God. Thus, the idea that these understandings change and evolve over time, and are not fixed in time, seems to be inescapable. Such historical, evolutionary processes support the proposal that it is unlikely that one single understanding of God and faith overrides all others.

We have also looked into the nature of the "being" we call God and noted the factors that may influence belief in his existence, given that he seems to have set things up in a way that does not program human beings to automatically believe. We have highlighted the faith and courage required to believe in God. Particularly, in the light of a number of paradoxes and difficult questions and uncertainties, for which we do not have satisfactory answers.

We have looked at the various ways by which God does communicate with each and everyone of us. Using all manner of means, situations, people and methodologies, and that he probably has done from when the earliest Hominids first had the ability to think and communicate between themselves. His ultimate revelation, for Christians, was coming to earth himself as a human being in the form of Jesus Christ. Then followed, Christ's crucifixion, resurrection and ascension into heaven (however we understand the manner by which this all occurred). We have suggested a "mechanism" whereby we may become aware of his personal communication. We have noted that such communication only makes sense, because God has given us freedom to choose. To choose whether to believe in him at all, and if so, whether to respond to what we may discern as his communications, or not.

We have noted that the limited capacity and capability to fully understand God's nature and purpose, follows from the way God has set things up and by the nature of the "created being unable to fully comprehend the creator." This concept, coupled with the freedom to choose, inevitably leads to different understandings of religion, faith and God.

Sadly, our human nature, often leads us to respond to these differences in ways that has and does cause harm and distress both to individuals, and through wars and violence to communities and nations. This harm and distress has occurred throughout history and still continues.

If having decided that having a faith makes sense, and our concept of God is of a 'being' that is all knowing and all powerful and capable of creating the world and the universe, then, on the face of it and in the light for the foregoing, it would seem unlikely that God does have a communication problem. Certainly, not in the way we would understand a problem, as a difficulty to be overcome. The harmful consequences result, not from a failure in communication on God's behalf, but as a result of the choices we make as human beings.

So all in all, it seems that God does not have a communication problem. He has chosen to set things up the way they are. He has willingly relinquished control over his creation by giving us free–will, and has chosen to work in partnership with human beings, by accepting the consequences of our freedom to choose. The cost to him of this decision was the crucifixion and perhaps an ongoing "frustration" with some of the decisions we make. There is a sense expressed within Christianity that the only hands, feet, eyes and ears God has, are ours (ascribed to St. Teresa of Avila [1515 to 1582 CE]— a Spanish noblewoman who chose a monastic life in the Catholic Church). That is, the notion that God works in the world by and through the actions of human beings, consequent on the "good" decisions that we make, in response to his communications. But ultimately, through his promised second coming (however we may understand this), he will re–take control. Faith is the means whereby, in the interim, we continue to hold on to this promise and our belief in this ultimate good.

So, having noted there are different understandings of religion, faith and God (both within and between different faith communities) and that some faith traditions have a propensity to claim the "correctness" of their own position compared to others and the harms that follow from our reactions to all of these differences, we come back to the title – "does it matter ?" If it does, we face the $64,000 question: what should our response be ?

What Is Our Response ?

It seems to me that all people of faith should be concerned wherever harm or distress are found. Particularly, if it seems that it is our responses to the differences in matters of faith and belief in God that is the cause. Therefore, I believe that people of faith have a duty and responsibility to minimise such harms on behalf of everyone. That question now of course, is how ?

The differences, as we have seen, have arisen through the evolutionary process of human development and human thought. We have seen how ideas relating to religion and its practices and rituals emerged as people began to think about their environment and why and how things were the way they were. Over the centuries, with increasing knowledge and the evolution of more structured ways of thinking and analysis, philosophers and later theologians devoted more time to such matters. Ideas were written down, some becoming designated as scriptures, and with the emergence of the concept of monotheism and the birth of Jesus, other complex theological ideas were generated.

The development of Christianity, as a religion and the variety of institutions and belief systems created within it, partly represent the struggle of human beings to make sense of the momentous occurrence of Christ's birth, crucifixion and resurrection. Also influenced by our limited capability and capacity to understand fully the nature of these events and by our human imperfections and characteristics (our greed and self interest as well as our altruism and capacity for self–sacrifice) – for good or ill. These struggles will also have, no doubt, been influenced by God as Holy Spirit, as outlined earlier.

As humans, it is more than likely that for some aspects of God and the scriptures, our speculations and interpretations will have both misunderstood and misrepresented the nature of God and his purposes. As our worldly experience tells us, it is highly likely that the powerful, however they may have acquired such power, use their power to get what they want, and as such, may have had a disproportionate influence on Christianity's evolution. Some like to think that, for example, the Bible was created by the Holy Spirit leading the process of decision making, resulting in an infallible record of God's word. But having imperfect human beings involved in the process, it is highly unlikely to have been quite as "pure" as that.

That is not to say the Holy Spirit was not at work or was not able to shape the nature of the collection, but how and to what extent is not possible to know. If free will is to mean anything, it has to extend to all aspects of life. So inevitably, different understandings and interpretations have arisen. With this added uncertainty, it is possible that some of our beliefs could be based on things that have been misunderstood. Who is to know ?

I rather like the story of the monk who, as he and his colleagues were sitting at their desks copying out the scriptures, asked the supervisor about "quality control." So the supervisor went off to the basement to check on the originals and after several hours had not returned. So the monks went to find him and discovered him banging his head against the wall, muttering – "the word was celebrate, celebrate "!

Although not perhaps in the specific, the story illustrates a general principal and probable reality.

Thus, Christianity, and the other major religious faiths, have, over the centuries, continued to develop understandings and insights into the nature of God through their experiences and new perspectives on the scriptures. Although some of these "insights" may have been misunderstandings or misrepresentations of what God intended, many have helped take our understanding forward. Sadly, many of the religious institutions failed and still fail to embrace such insights, or at least seem reluctant to share them with all their members. Not that all new insights are necessarily worthy of adoption, but most will be worthy of acknowledgement and discussion. New ideas that challenge the status quo are often at best dismissed by the leaders of the Church or at worst, the purveyors of the ideas are personally verbally abused and discredited (we only have to think of Copernicus and Galileo). On occasions they have also been attacked physically and even murdered. (Anglican bishops Hugh Latimer, Nicholas Ridley and Thomas Cranmer). Although unlikely to be murdered for his views, Bishop John Selby Spong and other theologians have received much criticism for their speculations on the nature of God.[502]

If the Church hierarchies do acknowledge new ideas, they may fail to disseminate them to all their followers. This maybe for a variety of reasons, not least of which may be the desire to maintain the status quo, in terms of their own power and status. Also, from a paternalistic (and given that most Church hierarchies are dominated by men, then it is correct to use the term paternalistic) desire to protect people from the risk of having their faith undermined. In thinking for themselves, people may come to conclusions that in turn could risk undermining the religious institutions ! Thus, beliefs and religious practices may remain entrenched in historical patterns of thinking, far longer than they need or even should. Hence, the negative aspects of religion are maintained and perpetuated. Trusting "ordinary" people with their own faith is surely no more of a risk than trusting Church leadership to other fallible human beings and then having to explain situations in which this trust has been abused.

This book is not about encouraging dissent. But rather, promoting a recognition that questioning demonstrates a healthy engagement with complex and difficult concepts and is a means of strengthening belief rather than leading to its rejection. If a person's faith is built on a firm basis of what makes sense to them, then questioning specific doctrines is about continuing the struggle to find ways in which these and other aspects of scripture can provide new insights and so be better understood and integrated into that faith. New information and knowledge from all disciplines can be treated in the same way and be seen as enhancing the richness of our understanding of God's creation, rather than as a threat to one's own faith and belief system.

Theologians, philosophers and other academic thinkers who spend their lives struggling with scriptures, doctrines, seemingly impossible concepts, questions of an ethical and moral nature and the new knowledge and information from other disciplines, aim to create new ways in which these concepts may be understood. Such new ways and new knowledge may or may not always make sense to others. But rather than judge them as being either right or wrong, they should be seen more as representing one person's personal struggle to understand. They can provide useful insights for others to build on. In this way our understanding of scripture evolves and may be seen as a work in progress. Such study provides new ways of understanding scripture, rather than undermining faith.

These insights do not need to be hidden, rejected or ridiculed or their proponents killed or tortured ! Some readers may remember the furore over remarks made by the Reverend David Jenkins as Bishop of Durham[503] or the publication of Honest to God by John Robinson.[504] Not forgetting the way in which Christ himself was treated, because he spoke of new insights of God, deemed unacceptable to the religious authorities.

New insights and ideas should be welcomed and encouraged and openly discussed. Some ideas may be accepted, others rejected, but all are stepping stones to further enquiry and understanding. By acknowledging, discussing and questioning such new insights, in a climate of honest love and respect, it seems to me that faith can be strengthened and understanding fostered, so reducing the risk of both personal distress and future conflict.

The preceding remarks primarily refer to those differences in understanding within the Christian faith and within and between the different denominations. Such differences tend to focus on interpretation of the scriptures and the theological basis for such interpretation. The conversations are often rather combative with each side trading scriptural quotes in support of their position, even though most would accept the common underlying foundation of their beliefs as Jesus Christ, as the Son of God.

If we broaden our consideration to include other faiths, the parameters of the discussion are different. There are generally still some common themes which most will share – such as showing care and love towards one's neighbours and the vulnerable in society. The three major Abrahamic faiths – Judaism, Islam and Christianity – also have many of the prophets in common. However, the underlying theology and practices are quite different. Thus, the conversations tend to be more comparative in nature, rather than combative.

But all faiths will, understandably from a human perspective, feel that their histories, traditions, understanding and insights are unique and special. So each will have particular insights they feel it important to share for the benefit of all. This idea is embodied in the concept known as Receptive Ecumenism.[505] This comes from the perspective of "what can we learn from others, NOT what can we teach others". For instance, if, as Christians, we see Jesus as the human manifestation of God, of the divine, and hence God's ultimate revelation of himself (that is; his ultimate form of communication with human beings) then Christianity has a duty to share this insight with everyone else. However, not in a way that implies Christianity, as a religious movement, is necessarily the one, true and only way to understand God, but in humility, recognising that no–one knows the mind of God. So for example, in John's Gospel chapter 14: verses 6 and 7, Jesus said:

"I am the way and the truth and the life. No one comes to the Father except through me. If you really know me, you will know my Father as well. From now on, you do know him and have seen him."

This passage has been interpreted to mean that the only way to be in relationship with God is by believing in Jesus and so becoming a Christian. This then provides justification for the idea that Christians should proselytise (try to convert) those of other faiths. Indeed, it may be said that aiming to "convert" those of other faiths is the best way of loving them, as it means they can share in God's kingdom, through Christ. But experience tells us that this has not happened in 2000 years. In fact, Islam emerged 600 years after Christ and has spread widely since. Force and violence (of the Crusades and subsequently) has not worked and whilst individuals may "convert' both from and to Christianity, wholesale shift has not happened yet, and seems unlikely to do so. Human nature tells us that, by informing large–scale movements that their traditions, beliefs and practices are wrong and that "we" know best, is not an approach that is likely to meet with much success. Quite the reverse ! It entrenches positions, increases resistance and hostility and potentially leads to violence. Such actions therefore may not be

439

following the "general drift", as the process and outcomes are unlikely to be loving, even if one does argue that the motive is.

Accepting that God has set things up the way they are, and that we should therefore be embracing and celebrating difference, in a loving and caring way, can we see this biblical passage in the light of the 'bigger picture' of God's love ? Is there an interpretation of this verse that does follow the "general drift". It seems to me that there is such an interpretation that, dare I say, is more "Christian." If we say that Jesus's 'way' is the way of love, then all love must be of God. Wherever love is found, there is God also (whether associated with another faith or none). So actions, beliefs and practices that demonstrate love and caring for others, as Jesus himself did, under whatever religious label or none it is exhibited, must be of God. Then, again:

> *Ubi caritas et amor, Deus ibi est.*
>
> *Where charity and love are, God is there.*

By showing consideration and respect to the beliefs of others, we may, as it says in the Letter to the Hebrews : *"Do not neglect to show hospitality to strangers, for thereby some have entertained angels unawares"* (The Letter to the Hebrews chapter 13: verse 2). In other words, we should accept that the ideas and practices of other faiths may well provide valuable insights into God's nature and purposes. By the nature of things and their evolution, no–one has a monopoly on the truth. If God is God, then he does not require groups of human beings, by claiming that only they know God's mind, to defend "his teachings" or promote them in violent, exclusive and repressive ways.

Such sharing and interchange however, needs to be undertaken within a climate of respect and with a recognition that we all may have just as much to learn, as to offer. Working out together what God means, through dialogue, is more consistent with the belief systems of the major religions, than is suspicion, threats, violence and fear. There are numerous examples of the main world religions engaging in dialogue,[506] [507] [508] and of ecumenism [509] (defined by The Oxford English Dictionary as: the principle or aim of promoting unity among the world's Christian Churches) being promoted by a wide range of organisations.[510] But often there seems to be a disconnect between the aspirations of such international initiatives and the practical outputs.

There is also a gulf between these initiatives and local church communities. Though this gulf can act both ways as local communities are often ahead of the Church hierarchies in terms of local inter–faith working and co–operation. But such disconnections still apply even where joint working is the everyday experience of individual members of these faiths.[511] [512] [513] Any initiatives that promote dialogue are to be welcomed but clearly much more is required if true co–operation and respect are to be the norm. Often, the only people getting to know each other are the people in the room ! Communicating the outcomes and outputs of such dialogue, to the wider Church and faith groups is essential, if we are to reduce suspicion and the harmful consequences of our response to different understandings. Different understandings reveal different facets of God's nature. By affirming such differences, and accepting the legitimacy of different view points, we can all benefit from glimpsing a bigger and more comprehensive picture of the nature of God.

Acknowledging the legitimacy of another point of view is not about necessarily agreeing with it or having to adopt the same belief system

441

and practices. It is the actions that are of God, and the religious framework is secondary. A sentiment echoed in Hinduism. But clearly, if such a framework facilitates its followers to do God's work then we should rejoice in that. In the early days of the Church, it was necessary for the members to agree what they believed. To try and write down the essence of their faith which they could share with others, as they were assailed by all sorts of other groups making various claims about Jesus. Paul and James felt the need to write letters to the Churches, warning them to beware of false teachers. So statements or creeds (we have referred to these in earlier chapters) were derived to try and clarify the basics of the faith. These took years of argument before final versions were reached, and even then, not everyone agreed.

However, the problem with writing things down is that one cannot capture every possible nuance or eventuality and so inevitably some things are excluded. Thus, there is a danger of stating that "this is what we believe" and so automatically, those ideas and understandings that are not written down are excluded. This automatically lessens the potential richness of understanding and leads to exclusion of people who perhaps understand things in as genuine, but different way. It can thus create barriers and cause harm to those who feel excluded. We are unlikely to be judged on our biblical theology but on our responses to each other, as per the parable of the sheep and the goats.[514] A similar sentiment is expressed in the Hadith.[515]

Where violence, self–interest, manipulation of power, injustice and other forms of harm and distress are found, it is those actions that are not of God and are to be condemned by everyone. The religious framework itself and those that follow it for good, should not be "all lumped together" within the condemnation. The use of a faith position to justify harmful or violent actions is yet another human violation and manipulation of the scriptures. Believing that one's own understanding is correct does not make it OK to say things that harm and hurt others.

However, not all different understandings represent just a difference of opinion. There are situations where individual harm follows from either a particular interpretation of scripture such as homophobia, discrimination against women for the priesthood for example, or from the considered implementation and interpretation of a religious framework, such as female genital mutilation (FGM), circumcision, or punishments like amputation or stoning. In such circumstances, people with different views, whether they be of faith or not (for example, through the judiciary or governmental legislature, where appropriate), need to engage with the appropriate religious leaders to share other insights that address such matters.

SUMMARY SO FAR 2

Such sentiments regarding respect and the toleration of other views are not new. There is a whole literature on toleration, stretching from Socrates, through Augustine and Aquinas and philosophers such as Erasmus, Kant, Locke and John Stuart Mill and others.[516][517] The focus of these writings was initially in support of religious tolerance, but broadened to include political tolerance, freedom of speech, freedom of the press and was really a precursor for the ideas that led to human rights legislation.

But toleration, whilst important, often fails to affect the underlying mind–set producing the tensions that our human response to differences frequently seems to generate. Human beings tend to be naturally suspicious of people who exhibit differences of any kind and often respond to these situations with fear, mistrust and not infrequently, violence. When life becomes difficult and the chips are down, particularly economically, people look round to find scapegoats. Whether it be individuals or groups as collectives. One can readily think of the Jews, travellers, refugees, people of colour and so on. The evolutionary benefits of such reactions have perhaps now been overtaken by the societal dis–benefits.

It has taken 200,000 years or so to get to where we are now, and we are still causing harm by how we respond to difference. Talks between faith groups have gone on for years and whilst progress has been made in some relationships, the fundamental differences remain. There are many initiatives of local inter–faith groups working together on humanitarian projects with many examples of solidarity shown between local churches at times of adversity. For example, the American Baptist Church opening its doors to local Muslims when they were building a mosque next door,[518] letters and shows of support from Christian Churches when Muslims were murdered by a so called "white supremacist" in New Zealand.[519] [520] All these things are important and necessary, but making a global impact on fundamental attitudes, changing the big picture, changing the paradigm, needs something more radical.

Jesus was a radical. He was aware of the challenges he was posing for both the religious and political systems. By doing what ? By demonstrating love and concern for everyone, but particularly for people who needed help themselves or needed to be shown how to help, rather than exploit. Who could argue with that ? Lots of people it seems. Why ? There were probably many reasons. There was implied criticism of the system in that it allowed people to be in need and which was not doing enough to meet those needs. There was implied criticism of a system that exploited the vulnerable, in terms of unfair taxes and temple practices (The story of Zacchaeus in the Gospel according to Luke chapter 19:

verses 1–10 and the story of Jesus overturning the tables in the Temple in the Gospel according to Matthew chapter 21: verse 12). Tackling these issues required the systems to be overhauled, which would inevitably affect the income and status of some of those in charge. So recognition and acknowledgement of inequality required wholesale changes to the traditional ways of running things, if things were to change. No can do ! We know that this "clash" of values and Jesus' single–mindedness in proclaiming his message, led him to the cross.

As people of faith we inherit God's mission. There are countless examples of people of faith, individually and as organisations, working to meet the needs of the oppressed and vulnerable all around the world. But, for Christians, Jesus is also about relationships. One of the central themes of this book is the harm and distress to people and relationships caused by our response to differences in understandings of faith, God and scripture. So just as Jesus required people to change things round in his day, he requires people of faith to turn things round today.

I have no blue print to offer. It may be thought naive, and that this is more than being just about differences in theology and just as much about politics. Of course it is. Jesus was acutely aware of the radical politics he was engaged with. But the best politics is about having a vision of the Common Good and developing actions and an implementation plan that delivers that vision for the benefit of everyone.

It will undoubtedly involve some self sacrifice and consideration of others' needs before our own – hence the radical nature of Jesus' message.[521] [522] The emphasis in these publications, concerning the radical nature of Jesus, relates to Jesus' relationship with the poor and marginalised and what this means for our response as disciples of Jesus.

445

However, I wish to appropriate this same notion and apply it to our relationships with those of different understandings of God, both within the spread of Christian denominations and between the major faiths and others. As Jesus' message then required people to let go of their cherished understandings of faith and practice if they were to serve God's greater purpose, so he requires of us a willingness to let go of those things that place barriers in the way of our relationships.

So, if this listening to each other and respecting each others legitimacy is to really lead to a change in relationship, it seems to me that a paradigm shift in mind–set is required. Whilst any one faith system believes it has the "correct" understanding of God and his purposes, little progress will be made. If, "I am right, and therefore you are automatically wrong" is the prevailing climate, what hope is there of a meeting of minds? As Chris Hewer says in his article – "The importance of a paradigm shift in understanding Christianity and Islam"

"Our quest for mutual understanding cannot demand a denial of the essence of the faith of either party. Questions of ultimate truth must be left 'in the hands of God' to be made clear on the Day of Judgement."[523]

Such a sentiment can be applied to the whole of our discussion hitherto.

A RADICAL FUTURE

Let us see differences in understanding as a gift of God, except where they either directly or through distortion, lead to harm. Let us regard them as an inevitable part of the evolutionary nature of God's creation. If we can do this, then these differences can be embraced, welcomed and accepted as having equal worth as pathways to understanding God. This removes the need to think in terms of proselytisation and conversion. It smooths the move between religions without condemnation or fear. It removes the need for competition and warring confrontation and the persecution of those of other faiths. It allows for the incorporation and adoption of rites and practices between faiths, where this is seen to enhance the understanding of God and his purposes. There would be no need for Churches to split because of different interpretations of doctrine or scripture. Differences would be accepted on both sides. Solutions devised to reduce harm, would be accepted by those who disagree. All these things would be seen as being of God.

Let us celebrate difference as a God given gift that allows us to see other facets of God through different eyes. Let us embrace the variety, as it helps us get a better picture of God, with lots of people putting their minds to trying to understand all this stuff. Not with the intention of combining all these differences into one religious amalgam, or moving towards believing the same things, but by living and worshiping side by side. Knowing that the person next to you may have a different take on some doctrinal, theological point or concept of God, but trusting that the bottom line is that it follows the general drift of God's love. Such is the legitimacy of other understandings. Acceptance of the fact that no one knows, that we are all in it together and that questioning and sharing are healthy and enhances our understanding of God and his purposes.

It is the responsibility of the faith groups to bring about such a situation. It cannot happen through legislation or secular channels. It is not simply about toleration of other points of view, it has to be a genuine mutual recognition of inter–denominational and inter–faith legitimacy. The study in chapter nine suggests that exposure to faith experiences are a major factor in people coming to their own personal faith. Perhaps then, the more we share our understandings of faith with each other, the more likely we are to cement trust and mutual respect, within and between faiths.

So how could such a situation come about ?

There are no "pat" answers to this question or pre–determined pathways. In fact, in keeping with the patterns identified throughout this book, it is likely there will be many different answers. But there can be a shared and common goal. That of genuinely accepting the legitimacy of other understandings and of other faiths and a readiness to share their insights into God and his purposes.

It seems to me that action is required on two levels; that of the institution and that of the individual/local faith organisation. The question is not whether it can be done or achieved, but whether there are sufficient people of influence who wish it to be done. However, irrespective of whether institutions are prepared for the challenge, at an individual level, each of us can begin the process of changing our own attitudes to difference and acting on the outcomes. There is precedence, for where individuals lead, hierarchies may follow !

The solution or end game is that of maintaining and celebrating the differences but behaving as though there are none !

I can only offer some questions to kick start any conversations as to how this might come about.

Questions For Institutions

- Is it possible to accept the legitimacy of another faith and so acknowledge that no one knows the mind of God and our way is one of many ?

- What would accepting the legitimacy of another faith look and feel like ?

- Will it include the interchange of ordained / authorised personnel ?

- Will it involve full sharing of sacraments or sacred rites and rituals ?

- Will it involve shared rites and practices, knowing that attendees could perceive such rites and sacraments differently ?

- What steps are required to get there ?

- Who needs to be involved ?

- How can appropriate discussions begin ?

- How can such change be communicated to their members ?

- How can the current theological thinking of theologians be disseminated ?

- How can members be supported in engaging with it ?

- How can honest questioning and engagement be supported and shared ?

- What day to day changes would be evident ?

- How do we discern whether something "follows the general drift " ?

449

- How can God's prompts and nudges in decision making be recognised?

Questions For Individuals / Local Churches

- How do I know what to believe ?
- How do I question aspects of my faith ? – Who do I talk to ?
- How do I recognise God's prompts and nudges in my life ?
- How do I know when God is talking to me ?
- How do I know what God is wanting me to do ?
- How do I "let go" of my need to be in control and align myself with God's will for my life ?
- How do I know if something "follows the general drift " ?
- What does accepting the legitimacy of another faith or other biblical interpretation mean for me ?
- How do I respond to views or biblical interpretations that differ from my own ?
- Do I accept and incorporate them into my faith view ?
- If I still disagree, how do I respond to the person who still holds them ?
- How do we have open and honest discussions over matters of faith and belief ?
- How do I or we as a local church, respond to other views and other faiths ? – Is this dependant on geographical proximity ?
- Can we communicate, visit, meet, write to other local faith groups ?

- Are their joint responses to local needs or international events we could co–ordinate ?

EPILOGUE

Why Does God Communicate ?

According to Ted Grimsrud,[524] *"because God wants us to grow in love, wants us to learn his truth and wants us to find the power and hope we need to help transform the world. God also wants us to join in his work of healing. Through studying the scriptures he will reveal himself as an interpretation of how to live the life of love in Jesus in the world each day."*

God does communicate with all of us. Though in the way he has set things up with the evolutionary patterns of development, the imperfections and limitations of human nature and by giving us free will, different ways of understanding his nature and purposes are inevitable. The key, is that by recognising God's prompts or nudges in our lives and following the way of love that makes sense to us in our culture, we have an opportunity to embrace these differences and gain broader insights into his nature and purposes.

I am not advocating an "anything goes" type of personal religion or belief system, although I do believe we all need to work out our own theology and find what makes sense to us. Taking account of new theological insights and new scientific discoveries is part of that working out. Within Christianity certain aspects and characteristics of God, as indicated by Christ, will guide the resulting understanding that any individual may reach. The test will be "whether the result is loving and just and Christ–like" in respect of its impact on themselves and other people. Do any conclusions "follow the general drift" ? There are perhaps many examples of "main stream" beliefs that may fall at this hurdle – the so called "prosperity gospel" for one [525] [526] (this is a religious belief among some Christians, who hold that financial blessing and physical well–being are always the will of God for them, and that faith, positive speech, and donations to religious causes will increase one's material wealth).

Other faiths will have their own criteria by which a particular view may be measured, and so decide whether such a view is compatible with their mainstream belief. But this requires openness and vision and prayerful handling from those charged with establishing and preserving the doctrines of the faith organisation.

So for all our sakes, let us accept the legitimacy of different understandings and in a climate of love and mutual respect, reduce the harm our past approaches have caused. In this way we can become true disciples and servants of God, working for the Common Good of all peoples. Let us start now and not wait for another 200,000 years !

INDEX

Page numbers (Pg) highlighted, indicate main sections relating to the topic

APPENDIX 1   THE CREEDS

NICENE

NICENO-CONSTANTINOPOLITAN

First Council of Nicaea (325 CE)	First Council of Constantinople (381CE)
We believe in one God, the Father Almighty, Maker of all things visible and invisible.	We believe in one God, the Father Almighty, Maker of heaven and earth, and of all things visible and invisible.
And in one Lord Jesus Christ, the Son of God, begotten of the Father [the only–begotten; that is, of the essence of the Father, God of God,] Light of Light, very God of very God, begotten, not made, being of one substance with the Father;	And in one Lord Jesus Christ, the only–begotten Son of God, begotten of the Father before all worlds (æons), Light of Light, very God of very God, begotten, not made, being of one substance with the Father;
By whom all things were made [both in heaven and on earth];	by whom all things were made;
Who for us men, and for our salvation, came down and was incarnate and was made man;	who for us men, and for our salvation, came down from heaven, and was incarnate by the Holy Ghost and of the Virgin Mary, and was made man;
He suffered, and the third day he rose again, ascended into heaven;	he was crucified for us under Pontius Pilate, and suffered, and was buried, and the third day he rose again, according to the Scriptures, and ascended into heaven, and sitteth on the right hand of the Father;
From thence he shall come to judge the quick and the dead.	from thence he shall come again, with glory, to judge the quick and the dead. ;
	whose kingdom shall have no end.
And in the Holy Ghost.	And in the Holy Ghost, the Lord and Giver of life, who proceedeth from the Father, who with the Father and the Son together is worshiped and glorified, who spake by the prophets.
	In one holy catholic and apostolic Church; we acknowledge one baptism for the remission of sins; we look for the resurrection of the dead, and the life of the world to come. Amen.

[But those who say: 'There was a time when he was not;' and 'He was not before he was made;' and 'He was made out of nothing,' or 'He is of another substance' or 'essence,' or 'The Son of God is created,' or 'changeable,' or 'alterable'— they are condemned by the holy catholic and apostolic Church.]

The Apostles' Creed ✣

A version of the Apostles Creed, said to have been written with contributions by all the apostles was in circulation around the third century. It reached something like its present form around the eighth century:

I believe in God, the Father almighty,
creator of heaven and earth.
I believe in Jesus Christ, his only Son, our Lord,
who was conceived by the Holy Spirit,
born of the Virgin Mary,
suffered under Pontius Pilate,
was crucified, died, and was buried;
he descended to the dead.
On the third day he rose again;
he ascended into heaven,
he is seated at the right hand of the Father,
and he will come to judge the living and the dead.

I believe in the Holy Spirit,
the Holy Catholic Church,
the communion of saints,
the forgiveness of sins,
the resurrection of the body,
and the life everlasting.
Amen.

APPENDIX 2 (CH. 9) FACTORS IN FAITH

IF completing THE QUESTIONNAIRE ELECTRONICALLY :
 PLEASE DELETE THE BOXES AND ANSWERS THAT DO NOT APPLY
if you are completing a paper copy:
 Please tick the boxes THAT DO APPLY (D/K = don't know)

SECTION 1 YOUR BELIEFS
Q1. Would you describe your position on faith / belief as being:
CHRISTIAN ☐ OTHER (includes other faiths and none and D/K) ☐

IF CHRISTIAN GO TO Q2 IF OTHER – PLEASE DESCRIBE:
Atheist ☐ Agnostic (not decided) ☐ not made any specific decision ☐ D/K ☐ OTHER FAITH / RELIGION Please describe
……………………………………………………………
NOW PLEASE GO TO SECTION 2 – Q8.

Q 2. Have you become a Christian through a conscious deliberate decision after thought and consideration ? YES ☐ NO ☐ D/K ☐
IF YES :GO TO Q3 IF NO: GO TO Q4

Q 3 (a) is there a definite time when you made this decision ? YES ☐ NO ☐ D/K ☐
IF YES : please explain what happened
……………………………………………………………
IF NO: is your position one you have:

Q 4 found you have just grown intoYES ☐ NO ☐ D/K ☐
Q 5 "acquired" / just adopted / not really thought YES ☐ NO ☐ D/K ☐

Q 6 What are the main things that have influenced your position

....

Q 7 Do you express your faith Regularly (i.e., it is generally part of your life rather than just ad hoc or now and then) by:

Going to church YES ☐ NO ☐ Praying YES ☐ NO ☐ Bible reading YES ☐ NO ☐ Helping others YES ☐ NO ☐ Speaking about your faith to others YES ☐ NO ☐

OTHER please describe

....

SECTION 2 CHRISTIAN INFLUENCES

At home, my :

Q8.(a) Mother would describe herself as Christian YES ☐ NO ☐ D/K ☐ OTHER please describe

....

Q8(b) Father would describe himself as Christian YES ☐ NO ☐ D/K ☐ OTHER please describe

....

FACTORS IN FAITH

Q9 (a) Mother went to churchREGULARLY ☐ RARELY ☐ NEVER ☐

IF REGULARLY – how often: most weeks ☐quite often ☐

IF RARELY – just special church occasions (christmas, easter etc)☐ just family occasions (funerals, weddings, baptisms etc.) ☐

OTHER – please describe

Q9 (b).Father went to churchREGULARLY ☐ RARELY ☐ NEVER ☐

IF REGULARLY – how often: most weeks ☐quite often ☐

IF RARELY – just special church occasions (christmas, easter etc.) ☐just family occasions (funerals, weddings, baptisms etc.) ☐.

OTHER – please describe

……………………………………………………………………………

Q.10 I went to Sunday School as a child YES ☐ NO ☐ D/K ☐

If YES – for how long (approximately) ………………yrs

Q11.(a) I went to church as a child YES ☐ NO ☐ D/K ☐

If YES – for how long (approximately) ………………yrs

Q 11 (b) I went to a church related youth group as a teenager YES ☐ NO ☐ D/K ☐

If YES – for how long (approximately) ………………yrs

Q12. Do you go to church now YES ☐ RARELY ☐ NEVER ☐

IF YES / RARELY: most weeks ☐quite often ☐ just special church occasions (christmas, easter etc.) ☐just family occasions (funerals, weddings, baptisms etc.) ☐.

OTHER – please describe

………………………………………………………………………

IF NEVER can you say why

…………………………………………………………………………..

Q 13 (a)Do you have bothers or sisters YES ☐ NO ☐

Q 13 (b) How many brothers ……… How many sisters………..

Q 13 (c) PLEASE INDICATE HOW MANY:

Share your belief ….. Don't share your belief …. D/K……..

At school, I :

Q14. Went to a Christian Faith School YES ☐ NO ☐D/K ☐

If YES – To Infant/Junior school YES ☐ NO ☐To Senior School YES ☐ NO ☐

Q15 (a) Was taught about Christianity YES ☐ NO ☐D/K ☐

If YES – At Infant/Junior school YES ☐ NO ☐ At Senior School YES ☐ NO ☐

Q15. (b) Was taught about other faiths / religionsYES ☐ NO ☐D/K ☐

If YES – At Infant/Junior school YES ☐ NO ☐ At Senior School YES ☐ NO ☐

Friends

Q16. I have close friends who are ChristiansYES ☐ NO ☐ D/K ☐

Q17. I know people who are Christians YES ☐ NO ☐ D/K ☐

Q18. Do you discuss matters of faith or religion with themYES ☐ NO ☐ D/K ☐

Q19. Have your friends ever asked you to church with them YES ☐ NO ☐ D/K ☐

IF YES Have you been with them YES ☐ NO ☐ D/K ☐

Q20. Have your friends influenced your beliefYES ☐ NO ☐ D/K ☐

IF YES – please explain how and in what ways...

Media

Q 21 Do you find any value in religion related items in the media :
(newspapers, internet, magazines, social networking sites etc.)

TV, radio,YES ☐ NO ☐ D/K ☐

newspapers, internet, magazines YES ☐ NO ☐ D/K ☐

internet, social networking sites YES ☐ NO ☐ D/K ☐

IF YES to any – please explain how and in what ways

…………………………………………………………………………….

Q21 (d) Are "catchy/funny" religious signs outside churches useful YES □ NO □ D/K □

IF YES – please explain how and in what ways

……………………………………………………………..

Reading the Bible:

Q22. Have you ever read ANY of the Bible at allYES □ NO □ D/K □

IF NO GO TO Q 26 IF YES, GO TO Q 23 – IS THIS FOR:

Q23. (a) pleasure YES □ NO □

(b) study purposes at school YES □ NO □

(c) study purposes to find out more about it YES □ NO □

(d) OTHER REASONS Please describe

………………………………………………………….

Estimate how much YOU MAY have read:

Q24 (a). All the Bible more or less YES □NO □ (b). Parts of it YES □ NO □

(c). Old Testament YES □ NO □If YES More than ½ □ ½ to ¼ □ Less than ¼ □

(d). New TestamentYES □ NO □ If YES More than ½ □ ½ to ¼ □ Less than ¼ □

(e). The Gospels (Matthew, Mark, Luke, John)YES □ NO □

If YES More than ½ □ ½ to ¼ □ Less than ¼ □

(f). OTHER bits specifically – please describe

………………………………………………………….

Q 25. Have you read the Bible for OTHER REASONS YES □ NO □
IF YES Please describe

....

Q26 IF NO Please describe if there are any particular reasons why you have not read any of the Bible

....

Reading other books about Christianity, religions, faith, etc:
Q27. Have you read any other books about religion etc. YES □ NO □
IF NO GO TO Q 31 IF YES – tick box that applies : more than 10 □ 5–10 □
less than 5 □ occasional one □

Q 28(a) Have you read these to really find out about faith / religion YES □ NO □ D/K □
(b) just out of curiosity YES □ NO □ D/K □
(c) OTHER please describe

....

Q29 – Has particular person influenced beliefs YES □ NO □ D/K □

Q30 IF YES Please describe
Q31 PLEASE ADD ANY OTHER COMMENTS THAT YOU THINK RELEVANT FOR THE AUTHOR TO KNOW ABOUT YOUR VIEWS ON FAITH AND BELIEF AND THE THINGS THAT HAVE INFLUENCED YOU

....

Q32 PLEASE ADD ANY OTHER COMMENTS THAT YOU THINK could INFLUENCE YOU in the future to change your position...

SECTION 3 ABOUT YOU

Q33. AGE : under 20 □ 20–25 □ 25–30 □ 30–40 □40–50 □50–60 □60–70 □over 70 □

Q34.SEX: M □F □ OTHER □

INDICATE HIGHEST LEVEL OF ATTAINMENT THAT APPLIES BY DELETING OTHERS :

Q35. LEVEL OF EDUCATIONAL ATTAINMENT : still at school □ left school with less than 5 GCSEs Grade C or above □ left school with 5 or more GCSEs Grade C or above □
post school college qualification □ University Degree □ Postgraduate qualification □

Q36. Would you say you :
(a) are more introvert than extrovert YES □ NO □ D/K □
(b) are more extrovert than introvert YES □ NO □ D/K □
(c) act more on impulse than considered thought YES □ NO □ D/K □
(d) act more after considered thought than on impulse YES □ NO □ D/K □

Q37. Would you say your approach to life is a more:
scientifically centred oneYES □ NO □ D/K □

Please describe why you answered as you did

....

(b) artistically centred one YES ☐ NO ☐ D/K ☐

Please describe why you answered as you did

....

THANK YOU VERY MUCH FOR YOUR TIME

BIBLIOGRAPHY

Bryson, B. (2003). A Short History of Nearly Everything (1st ed.). Broadway Books.

Van Doren, C. (1991). A History of Knowledge (1992nd ed.). New York: First Ballentyne Books, Random House.

Diamond, J. (1999). Guns,Germs and Steel (1st ed.). London: W.W.Norton & Company.

Life on the Front line, Fruitfullness on the Front Line – Resources from The London Institute for Contemporary Christianity – https://www.licc.org.uk/ourresources/lifeonthefrontline/

COPYRIGHT ACKNOWLEDGEMENTS

FIG. 1: INTER–RELATIONS OF KNOWLEDGE CATEGORIES
Reproduced with personal permission :Rafols, I., Porter, A. L., &
Leydesdorff, L. (2010). Science overlay maps: A new tool for research
policy and library management. Journal of the American Society for
information Science and Technology, 61(9), 1871–1887.

171 ISI (Institute for Scientific Information) subject categories in the
citing dimension; cosine > 0.2. Node sizes set proportional to the
logarithm of the number of citations given by each category. Whereas the
traditional disciplines are represented by clear factors (e.g., Physics or
Chemistry), specific fields of application in mathematics or engineering
do not fall in the disciplinary classification, but in the factor representing
their topic. For example, Mathematical Physics is classified as Physics,
while Mathematics is classified as part of Engineering and with a second
factor loading on the Computer Sciences. Chemical Engineering loads on
the Chemistry factor more than on the one representing Engineering.

FIGURE 2 TIME–LINE OF THE HISTORY OF THE UNIVERSE
Reproduced with personal permission of Particle Data Group, LBNL

**FIGURE 3 RODINIA SUPERCONTINENT – A COALITION OF
SEVERAL CONTINENTAL LAND MASSES.** Reproduced under
permission of the public domain regulations of work produced by / for
the us government under the terms of Title 17, Chapter 1, Section 105 of
the US Code. Wikimedia Commons contributors, Wikimedia Commons,
the free media repository, 22 September 2019, 21:18 UTC,
<https://commons.wikimedia.org/w/index.php?
title=File:Rodinia_reconstruction.jpg&oldid=367720141> [accessed 5
May 2020] source and author: http://antarcticsun.usap.gov/AntarcticSun/
science/images2/rodinia_map.jpg John Goodge reproduced under
permission of the public domain regulations of work produced by/for the
us government under the terms of Title 17, Chapter 1, Section 105 of the

US Code.

FIGURE 4 CHANGING CONFIGURATION OF THE CONTINENTS. Reproduced under permission of the Creative Commons Attribution–Share Alike 4.0 International license. (CC BY–SA 4.0). https://creativecommons.org/licenses/by–sa/4.0/ public domain regulations of work produced by / for the us government under the terms of Title 17, Chapter 1, Section 105 of the US Code.Wikimedia Commons contributors, 'File:Pangaea to present cy.svg',Wikimedia Commons, the free media repository, 23 February 2018, 06:05 UTC, <https://commons.wikimedia.org/w/index.php?title=File:Pangaea_to_present_cy.svg&oldid=288784785> [accessed 5 May 2020]

author Llywelyn20

FIGURE 5 DESIGNATED GEOLOGICAL TIMES IN MILLIONS OF YEARS AGO (MYA) Reproduced under permission : wikipedia:this image is in the public domain in the United States because it only contains materials that originally came from the United States Geological Survey, an agency of the United States Department of the Interior. Source: U.S. Geological Survey Department of the Interior/USGS. https://commons.wikimedia.org/wiki/File:Geologic_time_scale.gif

http://geomaps.wr.usgs.gov/socal/geology/geologic_history/images/timescale.jpg.

FIGURE 6A DIAGRAMMATIC REPRESENTATIONS OF THE STRUCTURE OF THE DNA DOUBLE HELIX Reproduced with personal permission: Shannan Muskopf biologycorner.com

FIGURE 6B

Reproduced under permission:Wikimedia Commons contributors, 'File:DNA simple2.svg', Wikimedia Commons, the free media repository,

29 October 2019, 10:57 UTC, <https://commons.wikimedia.org/w/index.php?title=File:DNA_simple2.svg&oldid=372423171>[accessed 5 May 2020] author Forluvoft I, the copyright holder of this work, release this work into the public domain. This applies worldwide.In some countries this may not be legally possible; if so: I grant anyone the right to use this work for any purpose, without any conditions, unless such conditions are required by law.

FIGURE 7 THE CHEMICAL STRUCTURES OF NUCLEOTIDES, NUCLEIC ACIDS, AMINO ACIDS: WITH PROTEINS BEING MADE UP OF A SEQUENCE OF AMINO ACIDS Reproduced with personal permission : A Tool for Structure Alignment of Molecules Authors: Pei–Ken Chang, Chien–Cheng Chen and Ming Ouhyoung Department of Computer Science and Information Engineering, National Taiwan University Proc. of IEEE Sixth International Symposium on Multimedia Software Engineering (IEEE–MSE2004) Special Session on Bioinformatics, pp. 354–361, Miami, USA, Dec. 2004. (ISBN 0–7695–2217–3)

FIGURE 8 THE TREE OF LIFE – A DIAGRAMMATIC REPRESENTATION OF THE EVOLUTIONARY PATHWAYS OF SPECIES DEVELOPMENT. Reproduced under: Permission creative commons attribution–share alike 3.0 unported licence. https://creativecommons.org/licenses/by–sa/3.0/ Wikimedia Commons contributors, 'File:Pedigree of Man English.jpg', Wikimedia Commons, the free media repository, 8 March 2016, 09:53 UTC, <https://commons.wikimedia.org/w/index.php?title=File:Pedigree_of_Man_English.jpg&oldid=189760056> [accessed 5 May 2020]. Ernst Haeckel's "Pedigree of Man" source and author :Haeckel, Ernst. The Evolution of Man: A Popular Exposition of the Principal Points of Human Ontogeny and Phylogeny. New York: Appleton & Co., 1897, Plate xv

FIGURE 9 TIME LINE OF HUMAN EVOLUTION Reproduced with permission under licence: Creative Commons Attribution-Share Alike 4.0 International license. Wikimedia Commons contributors, 'File:Hominin evolution.jpg', Wikimedia Commons, the free media repository, 5 June 2018, 02:16 UTC, <https://commons.wikimedia.org/w/index.php? title=File:Hominin_evolution.jpg&oldid=304659580> [accessed 12 May 2020] author: Cruithne9. Modified by selecting part of the image.

FIGURE 10 INDICATIVE MIGRATION ROUTES OF HOMO SAPIENS WITH ESTIMATED TIMINGS (in thousands years ago) Reproduced under permission : Based on Wikipedia under the following license: Creative Commons Attribution-Share Alike 2.5 Generic. This item is in the public domain, and can be used, copied, and modified without any restrictions. Source: File:Spreading homo sapiens.jpg (2006). Author: Altaileopard SVG by Magasjukur2. https://commons.wikimedia.org/wiki/ File:Spreading_homo_sapiens.svg

FIGURE 11 CHRONOLOGICAL TIMELINE FOR OCCUPATION OF MESOPOTAMIA Reproduced under licence: creative commons attribution – share alike 3.0 unported (cc sa–by 3.0) https:// commons.wikimedia.org/wiki/File:Mesopotamian_Chronology_2–2011– 29–03.png.https://creativecommons.org/licenses/by–sa/3.0/ source and author: Chronologie_Mesopotamie_2.png: Venal. derivative work: SimonTrew (talk)

CHART 1 WORLD MAP OF THE DEGREE OF RELIGIOUS INTENSITY ACROSS THE WORLD Reproduced under permission Creative Commons Attribution 3.0 Unported licence source and author: File: https://en.wikipedia.org/wiki/Major_religious_groups%25/media/ File:World_religions_map_en.svg Source: Self made, data from 2009 Gallup poll Author: Sbw01f https://commons.wikimedia.org/wiki/ File:Religion_in_the_world.png

CHART 2 SPLITS DURING THE EVOLUTION OF CHRISTIANITY. The chart summarises the major splits that have accompanied the evolution of Christianity, https://en.m.wikipedia.org/wiki/File:Christianity_major_branches.svg Reproduced under Permission which is granted to copy, distribute and/or modify this document under the terms of the GNU Free Documentation License, Version 1.2. https://commons.wikimedia.orgwikiCommons:GNU_Free_Documentation_License,_version_1.2 taken from Wikipedia. Source and Author : Christian–lineage.png – unknown.

LIST OF FIGURES, CHARTS AND TABLES

END NOTES: REVIEWS, ABOUT THE AUTHOR

Thank you for reading my book. If you enjoyed it, won't you please take a moment to leave me a review at your favourite retailer?

Thanks! John Cornell

※∩ ※∩ ※∩※

※∩ ※∩ ※∩※

ABOUT THE AUTHOR:

I qualified in medicine in 1975, and after training in surgery became a general practitioner (doctor of family medicine) In 1974 I left clinical practice and re-trained in Public Health Medicine I became a Consultant in Public Health and then a Director of Public Health in Doncaster. For the last six years before retiring, in 2011, I went back into general practice. Presently I am doing a certificate course in urban theology.

I am active ecumenically at both local and national church levels and volunteer with a number of groups within my local community.

I have published a number of papers, articles and letters in professionally peer reviewed journals and co-edited a book on public health.

Connect with me:

Email: sjohncornell@talktalk.net

⁂⌒ ⁂⌒ ⁂⌒⁂

⁂⌒ ⁂⌒ ⁂⌒⁂

REVIEWS

This book tackles a big question: how can my knowledge of God connect to yours, when we travel on different religious paths? John Cornell's experience is broad – a scientific training, long and responsible church involvement, and high respect for neighbours of other faiths. His outlook is broad too: he asks for greater mutual openness in our faith communities, as a step towards better relationships and more faithful practice. There is serious substance here – intelligent, organised, wide-ranging, accessible and very clear. Rev. John Proctor John Proctor was, before his retirement, General Secretary of the United Reformed Church.*The Rev. John Proctor General Secretary to the The United Reformed Church*

⁂⌒ ⁂⌒ ⁂⌒⁂

John Cornell has developed a structured and frank discussion on religion and the faith it has engendered throughout human history. With sensitivity and thoroughness he tackles the attempts of the faithful to understand the nature and the will of the God in which they believe. A well written, interesting and thoughtful book. *Dr. Ray Chesworth*

⁂⌒ ⁂⌒ ⁂⌒⁂

Dr Cornell has written a book reflecting on our understanding of religion that even fascinates a confirmed atheist like me. Whilst acknowledging that the existence of a God is immutable for people of faith he argues that religion like any bother branch of knowledge should be open to challenge argument and change. By examining the history of religion, science and philosophy he shows how mankind at different times has interpreted the

word of God differently and concludes that this is a right and proper way to ensure communication from God remains relevant."
Dr. Richard Watton

"a brave, honest and provisional undertaking. Here is a Christian layman, open and committed to the Jesus' way, daring to step-out and deploy interdisciplinary 'takes' to wrestle with tough, transgressive, and uncomfortable questions." *Revd Dr Michael N. Jagessar, blogger, writer and theologian.*

※∩ ※∩ ※∩※

REFERENCES

[1]From the outset I would like to draw a distinction between faith and religion. By faith, I mean ones' personal spirituality and belief and relationship with God. By religion/ Church I refer to the organised structures and doctrines of the institutional Church (rather than any local congregation). Ideally, the two are aligned but not all members of a religion subscribe to all the doctrines of their chosen religion and sometimes the Church acts in ways that not all its members would support. Sometimes too, the Church acts in ways that appear contrary to the faith it professes to represent.

[2] Where I use the word Church, it most naturally refers to the Christian Church (which of course is itself not just one unified organisation). Church usually refers to the leadership of the denominational (defined later) faith organisation as an Institution. Thus, it should also be read, where appropriate, to include the institution of any faith organisation. When i use 'church' I usually refer to the building and members of a local community worshiping group. I realise that neither of these definitions have direct equivalents in other faiths, which are organised and operate differently. I do this for ease of writing – without I hope, giving any offence or wishing to exclude any faith or religious group.

[3] I use the term "understanding" to indicate that there are different views, beliefs, ways of knowing or different thoughts and opinions on how to interpret any particular portion of the Bible, scripture or aspect of faith or religious doctrine.

[4] Denomination refers to those different branches of Christianity, defining themselves by having distinct beliefs about aspects of the faith and the Church, whilst still claiming a Christian basis for such beliefs. They include: Roman Catholicism, Church of England – Anglicanism, Methodism, United Reformed Church, Baptists and others within the UK and many other groupings such as the Orthodox, Lutherans, Moravians, Seventh Day Adventist and so on across the world. There are too many variations to name them all. That particular groups are not named is purely down to lack of space.

[5] I will use the term "describe" to apply to the external physical appearance of an object and the term "definition" to apply to the character, nature or essence of an object. So using an apple as an exemplar, a description would include : its size, dimensions, shape and colour but a definition would refer to its nature or essence – as a fruit, its taste and texture. We return to this in chapter 10

[6] The premise of this book is that God exists. The following definition is what I mean when referring to God. I refer to the concept of a "something that is other than and outside / beyond ourselves and the perception of our senses i.e., that God is "supernatural". So it can be assumed that each time I use the word "God", this is what I mean. This avoids the need to continually repeat the phrase in parentheses, as a caveat when discussing ideas in connection with God. In this sense, God "exists" as an independent and separate 'being' / entity that really does things of its own accord and can act, interact and communicate with humankind, as it sees fit. This is therefore a definition of a God that is presumed to exist. It is not a construct or creation of man's imagination, only in so far as we have to use human language to try and convey his nature. This is in contrast to "god" as a human construct, an assumption made by many academic researchers that we will refer to in the coming chapters. It may be attributed all the characteristics of God, but which, being a creation of man's imagination, cannot have any independent interaction with human kind. I do not ascribe any particular gender, colour or creed to God, though may use pronouns in the male form to make the writing flow more easily, and to avoid using the word God too frequently. However, this is not to impute any other particular quality or anthropomorphic (human–like) characteristics to God, except where specifically stated. The reader should not assume, therefore, any automatic similarity to their own concept of God. An Atheist may assume these two definitions of "God"/"god" are actually one and the same. It comes down to a matter of faith !

[7] The Congregational movement originated within the protestant Separatist movements of the 16th and 17th Century. It joined with the Presbyterian Church in 1972 to become the United Reformed Church which was later by the Churches of Christ in 1981 and Scottish Congregationalists in 2000.

[8] Wikipedia contributors, 'Congregational church', *Wikipedia, The Free Encyclopedia,* Last modified 24 July 2018, https://en.wikipedia.org/w/index.php?title=Congregational_church&oldid=851723764 [Accessed 4 August 2018]

[9]. Shia, Sunni, Suffi and others

[10] Orthodox, Conservative, Reformists and others

[11] Ehrman, B. D. (2008) God's problem : how the Bible fails to answer our most important question—why we suffer. HarperOne.

12 Wikipedia Contributors, 'Crusades', *Wikipedia, The Free Encyclopedia,* Last modified 3 July 2019, https://en.wikipedia.org/w/index.php?title=Crusades&oldid=904573074 [Accessed 3 July 2019]

[13] Wikipedia Contributors, 'Inquisition', *Wikipedia, The Free Encyclopedia,* Last modified 17 June 2019, https://en.wikipedia.org/w/index.php?title=Inquisition&oldid=902183710 [Accessed 3 July 2019]

[14] Wikipedia Contributors, 'Ethnic cleansing', *Wikipedia, The Free Encyclopedia,* Last modified 4 July 2019, https://en.wikipedia.org/w/index.php?title=Ethnic_cleansing&oldid=904746450 [Accessed 6 July 2019]

[15] Stark, R., & Bainbridge, W. S. (2112). Religion, Deviance, and Social Control. (Pg. 1–2) New York: Routledge – Taylor Francis Group.

[16] Ruth Benedict, Patterns of Culture (Boston: Houghton Mifflin Company, 1934).

[17] Ruth Benedict, The Chrysanthemum and the Sword: Patterns of Japanese Culture (Boston: Houghton Mifflin Company, 1946), p. 1

[18] Tangney, J. P. (1991). Moral Affect: The good, the bad, and the ugly. Journal of Personality and Social Psychology, 61(4), 598–607.

[19] Lutwak, N., Panish, J., Ferrari, J., & Razzino, B. (2001). Shame and guilt and their relationship to positive expectations and anger expressiveness. Adolescence, vol: 36 (1), pp: 641–53

[20] Leith, K. P., & Baumeister, R. F. (1998). Empathy, Shame, Guilt, and Narratives of Interpersonal Conflicts: Guilt–Prone People Are Better at Perspective Taking. Journal of Personality, 66(1), 1–37. doi.10.1111/1467–6494.00001

[21] Muller, R. (2000). Honor and Shame In A Middle Eastern Setting. Retrieved July 28, 2016, from http://nabataea.net/h %26s.html

[22] Tangney, J. P., Stuewig, J., & Mashek, D. J. (2007). Moral Emotions and Moral Behaviour. Annual Review of Psychology, 58, 345–72. http://doi.org/10.1146/annurev.psych.56.091103.070145

[23] Georges, J. (n.d.) Honor and Shame based Societies: 9 Keys to Working With Muslims. Retrieved July 27, 2016, from http://www.zwemercenter.com/guide/honor–and–shame–9–keys/

[24] Khan, S. (2010 September 1st). Are Islamic Cultures Really Shame–Based? Part 1. Retrieved July 27, 2016, from https://iecrcnewsletter.wordpress.com/2010/09/01/are–islamic–cultures–really–shame–based–part–1/

[25] Ed West (2015). Guilt vs Shame cultures: the silent triumph of Christianity. The Spectator, Coffee House, September the 5th.

[26] Boschen, Timothy L.. "Shame and Guilt In Religious Fundamentalism" ChristianEthicsToday. The Christian Ethics Today Foundation. December 2001 (Issue 37 Page 21) pastarticles.christianethicstoday.com/cetart/index.cfm?fuseaction=Articles.main&ArtID=613

[27] Bierbrauer, G. (1992). Reactions to Violation of Normative Standards: A Cross–Cultural Analysis of Shame and Guilt. International Journal of Psychiatry, 27(2), pages 181–193.

[28] Muller, R. (2000). Honor and Shame In A Middle Eastern Setting. Retrieved July 28, 2016, from http://nabataea.net/h %26s.html

[29] Witness the fighting between Shia and Sunni Muslims. The differences between Jewish factions – orthodox, liberal, American and other groups. The Christian Church's record, including – the Spanish Inquisition, the imprisonment of Galileo, the excommunication of Martin Luther, the execution of William Tyndale and numerous other examples. Disagreements still continue within and between the Christian denominations over the role of women in the Church, homosexuality, same sex marriage and the understanding of the nature of the eucharist (holy communion) in particular.

[30] Winell, M. Religious Trauma Syndrome. Retrieved July 27, 2016, from http://www.babcp.com/Review/RTS–Trauma–from–Leaving–Religion.aspx

[31] Winell, M. (2006). Leaving the Fold: A Guide for Former Fundamentalists and Others Leaving Their Religion. Canada: Raincoat Books.

[32] Richard P. Feynman. 1990. QED : The Strange Theory of Light and Matter. Introduction Page 9. London: Penguin Books.

[33] Sherwood, C. C., Subiaul, F., & Zawidzki, T. W. (2008). A natural history of the human mind: tracing evolutionary changes in brain and cognition. Journal of Anatomy, 212(4), 426–54. http://doi.org/10.1111/j.1469–7580.2008.00868.x

[34] Hofman, M. A. (2014). Evolution of the human brain: when bigger is better. Frontiers in Neuroanatomy, 8:15. http://doi.org/10.3389/fnana.2014.00015

[35] Paul Tillich, Dynamics of Faith, First Published 1957 (New York: Harperone, 2009)

[36] Wikipedia Contributors, "Pluto." Wikipedia, The Free Encyclopedia, Last Modified 10 January 2020. Accessed January 14, 2020. https://en.wikipedia.org/w/index.php?title=Pluto&oldid=935110950.

[37] Wikipedia Contributors. 2018. "Ignaz Semmelweis." In Wikipedia, The Free Encyclopaedia, Last Modified 10th April 2018. Accessed January 14, 2020. https://en.wikipedia.org/w/index.php?title=Ignaz_Semmelweis&oldid=734786976.

[38] Dobbins, M. et al., (2004) A Framework for the Dissemination and Utilization of Research for Health-Care Policy and Practice. Worldviews on Evidence-based Nursing presents the archives of Online Journal of Knowledge Synthesis for NursingVolume E9, Issue 1.https://doi.org/10.1111/j.1524-475X.2002.00149.x Accessed 15th January 2009

[39] Wikipedia Contributors. "Knowledge Transfer." Wikipedia, The Free Encyclopedia, Last Modified 16 September 2017. Accessed October 7, 2017.https://en.wikipedia.org/w/index.php?title=Knowledge_transfer&oldid=935123253

[40] Wikipedia Contributors. "Criticism of Wikipedia." Wikipedia, The Free Encyclopedia, Last Modified 2016, November 24. Accessed November 24, 2016. https://en.wikipedia.org/w/index.php?title=Criticism_of_Wikipedia&oldid=751244851

[41] Wikipedia, Contributors."Prehistory". Last modified September 26, 2016, from https://en.wikipedia.org/w/index.php?title=Prehistory&oldid=741258813.

[42] William A. Haviland, Harald E.L. Prins, Dana Walrath, and Bunny McBride. 2012. The Essence of Anthropology. 3rd ed. Cengage Learning. Pg.83

43 "Prehistoric Art: Origins, Types, Characteristics, Chronology." Encyclopaedia of Art. Accessed September 14, 2016. http://www.visual-arts-cork.com/prehistoric-art.htm.

44 Kelly, L., 2015. *Knowledge and Power in Prehistoric Societies: Orality, Memory and the Transmission of Culture*, Cambridge: Cambridge University Press. ISBN: 978-1-107-05937-5. Available at: http://www.cambridge.org/us/academic/subjects/archaeology/prehistory/knowledge-and-power-prehistoric-societies-orality-memory-and-transmission-culture.

45 Kelly, L., 2015. *Knowledge and Power in Prehistoric Societies: Orality, Memory and the Transmission of Culture*, Pg. 95 ISBN: 978-1-107-05937-5.

46 Wade N (2006). Before the Dawn : Recovering the Lost History of Our Ancestors. Penguin Press HC

47 Ellwood P. Cubberley. 1920. The History of Education:Educational Practice and Progress Considered as a Phase of the Development and Spread of Western Civilization. Riverside Books in Education. 2nd ed. Boston: Houghton Mifflin, Riverside Books Cambridge.

48 Pérez, Rolando, "Francisco Sanches", The Stanford Encyclopedia of Philosophy (Summer 2020 Edition), Edward N. Zalta (ed.), forthcoming URL = <https://plato.stanford.edu/archives/sum2020/entries/francisco-sanches/>.

49 Gettier, E. L. (1963). Is Justified True Belief Knowledge? Analysis, 23, 121–123.

50 Shope, R. K. (1963). The Analysis of Knowing: A Decade of Research. Princeton, New Jersey: Princeton University Press.

51 Glanzberg, Michael, "Truth", The Stanford Encyclopedia of Philosophy (Winter 2016 Edition), Edward N. Zalta (Ed.) Metaphysics Research Lab, Stanford University, forthcoming URL = <https://plato.stanford.edu/archives/win2016/entries/truth/

52 Wikipedia Contributors, 'Truth', In Wikipedia, The Free Encyclopaedia, Last Modified 1st December 2016, 2016 <https://en.wikipedia.org/w/index.php?title=Truth&oldid=752442805> [Accessed 14 January 2020]

53 Truth : Stanford University., & Center for the Study of Language and Information (U.S.). (1997). *Stanford encyclopedia of philosophy.* Stanford University. Retrieved from https://plato.stanford.edu/entries/truth/

54 Wikipedia Contributors, 'Epistemology', Wikipedia, The Free Encyclopedia, Last Modified 22nd March 2017 <https://en.wikipedia.org/w/index.php?title=Epistemology&oldid=771680921> [Accessed 29 March 2017]

55 Klein, Peter, "Skepticism", The Stanford Encyclopedia of Philosophy (Summer 2015 Edition), Edward N. Zalta (Ed.) Metaphysics Research Lab, Stanford University, https://plato.stanford.edu/archives/sum2015/entries/skepticism/

56 Peirce, C. S. (1955). The Scientific Attitude and Fallibilism. (J. Buchler, Ed.) (Philosophical Writings of Peirce). New York: Dover.

57 Wikipedia Contributors. "Fallibilism." Wikipedia, The Free Encyclopedia, last modified 28th March 2017. Accessed January 15th 202o. https://en.wikipedia.org/w/index.php?title=Fallibilism&oldid=772599762.

58 Lewis, D. (1996). Elusive Knowledge. *Australasian Journal of Philosophy*, *74*(4), 549

59 Andrea, A., & Overfield, J. (2015). Plato, Apology 21d. Human Record: Sources of Global History, Volume I: To 1500. p. 116 Cengage Learning.

60 Wikipedia Contributors, 'René Descartes', Wikipedia, The Free Encyclopedia, Last Modified November 21, 2016 <https://en.wikipedia.org/w/index.php?title=René_Descartes&oldid=750828160> [Accessed 21 November 2016]

61 de Jong, T., & Ferguson-Hessler, M. G. M. (1996). Types and qualities of knowledge. Educational Psychologist, 31(2), 105–113. Retrieved from http://doc.utwente.nl/26717/1/types.pdf

62 Pritchard, D. & Turri, J. (2014). 'The Value of Knowledge', The Stanford Encyclopedia of Philosophy, Edward N. Zalta (Ed.) Metaphysics Research Lab, Stanford University, 2014, <https://plato.stanford.edu/archives/spr2014/entries/knowledge-value/> [Accessed 7 December 2016]

63 Ichikawa, J. J., & Steup, M. (2014). The Analysis of Knowledge. The Stanford Encyclopedia of Philosophy, Edward N. Zalta (Ed.) Metaphysics Research Lab, Stanford University (Spring 2014). Retrieved from http://plato.stanford.edu/archives/spr2014/entries/knowledge-analysis/

64 64 Glanzberg, Michael, "Truth", The Stanford Encyclopedia of Philosophy (Winter 2016 Edition), Edward N. Zalta (Ed.) Metaphysics Research Lab, Stanford University, forthcoming URL = <https://plato.stanford.edu/archives/win2016/entries/truth/

65 de Jong, T., & Ferguson-Hessler, M. G. M. (1996). Types and qualities of knowledge. Educational Psychologist, 31(2), 105–113. Retrieved from http://doc.utwente.nl/26717/1/types.pdf

66 Reif, F., & Allen S. (1992). Cognition of interpreting scientific concepts: A study of acceleration. Cognition and Instruction, 9, 1-44.

67 Loet Leydesdorff [1] & Ismael Rafols. (n.d.). A global map of science based on the ISI subject categories. Retrieved from http://www.leydesdorff.net/map06/texts/index.htm

68 Kim, Y., & Stanton, J. M. (2012). Institutional and Individual Influences on Scientists' Data Sharing Practices. Journal of Computational Science Education, Vol. 3(Issue 1), 47–56

69 Benowitz, S. (2002). When scientists don't share: is secrecy a necessary evil? Journal of the National Cancer Institute, 94(10), 712–3. http://doi.org/10.1093/JNCI/94.10.712

70 Benowitz, S. (2002). When scientists don't share: is secrecy a necessary evil? 712-713

[71] Blumenthal, D. et al., 1997. Withholding Research Results in Academic Life Science. *JAMA*, 277(15), p.1224. Available at: http://jama.jamanetwork.com/article.aspx?doi=10.1001/jama.1997.03540390054035 [Accessed August 30, 2017].

[72] Campbell, E.G. et al., 2002. Data Withholding in Academic Genetics. *JAMA*, 287(4), p.473. Available at: http://jama.jamanetwork.com/article.aspx?doi=10.1001/jama.287.4.473 [Accessed August 30, 2017].

[73] Wikipedia Contributors, 'Scientia Potentia Est', Wikipedia, The Free Encyclopedia, Last Modified 17 December 2017 <https://en.wikipedia.org/w/index.php?title=Scientia_potentia_est&oldid=931149031> [Accessed 15 January 2020]

[74] Kelly, L., 2015. *Knowledge and Power in Prehistoric Socie*ties: Orality, Memory and the Transmission of Culture, Cambridge: Cambridge University Press. ISBN: 978-1-107-05937-5.

[75] International Knowledge Ecology, 'Access to Knowledge' <https://www.keionline.org/category/access-to-knowledge> [Accessed 15 January 2020]

[76] Wikipedia Contributors, 'Geneva Declaration on the Future of the World Intellectual Property Organisation', Wikipedia, The Free Encyclopedia, Last Modified 23 October 2017 <https://en.wikipedia.org/w/index.php?title=Geneva_Declaration_on_the_Future_of_the_World_Intellectual_Property_Organization&oldid=806638783> [Accessed 14 January 2020]

[77] Shaver, L.B., 2010. The Right to Science and Culture. *Wisconsin Law Review*, (1), pp. 121–184. Available at: http://www.ssrn.com/abstract=1354788 [Accessed October 6, 2017].

[78] The International Consortium of Investigators for Fairness in Trial Data Sharing, 'Toward Fairness in Data Sharing', New England Journal of Medicine, 375 (2016), 405–7 <https://doi.org/10.1056/NEJMp1605654>

[79] Popkin, R. 2003). The History of Scepticism : From Savonarola to Bayle: Oxford University Press. USA. SBN:978-0195107685

[80] Kvanvig, J. L. (2003). The Value of Knowledge and the Pursuit of Understanding. Canbridge: Cambridge University Press.

[81] Baghramian, Maria and Carter, J. Adam, "Relativism", *The Stanford Encyclopedia of Philosophy* (Winter 2019 Edition), Edward N. Zalta (ed.), URL = <https://plato.stanford.edu/archives/win2019/entries/relativism/>

[82] Quote from Jim Croft, a neighbour and friend.

[83] Wikipedia contributors, 'Fake news website', *Wikipedia, The Free Encyclopedia,* 31 March 2017, 07:54 UTC, <https://en.wikipedia.org/w/index.php?title=Fake_news_website&oldid=773096812> [Accessed 15 January 2020]

[84] Little, D., 2017. History of Philosophy. *Stanford Encyclopedia of Philosophy Edward N. Zalta (ed.)* Metaphysics Research Lab, Stanford University. https://plato.stanford.edu/archives/sum2017/entries/history/ [Accessed January 10, 2018].

85 'Different American Wests: Introduction to Historical Controversies', A2 History: Introduction to Histiography <http://www.andallthat.co.uk/uploads/2/3/8/9/2389220/introduction_to_historiography.pdf> [Accessed 10 January 2018]

86 Wikipedia Contributors, 'Recorded History', Wikipedia, The Free Encyclopedia, Last Modified 17 December 2017 <https://en.wikipedia.org/w/index.php?title=Recorded_history&oldid=815896160> [Accessed 10 January 2018]

87 Dalton, M.S. & Charnigo, L., 2004. Historians and Their Information Sources. *College & Research Libraries*, pp.400–25, at 416 no..3. Available at: https://crl.acrl.org/index.php/crl/article/view/15685/17131[Accessed January 15 2020].

88 Wikipedia contributors, 'I'm Sorry I Haven't a Clue', *Wikipedia, The Free Encyclopedia,* 2 July 2019, <https://en.wikipedia.org/w/index.php?title=I%27m_Sorry_I_Haven %27t_a_Clue&oldid=904440977> [Accessed 15 January 2020]

89 Andy Bodle, 'How New Words Are Born', The Guardian, 4 February 2016 <https://www.theguardian.com/media/mind-your-language/2016/feb/04/English-neologisms-new-words> [Accessed 30 October 2017]

90 Lloyd, G.E.., 1970. *Early Greek Science: Thales to Aristotle* First., New York: W.W. Norton and Company Ltd. ISBN:0-393-00583-6

91 Norman Kemp Smith, 'How Far Is Agreement Possible in Philosophy', The Journal of Philosophy, Psychology and Scientific Methods, 9 (1912), 701–11 <https://doi.org/10.2307/2013045>

92 Laudan, L. (1990). Science and relativism: Some key controversies in the philosophy of science. (D. L. Hull, Ed.), Science and Its Conceptual Foundations. The University of Chicago Press.

93 Kuhn, T.S. et al., 1970. *The Structure of Scientific Revolutions Second Edition, Enlarged The Structure of Scientific Revolutions* Second., Chicago. Available at: https://projektintegracija.pravo.hr/_download/repository/Kuhn_Structure_of_Scientific_Revolutions.pdf [Accessed December 1, 2017].

94 Kelly, L., 2015. *Knowledge and Power in Prehistoric Societies: Orality, Memory and the Transmission of Culture*, Cambridge: Pg. 35 Cambridge University Press. ISBN: 978-1-107-05937-5

95 Grant, E., 2007. *A History of Natural Philosophy - From the Ancient World to the Nineteenth C* 1st Ed., New York: Cambridge University Press. (P.17)

96 Wikipedia Contributors. "Aristotle." Wikipedia, The Free Encyclopedia, Last Modified 7th August 2017. Accessed January 15, 2020. https://en.wikipedia.org/w/index.php?title=Aristotle&oldid=931527798.

97 Grant, E., 2007. *A History of Natural Philosophy - From the Ancient World to the Nineteenth C* 1st Ed., New York: Cambridge University Press. (Pgs. 236-238)

98 Grant, E., 2007. *A History of Natural Philosophy - From the Ancient World to the Nineteenth* (Pg. 295).

[99] The Science Council https://sciencecouncil.org/

[100] Ross, S. (1962). Scientist: History of Word. Annals of Science, 18(2). Retrieved from http://www.tandfonline.com/doi/pdf/10.1080/00033796200202722

[101] Grant, E. (2007). A History of Natural Philosophy - From the Ancient World to the Nineteenth C (1st ed.). New York: Cambridge University Press. Pg. 156

[102] Ross, S. (1962). Scientist: History of Word. Annals of Science, 18(2). Retrieved from http://www.tandfonline.com/doi/pdf/10.1080/00033796200202722

[103] Wikipedia contributors, 'List of people considered father or mother of a scientific field', *Wikipedia, The Free Encyclopedia,* Last Modified 15 January 2018 <https://en.wikipedia.org/w/index.php?title=List_of_people_considered_father_or_mother_of_a_scientific_field&oldid=820540841> [Accessed 15th January 2020]

[104] Grant, E. (2007). A History of Natural Philosophy - From the Ancient World to the Nineteenth C (1st ed.). New York: Cambridge University Press. Pg. 276

[105] Hansson, Sven Ove, "Science and Pseudo-Science", The Stanford Encyclopedia of Philosophy (Summer 2017 Edition), Edward N. Zalta (ed.) Metaphysics Research Lab, Stanford University URL = <https://plato.stanford.edu/archives/sum2017/entries/pseudo-science/>. [Accessed 15th January 2020]

[106] In modern times, the demarcation dispute is important in some fields as there may be practical consequences for funding research, efficacy and reliability of medical treatments and so on if certain claims are not given sufficient scrutiny and accepted as face value because someone purports their work to be "scientifically" based. However, the purpose here is not to class something as science or non-science, but to establish the credibility of any claim by examining the means by which someone has come to their conclusion or derived their claim. This is Feyerabend's position.

[107] Fara, P. (2009). *Science: A Four Thousand Year History.* Pg. 51 Oxford: Oxford University Press. ISBN:978-0-19-958027-9.

[108] Feyerabend, P., 1987. *Farewell To Reason*, London: Verso.

[109] Das, A. C. (1952). Similarities in Eastern and Western Philosophy. *The Review of Metaphysics.* 5 (4): 631–38. Pub.Philosophy Education Society Inc. https://www.jstor.org/stable/20123296 https://doi.org/10.2307/20123296.

[110] Eastern Philosophy - General - The Basics of Philosophy. (n.d.). Retrieved February 23, 2018, from http://www.philosophybasics.com/general_eastern.html.

[111] Fara, P., 2009. *Science: A Four Thousand Year History.* Pg. 217 Oxford: Oxford University Press. ISBN:978-0-19-958027-9.

[112] Fara, P. (2009). *Science: A Four Thousand Year History.* Pg. 55. ISBN: 978-0-19-958027-9.

[113] Huff, T. E. (1993). The Rise of Early Modern Science: Islam, China and the West. Cambridge: Cambridge University Press.

[114] Grant, E. (2007). A History of Natural Philosophy - From the Ancient World to the Nineteenth C (1st ed.). New York: Cambridge University Press.

[115] Frankopan P. 2015 The Silk Roads. Bloomsbury Paperbacks. ISBN:HB: 978-1-4088-3997-3

[116] Freeman, K. & Diels, H., 1948. Ancilla to the pre-Socratic philosophers : a complete translation of the fragments in Diels, Fragmente der Vorsokratiker, Blackwell.

[117] Wikipedia Contributors. "Multiverse." Wikipedia, The Free Encyclopedia, Last Modified 9th September 2016. Accessed January 15, 2020. https://en.wikipedia.org/w/index.php?title=Multiverse&oldid=738594976.

[118] Wikipedia Contributors. "Theory of Everything." Wikipedia, The Free Encyclopedia, Last Modified 8th September 2016. Accessed January 15, 2020. https://en.wikipedia.org/w/index.php?title=Theory_of_everything&oldid=738292324.

[119] Charles Van Doren. 1992. A History of Knowledge. Reissue Ed. New York: First Ballentyne Books, Random House.

[120] Wikipedia Contributors, 'Anaximander', Wikipedia, The Free Encyclopedia, Last modified 26 August 2016, <https://en.wikipedia.org/w/index.php?title=Anaximander&oldid=736310481> [Accessed 15 January 2020]

[121] Hardy, A. (1960). Was Man More Aquatic in the Past? New Scientist, 7, Pgs. 642-645.

[122] Roede, M. (1991). The aquatic ape : fact or fiction? : the first scientific evaluation of a controversial theory of human evolution. Souvenir Press.

[123] John H. Langdon. 2012. "Review: Was Man More Aquatic in the Past? Fifty Years after Alister Hardy: Waterside Hypotheses of Human Evolution, Mario Vaneechoutte, Algis Kuliukas, Marc Verhaegen (Eds.). Bentham EBooks, (2011)." HOMO - Journal of Comparative Human Biology. 2012. https://doi.org/10.1016/j.jchb.2012.06.001.

[124] Wikipedia Contributors, 'Anaximenes of Miletus', *Wikipedia, The Free Encyclopedia,* Last Modified 16 August 2016, <https://en.wikipedia.org/w/index.php?title=Anaximenes_of_Miletus&oldid=734685295> [Accessed 15 January 2020]

[125] Wikipedia Contributors. "Parmenides." Wikipedia, The Free Encyclopedia, Last Modified 29th August 2016. Accessed January 15, 2020. https://en.wikipedia.org/w/index.php?title=Parmenides&oldid=736699867.

[126] Lindberg, D. C. (2007). The beginnings of western science : the European scientific tradition in philosophical, religious, and institutional context, prehistory to A.D. 1450. University of Chicago Press.

[127] Grant, E. (2007). A History of Natural Philosophy - From the Ancient World to the Nineteenth C (1st ed.). New York: Cambridge University Press. (Pgs. 37 - 38).

[128] Wikipedia Contributors. "'Philosophy.'" Wikipedia, The Free Encyclopedia, Last Modified 26th June 2017. Accessed January 15, 2020. https://en.wikipedia.org/w/index.php?title=Philosophy&oldid=787484943

[129] Grant, E. (2001). *God and Reason in the Middle Ages*. (Introduction) Cambridge: Cambridge University Press. https://doi.org/DOI: 10.1017/CBO9780511512155.

[130] Wikipedia contributors, 'Islamic Golden Age', *Wikipedia, The Free Encyclopedia,* last modified 16 March 2018, https://en.wikipedia.org/w/index.php?title=Islamic_Golden_Age&oldid=830653287 [Accessed 15th January 2020]

[131] Wikipedia contributors, 'Science in the medieval Islamic world', *Wikipedia, The Free Encyclopedia,* Last modified 27 January 2018, https://en.wikipedia.org/w/index.php?title=Science_in_the_medieval_Islamic_world&oldid=822688322 [Accessed 15th January 2020]

[132] Harold Dorn, and James E. McClellan. 2015. Science and Technology in World History. 3rd ed. John Hopkins Univeristy Press.

[133] Grant, E. (2001). (Pg.102).*God and Reason in the Middle Ages*. Cambridge: Cambridge University Press. https://doi.org/DOI: 10.1017/CBO9780511512155

[134] Grant, E. (2007). A History of Natural Philosophy - From the Ancient World to the Nineteenth C (1st ed.). New York: Cambridge University Press. (Pg. 28)

[135] Barr, S. M. (2006) (Pg. 26) *A Students Guide to Natural Science*. Wilmington, DE: Intercollegiate Studies Institute.

[136] Grant, E. (2007) (Page 305) A History of Natural Philosophy - From the Ancient World to the Nineteenth C. 1st edn. New York: Cambridge University Press.

[137] Grant, E. (2007) (Page 321) A History of Natural Philosophy - From the Ancient World to the Nineteenth C. 1st edn. New York: Cambridge University Press.

[138] Grant, E. (2001). (Pg.102).*God and Reason in the Middle Ages*. Cambridge: Cambridge University Press. https://doi.org/DOI: 10.1017/CBO9780511512155

[139] Grant, E. (2007). A History of Natural Philosophy - From the Ancient World to the Nineteenth century (1st ed.) (Pg. 307) New York: Cambridge University Press.

[140] Feynman, Richard (1965). *The Character of Physical Law*. Modern Library. ISBN 0-679-60127-9.

[141] Jeaneane D. Fowler (2002). Perspectives of Reality: An Introduction to the Philosophy of Hinduism. Sussex Academic Press. ISBN 978-1-898723-93-6.

[142] Schaffer, S. (1986). Pages (387-420) Scientific Discoveries and the End of Natural Philosophy. *Social Studies of Science*. Sage Publications, Ltd. https://doi.org/10.2307/285025

[143] Bryson, B. (2003). A Short History of Nearly Everything (1st ed.). Broadway Books.

[144] Charles Van Doren. 1992. A History of Knowledge. Reissue Ed. New York: First Ballentyne Books, Random House.

[145] Diamond, J. (1999). *Guns, Germs and Steel* (1st ed.). London: W.W.Norton & Company.

[146] Thomas G West. 1979. Plato's Apology of Socrates: An Interpretation, with a New Translation. Ithaca and London: Cornell University Press. (out of print but available at: https://www.academia.edu/27928809/Plato_s_Apolo)

[147] Plato, and Francis Macdonald Cornford. 1957. *Plato's cosmology; the Timaeus of Plato*. Indianapolis: Bobbs-Merrill.

[148] Shields, C. (2007) Aristotle (Routledge Philosophers). 1st edn. Routledge.

[149] Wikipedia Contributors, 'Neoplatonism', *Wikipedia, The Free Encyclopedia*, Last modified 30 November 2019, https://en.wikipedia.org/w/index.php?title=Neoplatonism&oldid=928682693 [Accessed 15 January 2020]

[150] Grant, E. (2007) (Pg 247) A History of Natural Philosophy - From the Ancient World to the Nineteenth C. 1st edn. New York: Cambridge University Press.

[151] Grant, E. (2007) (Pgs 247 - 248) A History of Natural Philosophy - From the Ancient World to the Nineteenth C. 1st edn. New York: Cambridge University Press.

[152] Wikipedia Contributors, 'Sentences', *Wikipedia, The Free Encyclopedia*, last modified 28 December 2019, https://en.wikipedia.org/w/index.php?title=Sentences&oldid=932852564 [Accessed 15 January 2020]

[153] Wikipedia Contributors, 'Hermeticism', *Wikipedia, The Free Encyclopedia*, Last modified 22 July 2019, https://en.wikipedia.org/w/index.php?title=Hermeticism&oldid=907333262 [Accessed 15 January 2020]

[154] The Complete Idiots' Guide to Theories of the Universe. (2002) (Chapter 3) Moring,Gary. Alpha. Penguin Group. USA. eISBN:978-1-440-69572-8

[155] Wikipedia Contributors, 'Esoteric Christianity', *Wikipedia, The Free Encyclopedia*, 2 May 2019, https://en.wikipedia.org/w/index.php?title=Esoteric_Christianity&oldid=895234329 [Accessed 2 August 2019]

[156] Magee, B. (1998) Confessions of a Philosopher - a journey through western philosophy. London: Phoenix.

[157] Mastin, L. (2009) *A Quick History of Philosophy - General - The Basics of Philosophy*. Available at: https://www.philosophybasics.com/general_quick_history.html (Accessed: 24 March 2018).

[158] Hanne Andersen, and Brian Hepburn. 2016. "Scientific Method." Stanford Encyclopedia of Philosophy Edward N. Zalta (Ed.) Metaphysics Research Lab, Stanford University. http://plato.stanford.edu/archives/sum2016/entries/scientific-method/

[159] Wikipedia Contributors, 'Scientism', *Wikipedia, The Free Encyclopedia*, last modified 24 May 2018, https://en.wikipedia.org/w/index.php?title=Scientism&oldid=842784244 [Accessed 15 January 2020]

[160] Feser, E. (2010) *Recovering Sight after Scientism, Public Discourse*. Available at: http://www.thepublicdiscourse.com/2010/03/1184/ (Accessed: 29 August 2016).

[161] Feser, E. (2010) *Blinded by Scientism, Public Discourse*. Available at: http://www.thepublicdiscourse.com/2010/03/1174/ (Accessed: 29 August 2016).

[162] Vogt, B. (n. d.) *The Science Delusion: An Interview with Atheist Curtis White, Strange Notions*. Available at: http://www.strangenotions.com/the-science-delusion/ (Accessed: 29 August 2016).

[163] Kreeft, D. P. (n. d.) *20 Arguments For God's Existence : Strange Notions*. Available at: https://strangenotions.com/god-exists/ (Accessed: 28 May 2018).

[164] Shah, M. and M. Shah (2009). "Modern science and scientism a perennialist appraisal." European Journal of Science and Theology **5**: 1-24.

[165] Principe, L. (2013) *Scientism and the Religion of Science - YouTube*. Available at: https://www.youtube.com/watch?v=XFVARio4pAk (Accessed: 29 May 2018).

[166] Definition of philosophy in English by Oxford Dictionaries (n.d.). The study of the fundamental nature of knowledge, reality, and existence, especially when considered as an academic discipline. The study of the theoretical basis of a particular branch of knowledge or experience.'the philosophy of science'Available at: https://en.oxforddictionaries.com/definition/philosophy (Accessed: 30 May 2018).

[167] Feyerabend, P. (1993) *Against Method*. 3rd ed. London: Verso.

[168] Karlo Broussard. n.d. "Is Real Knowledge Only Scientific Knowledge?" Strange Notions. Accessed August 29, 2016. http://www.strangenotions.com/is-real-knowledge-only-scientific-knowledge/.

[169] Hanne Andersen, and Brian Hepburn. 2016. "Scientific Method." Stanford Encyclopedia of Philosophy Edward N. Zalta (Ed.) Metaphysics Research Lab, Stanford University. http://plato.stanford.edu/archives/sum2016/entries/scientific-method/.

[170] Cat, Jordi, "The Unity of Science", The Stanford Encyclopedia of Philosophy (Fall 2017 Edition), Edward N. Zalta (Ed.). Metaphysics Research Lab, Stanford University <https://plato.stanford.edu/archives/fall2017/entries/scientific-unity/>.

[171] Philosophy 1. A Guide Through the Subject. 1998. Edited by A. C. Grayling. Oxford University Press, ISBN: 9780198752431

[172] Karl R. Popper. 2013. Realism and the Aim of Science. Edited by William Warren Bartley. Routledge.

[173] Kuhn, T. S. (1970) 'The Structure of Scientific Revolutions.', *Philosophical Review* , II. Available at: https://doi.org/10.1119/1.1969660.

[174] Debra Nails. 2018. Socrates. Edward N. Zalta (Ed). Metaphysics Research Lab, Stanford University. Stanford Encyclopedia of Philosophy. Spring 201. Stanford University.https://plato.stanford.edu/archives/spr2018/entries/socrates/

[175] Famous Scientists at: https://www.famousscientists.org Accessed 15th January 2020

[176] Fara, P. (2009) *Science: A Four Thousand Year History -*. Oxford: Oxford University Press. Available at: https://books.google.co.uk/books?

[177] Blair, M. (2014) 'Getting evidence into practice—implementation science for paediatricians', *Archives of Disease in Childhood*, 99(4), p. 307 LP-309. Available at: http://adc.bmj.com/content/99/4/307.abstract.

[178] Dobbins, M. et al., (2004) A Framework for the Dissemination and Utilization of Research for Health-Care Policy and Practice. Worldviews on Evidence-based Nursing presents the archives of Online Journal of Knowledge Synthesis for NursingVolume E9, Issue 1.https://doi.org/10.1111/j.1524-475X.2002.00149.x Accessed 15th January 2009.

[179] Slote Morris Z, Wooding S, G. J. (2011) 'The answer is 17 years, what is the question: understanding time lags in translational research', *J R Soc Med*, 104, pp. 510–520. doi: DOI 10.1258/jrsm.2011.110180.

[180] Wikipedia contributors, 'History of science', *Wikipedia, The Free Encyclopedia,* last modified 4 April 2018, https://en.wikipedia.org/w/index.php?title=History_of_science&oldid=834233301[Accessed 16 January 2020]

[181] Moran, Gordon (1998). Silencing Scientists and Scholars in Other Fields: Power, Paradigm Controls, Peer Review, and Scholarly Communication. Santa Barbara, California: Ablex. pp. (cited on page) 38. ISBN 978-1567503432

[182] Merton, Robert K. (1973). *The Sociology of Science*. Chicago: University of Chicago Press. ISBN:0-226-52092-7

[183] Wikipedia Contributors, 'Cecilia Payne-Gaposchkin', *Wikipedia, The Free Encyclopedia,* Last modified 14 April 2018, https://en.wikipedia.org/w/index.php?title=Cecilia_Payne-Gaposchkin&oldid=836322091 [Accessed 16 January 2020]

[184] Grant, E. (1996) (Pg. 204) *The Foundations of Modern Science in the Middle Ages*. Cambridge: Cambridge University Press. doi: 10.1017/CBO9780511817908.

[185] Hansson, Sven Ove, "Science and Pseudo-Science", The Stanford Encyclopedia of Philosophy (Summer 2017 Edition), Edward N. Zalta (Ed.). Metaphysics Research Lab, Stanford University. URL = https://plato.stanford.edu/entries/pseudo-science/ Accessed 16th January 2020

[186] "The 10 Greatest Cases of Fraud in University Research." Staff Writers OnlineUniversities.com, February 27, 2012. https://www.onlineuniversities.com/blog/2012/02/the-10-greatest-cases-of-fraud-in-university-research/

[187] "Scientific Misconduct." *Wikipedia*, April 20, 2018. https://en.wikipedia.org/w/index.php?title=Scientific_misconduct&oldid=837423677.

[188] Bailey, B.L. & Immerman, R.H., *Understanding the U.S. wars in Iraq and Afghanistan*, Available at: http://www.jstor.org/stable/j.ctt13x0q17?loggedin=true [Accessed December 1, 2017].

[189] Shahrani, N.M., 2002. War, Factionalism, and the State in Afghanistan. *American Anthropologist*, 104(3), pp.715–722. Available at: http://doi.wiley.com/10.1525/aa.2002.104.3.715 [Accessed December 1, 2017].

[190] Kuhn, T.S., 1970. The Structure of Scientific Revolutions. *Philosophical Review* , II. Available at: https://doi.org/10.1119/1.1969660.

[191] Wikipedia Contributors, 'List of creation myths', *Wikipedia, The Free Encyclopedia,* Last modified 20 March 2018, https://en.wikipedia.org/w/index.php? title=List_of_creation_myths&oldid=831506109 [Accessed 11 April 2018]

[192] Dalley, S. (1989) Myths from Mesopotamia - Creation, The Flood, Gilgamesh and Others. 1998th edn. Oxford: Oxford World's Classics.

[193] Gascoigne, Bamber. "History of Creation Stories" HistoryWorld. From 2001, ongoing. http://www.historyworld.net/wrldhis/PlainTextHistories.asp?ParagraphID=bjh

[194] Origin: (Robert Langdon Book 5) by Dan Brown 2017. Pg. 81. Transworld Publishers. London. Penguin Random House. ISBN:9780593078754.

[195] Wikipedia Contributors, 'Cosmology in medieval Islam', *Wikipedia, The Free Encyclopedia,* Last modified 2 April 2018, Accessed 14 April 2018] https:// en.wikipedia.org/w/index.php?title=Cosmology_in_medieval_Islam&oldid=833774503

[196] Wendy Doniger. 1981. The Rig Veda "Creation Hymn 10.129." Penguin Classics.

[197] Wikipedia Contributors, 'Chinese creation myths', *Wikipedia, The Free Encyclopedia,* last modified 3 March 2018, <https://en.wikipedia.org/w/index.php? title=Chinese_creation_myths&oldid=828573517> [Accessed 12 April 2018]

[198] Definition astronomy : The scientific study of the universe and the objects in it, including stars, planets, nebulae, and galaxies. Astronomy deals with the position, size, motion, composition, energy, and evolution of celestial objects.

[199] Definition of cosmology : Cosmos is just another word for universe, and "cosmology" is the study of the origin, evolution and fate of the universe.

[200] Hellman, H. (1998). Two Views of the Universe: Galileo vs. the Pope. Retrieved August 26, 2016, from http://www.washingtonpost.com/wp-srv/national/horizon/sept98/ galileo.htm

[201] Wikipedia contributors, 'Ignaz Semmelweis', *Wikipedia, The Free Encyclopedia,* last modified 10 April 2018, <https://en.wikipedia.org/w/index.php? title=Ignaz_Semmelweis&oldid=835808906> [Accessed 20 April 2018]

[202] Wikipedia Contributors, 'Modern flat Earth societies', *Wikipedia, The Free Encyclopedia,* last modified 15 April 2018, <https://en.wikipedia.org/w/index.php? title=Modern_flat_Earth_societies&oldid=836569331> [Accessed 18 April 2018]

[203] Wikipedia Contributors, 'Sumer', *Wikipedia, The Free Encyclopedia,* Last modified 16 April 2018, https://en.wikipedia.org/w/index.php?title=Sumer&oldid=836740070 [Accessed 20 April 2018]

[204] Wikipedia Contributors."Babylonian Mathematics." In Wikipedia, The Free Encyclopedia Last Modified 15th September 2016. Accessed April 16, 2020. https:// en.wikipedia.org/w/index.php?title=Babylonian_mathematics&oldid=736675327.

205 Wikipedia Contributors. "Babylonian Astronomy." Wikipedia, The Free Encyclopedia Last Modified 30th August 2016. Accessed September 15, 2016. https://en.wikipedia.org/w/index.php?title=Babylonian_astronomy&oldid=736844336

206 Herodotus, with an English translation by A. D. Godley Cambridge Harvard University Press 1920 0674991303 0674991311 0674991338 0674991346 Perseus Project

207 Wikipedia Contributors. "Aryabhata." Wikipedia, The Free Encyclopedia Last Modified 9th March 2018. Accessed April 9, 2018. https://en.wikipedia.org/w/index.php?title=Aryabhata&oldid=829612981.

208 Wikipedia Contributors. "Antikythera Mechanism." Wikipedia, The Free Encyclopedia Last Modified 13th April 2018. Accessed April 19, 2018. https://en.wikipedia.org/w/index.php?title=Antikythera_mechanism&oldid=836275723.

209 Wikipedia Contributors. "Armillary Sphere." Wikipedia, The Free Encyclopedia Last Modified 14th April 2018. Accessed April 19, 2018. https://en.wikipedia.org/w/index.php?title=Armillary_sphere&oldid=836434857.

210 Wikipedia Contributors. n.d. "Nilakantha Somayaji." Wikipedia, The Free Encyclopedia Last Modified 30th March 2018. Accessed April 19, 2018. https://en.wikipedia.org/w/index.php?title=Nilakantha_Somayaji&oldid=833340084.

211 Patricia Curd. 2019. "Anaxagoras." Stanford Encyclopedia of Philosophy. Metaphysics Research Lab, Stanford University.

212 John Carl Villanueva (2015) *Structure of the Universe - Universe Today*. Available at: https://www.universetoday.com/37360/structure-of-the-universe/ (Accessed: 20 April 2018).

213 A light year is the distance light can travel in one Julian year {365.25 days} i.e. 9.46 trillion kilometres

214 Bang! : the complete history of the Universe. May, Brian, Moore, Patrick, Lintott, Chris. Carlton Books 3rd ed.2012. (Pg.11)

215 Bang! :the complete history of the Universe.2012. (Pg. 35)

216 Bang! : the complete history of the Universe..2012. (Pg. 92)

217 Cox, B. and Cohen, A. (2016) *Forces of Nature*. First. London: William Collins.

218 Wikipedia Contributors. "Discovery of Cosmic Microwave Background Radiation." Wikipedia, The Free Encyclopedia Last Modified 13th June 2018. Accessed July 18, 2018. https://en.wikipedia.org/w/index.php?title=Discovery_of_cosmic_microwave_background_radiation&oldid=845622255.

219 John Carl Villanueva (2015) *Oscillating Universe Theory - Universe Today*, *Universe Today*. Available at: https://www.universetoday.com/38195/oscillating-universe-theory/ (Accessed: 20 April 2018).

[220] Gerrit Smith, and Robert Weingard. "Quantum Cosmology and the Beginning of the Universe." *Philosophy of Science* 57, no. 4 (1990): 663–67.

[221] "Time and the "Big Bang" – Exactly What Is Time?" Accessed January 20, 2020. http://www.exactlywhatistime.com/physics-of-time/time-and-the-big-bang/.

[222] Wikipedia Contributors. "Flatness Problem." Wikipedia, The Free Encyclopedia Last Modified 25th March 2018. Accessed April 19, 2018. https://en.wikipedia.org/w/index.php?title=Flatness_problem&oldid=832419245.

[223] Wikipedia Contributors. "Inflation (Cosmology." Wikipedia, The Free Encyclopedia Last Modified 12nd April 2018. Accessed April 25, 2018. https://en.wikipedia.org/w/index.php?title=Inflation_(cosmology)&oldid=833856336.

[224] Wikipedia Contributors. "Cyclic Model." Wikipedia, The Free Encyclopedia Last Modified 15th April 2018. Accessed April 19, 2018. https://en.wikipedia.org/w/index.php?title=Cyclic_model&oldid=836523670.

[225] He, D., Gao, D., & Cai, Q. (2014). Spontaneous creation of the universe from nothing. Physical Review D, 89(8), 83510.

[226] Stephen Hawking. 2018. Brief Answers to the Big Questions. London: John Murray.

[227] Ethan Siegel. 2018. "There Was No "Big Bang" Singularity." https://www.forbes.com/sites/startswithabang/2018/07/27/there-was-no-big-bang-singularity/#209cfa6f7d81. [Accessed 16th January 2020]

[228] "Did The Universe Really Begin With a Singularity?" (2014) Accessed January 16, 2020. https://profmattstrassler.com/2014/03/21/did-the-universe-begin-with-a-singularity/.

[229] Bang! : The Complete History of the Universe.2012 Pg. 68

[230] Wikipedia Contributors. "Multiverse." Wikipedia, The Free Encyclopedia, Last Modified 9th September 2016. Accessed January 15, 2020. https://en.wikipedia.org/w/index.php?title=Multiverse&oldid=738594976.

[231] Wikipedia Contributors. n.d. "Albert Einstein." Wikipedia, The Free Encyclopedia Last Modified 19th April 2018. Accessed April 21, 2018. https://en.wikipedia.org/w/index.php?title=Albert_Einstein&oldid=837226582.

[232] Wikipedia Contributors. "Subatomic Particle." Wikipedia, The Free Encyclopedia Last Modified 8th April 2018. Accessed April 21, 2018. https://en.wikipedia.org/w/index.php?title=Subatomic_particle&oldid=835451568.

[233] Wikipedia Contributors. "List of Particles." Wikipedia, The Free Encyclopedia Last Modified 20th April 2018. Accessed April 25, 2018. https://en.wikipedia.org/w/index.php?title=List_of_particles&oldid=837414689.

[234] Wikipedia Contributors."Fundamental Interaction." Wikipedia, The Free Encyclopedia Last Modified 20th April 2018. Accessed April 21, 2018. https://en.wikipedia.org/w/index.php?title=Fundamental_interaction&oldid=837459849.

[235] Wikipedia Contributors."Grand Unified Theory." Wikipedia, The Free Encyclopedia Last Modified 11th April 2018. Accessed April 21, 2018. https://en.wikipedia.org/w/index.php?title=Grand_Unified_Theory&oldid=835984373.

[236] Wikipedia Contributors."Standard Model." Wikipedia, The Free Encyclopedia Last Modified 28th August 2018. Accessed August 29, 2018. https://en.wikipedia.org/w/index.php?title=Standard_Model&oldid=856977203.

[237] Wikipedia Contributors."Planck Time." Wikipedia, The Free Encyclopedia Last Modified 21st August 2018. Accessed August 29, 1BC. https://en.wikipedia.org/w/index.php?title=Planck_time&oldid=855863285.

[238] Wikipedia Contributors. "Matter." Wikipedia, The Free Encyclopedia Last Modified 28th August 2018. Accessed August 29, 2018. https://en.wikipedia.org/w/index.php?title=Matter&oldid=856988602.

[239] Bearing in mind that the speed of light is 186,000 miles per second and one light year is the distance light travels in one year, (if it was traveling in a vacuum) then if the diameter of a cloud is just 1 light year , this represents a distance of 5.88 trillion miles ! Then multiply this by 10s, 100s or even millions……..Mind blowing concepts.

[240] *KuiperBelt* (no date). Available at: http://www.kenpress.com/kuiperbelt.html (Accessed: 31 August 2018).

[241] Wikipedia Contributors. "James Lovelock." Wikipedia, The Free Encyclopedia Last Modified 11st August 2019. Accessed January 16, 2020. https://en.wikipedia.org/w/index.php?title=James_Lovelock&oldid=908830191.

[242] The Physics of the Universe. "The Early Earth and the Building Blocks of Life - The Beginnings of Life." Accessed January 16, 2020. https://www.physicsoftheuniverse.com/topics_life_early.html.

[243] J D Watson, and F H C Crick. 1953. "Molecular Structure of Nucleic Acids: A Structure for Deoxyribose Nucleic Acid." Nature 171 (April): 737. http://dx.doi.org/10.1038/171737a0.

[244] Wikipedia Contributors. "History of Evolutionary Thought." Wikipedia, The Free Encyclopedia Last Modified 27th July 2018. Accessed September 1, 2018. https://en.wikipedia.org/w/index.php?title=History_of_evolutionary_thought&oldid=852180068.

[245] Natural selection is the idea that any traits or mutations (spontaneous changes to genetic material) which are beneficial to an organism will be kept and passed on to subsequent generation. In this way, organisms become better equipped to cope with the environment they inhabit. So less useful characteristics gradually die out.

[246] Wikipedia Contributors. "Y-Chromosomal Adam." Wikipedia, The Free Encyclopedia Last Modified 4th July 2018. Accessed September 5, 2018. https://en.wikipedia.org/w/index.php?title=Y-chromosomal_Adam&oldid=848776559.

[247] Wikipedia Contributors."Mitochondrial Eve." Wikipedia, The Free Encyclopedia Last Modified 24th December 2019. Accessed January 16, 2020. https://en.wikipedia.org/w/index.php?title=Mitochondrial_Eve&oldid=932187087.

[248] Wikipedia Contributors.. "Cell (Biology)." Wikipedia, The Free Encyclopedia Last Modified 25th August 2018. Accessed September 6, 2018. https://en.wikipedia.org/w/index.php?title=Cell_(biology)&oldid=856423861.

[249] Wade N and N, W. (2006) Before the Dawn : Recovering the Lost History of Our Ancestors. Penguin Press HC,.

[250] Harari, Y. N. (2011) Sapiens A Brief History of Humankind. Vintage, Penguin Random House.

[251] Osborne, A. H. et al. (2008) A humid corridor across the Sahara for the migration of early modern humans out of Africa 120,000 years ago. Available at: www.pnas.orgcgidoi10.1073pnas.0804472105 (Accessed: 7 September 2018).

[252] Elton, S. (2008) 'The environmental context of human evolutionary history in Eurasia and Africa.', *Journal of anatomy*. Wiley-Blackwell, 212(4), pp. 377–93. doi: 10.1111/j.1469-7580.2008.00872.x.

[253] Intelligence is not just about brain size. Neanderthals possibly had brains about 10% larger than H. sapiens.The organisation of the internal neural connections and the development / size of certain areas of the brain also seem to be important.

[254] Wikipedia Contributors. "Brain-to-Body Mass Ratio." Wikipedia, The Free Encyclopedia Last Modified 28th August 2018. Accessed September 10, 2018. https://en.wikipedia.org/w/index.php?title=Brain-to-body_mass_ratio&oldid=856987369.

[255] Colom, R. *et al.* (2010) 'Human intelligence and brain networks.', *Dialogues in clinical neuroscience*. Les Laboratoires Servier, 12(4), pp. 489–501. Available at: http://www.ncbi.nlm.nih.gov/pubmed/21319494 (Accessed: 10 September 2018).

[256] Maclean, E. L. (2016) 'Unraveling the evolution of uniquely human cognition'. PNAS, 113(23), pp. 6348–6354. doi: 10.1073/pnas.1521270113.

[257] Sample, I. (2010) *First humans arrived in Britain 250,000 years earlier than thought | Science | The Guardian, The Guardian*. Available at: https://www.theguardian.com/science/2010/jul/07/first-humans-britain-stone-tools (Accessed: 9 September 2018).

[258] Encyclopedia.com. 2020. "Human Evolution." Last Modified 13th January 2020. 2020. https://www.encyclopedia.com/social-sciences-and-law/anthropology-and-archaeology/human-evolution/human-evolution.

[259] Michael Balter. 2005. "The Seeds of Civilization." Smithsonian Magazine. 2005. https://www.smithsonianmag.com/history/the-seeds-of-civilization-78015429/.

[260] Wonderwerk Cave, Northern Cape province, South Africa.', *Proceedings of the National Academy of Sciences of the United States of America*. National Academy of Sciences, 109(20), pp. E1215-20. doi: 10.1073/pnas.1117620109.

[261] Wikipedia Contributors. "Hunter-Gatherer." Wikipedia, The Free Encyclopedia Last Modified 6th September 2018. Accessed September 10, 2018. https://en.wikipedia.org/w/index.php?title=Hunter-gatherer&oldid=858392503.

[262] Weissbrod, L. *et al.* (2017) 'Origins of house mice in ecological niches created by settled hunter-gatherers in the Levant 15,000 y ago', *Proceedings of the National Academy of Sciences*, 114(16), pp. 4099–4104. doi: 10.1073/pnas.1619137114.

[263] Price, M. (2017) 'When did humans settle down? The house mouse may have the answer', *Science*. doi: 10.1126/science.aal0966.

[264] Layton, R., Foley, R., Williams, E. (1991) '"The Transition Between Hunting and Gathering and the Specialized Husbandry of Resources: A Socio-Ecological Approach', *Current Anthropology*, 32(3), pp. 255–274. Available at: https://www.jstor.org/stable/2743774?seq=1#metadata_info_tab_contents (Accessed: 8 September 2018).

[265] Wikipedia Contributors. "Code of Hammurabi." Wikipedia, The Free Encyclopedia Last Modified 4th October 2018. Accessed October 4, 2018. pedia.org/w/index.php?title=Code_of_Hammurabi&oldid=862387146.

[266] Harari, Y. N. (2011) Sapiens A Brief History of Humankind. Vintage, Penguin Random House.

[267] Wikipedia Contributors. "Cradle of Civilization." Wikipedia, The Free Encyclopedia Last Modified 14th September 2018. Accessed September 21, 2018. https://en.wikipedia.org/w/index.php?title=Cradle_of_civilization&oldid=859444892.

[268] Wikipedia Contributors. "Origin of Language." Wikipedia, The Free Encyclopedia Last Modified 8th September 2018. Accessed September 11, 2018. https://en.wikipedia.org/w/index.php?title=Origin_of_language&oldid=858660616.

[269] Gray, R. D., Atkinson, Q. D. and Greenhill, S. J. (2011) 'Language evolution and human history: what a difference a date makes.', *Philosophical transactions of the Royal Society of London. Series B, Biological sciences*. The Royal Society, 366(1567), pp. 1090–100. doi: 10.1098/rstb.2010.0378.

[270] Wade N (2006) Before the Dawn : Recovering the Lost History of Our Ancestors. Penguin Press HC.

[271] Wikipedia Contributors. "Evolutionary Origin of Religions." Wikipedia, The Free Encyclopedia Last Modified 16th January 2019. Accessed February 1, 2019. https://en.wikipedia.org/w/index.php?title=Evolutionary_origin_of_religions&oldid=878658646.

[272] Haselsberger, B. (2014) 'Decoding borders. Appreciating border impacts on space and people', *Planning Theory & Practice*. Routledge, 15(4), pp. 505–526. doi: 10.1080/14649357.2014.963652.

[273] Laine, J. P. (2015) 'A historical view on the studies of borders', *Introduction to border studies*, (September), pp. 15–33.

[274] Giuliano Bellezza. 2013. "On Borders: From Ancient to Postmodern Times." In The International Archives of the Photogrammetry, Remote Sensing and Spatial Information Sciences. ISPRS/IGU/ICA Joint Workshop on Borderlands Modelling and Understanding for Global Sustainability 2013, 5 – 6 December 2013, Beijing, China. Vol. XL-4/W3. https://doi.org/10.5194/isprsarchives-XL-4-W3-1-2013.

L

[275] "GeaCron Project." Accessed January 16, 2020. http://GeaCron.com/home-en/.

[276] BBC History. n.d. "Mesopotamia." Ancient History in Depth Last Modified 21st July 2017. Accessed January 16, 2020. http://www.bbc.co.uk/history/ancient/.

[277] DeMeo, James. 1998. "Origin and Diffusion of Patrism in Saharasia." http://www.orgonelab.org/saharasia_en.htm.

[278] Wikimedia Commons Distributors. "File:Mesopotamian Chronology 2-2011-29-03.Png." Wikimedia Commons Last Modified 16th March 2015. Accessed January 16, 2020. https://commons.wikimedia.org/w/index.php?title=File:Mesopotamian_Chronology_2-2011-29-03.png&oldid=153294985

[279] Frankopan, P. (2015) *The silk roads : a new history of the world.* Bloomsbury Publihing. Available at: https://books.google.co.in/books/about/The_Silk_Roads.html?id=M1FFCQAAQBAJ&printsec=frontcover&source=kp_read_button&redir_esc=y#v=onepage&q&f=false (Accessed: 29 September 2018).

[280] Chi Rho ☧ the first two letters of Christ's name in Greek

[281] Wikipedia Contributors. "List of New Religious Movements." Wikipedia, The Free Encyclopedia Last Modified 26th March 2019. Accessed April 4, 2019. https://en.wikipedia.org/w/index.php?title=List_of_new_religious_movements&oldid=889610977.

[282] David Hall. 1997. Lived Religion In America: Toward A History Of Practice. 1st ed. Princeton University Press.

[283] Robert Orsi. 2002. The Madonna of 115th Street: Faith And Community In Italian Harlem, 1880-1950. Yale University Press.

[284] Long, Charles H. (1987) *Popular Religion | Encyclopedia.com.* Available at: https://www.encyclopedia.com/environment/encyclopedias-almanacs-transcripts-and-maps/popular-religion (Accessed: 4 April 2019).

[285] Park, C. (2004) 'Religion and Geography', in Hinnells, J. (ed.) *Companion to the Study of Religion.* London: Routledge. Available at: https://www.lancaster.ac.uk/staff/gyaccp/geography and religion.pdf (Accessed: 4 April 2019).

[286] Harriet Sherwood (2018) 'The Briefing: Religion: why faith is becoming more and more popular', *The Guardian*, 27th August.

[287] Wikipedia contributors, 'List of religious populations', *Wikipedia, The Free Encyclopedia,* 28 April 2020, 01:38 UTC, <https://en.wikipedia.org/w/index.php?title=List_of_religious_populations&oldid=953595497> [accessed April 7th 2020]

[288] Wikipedia Contributors. "Definition of Religion." Wikipedia, The Free Encyclopedia Last Modified 10th October 2018. Accessed November 16, 2018. https://en.wikipedia.org/w/index.php?title=Definition_of_religion&oldid=863386436.

[289] Wikipedia Contributors. "History of Religion." Wikipedia, The Free Encyclopedia Last Modified 13th November 2018. Accessed November 16, 2018. https://en.wikipedia.org/w/index.php?title=History_of_religion&oldid=868563824

[290] Nongbri, Brent (2013). *Before Religion: A History of a Modern Concept.* Yale University Press. p. 26 ISBN 978-0300154160.

[291] Harrison, Peter (1990). *'Religion' and the Religions in the English Enlightenment.* Cambridge: Cambridge University Press. p. 1. ISBN 978-0521892933

[292] Nongbri, Brent (2013). *Before Religion: A History of a Modern Concept.* Yale University Press. p. 2. ISBN 978-0300154160

[293] Nongbri, Brent (2013). *Before Religion: A History of a Modern Concept.* p. 3.

[294] Wikipedia Contributors. "Renaissance." Wikipedia, The Free Encyclopedia Last Modified 14th March 2019. Accessed March 16, 2019. https://en.wikipedia.org/w/index.php?title=Renaissance&oldid=887729060.

[295] Fitzgerald, Timothy (2007). *Discourse on Civility and Barbarity.* Oxford University Press. pp. 45–46.

[296] Dubuisson, D. (2007) *The western construction of religion : myths, knowledge, and ideology.* Johns Hopkins University Press. Available at: https://jhupbooks.press.jhu.edu/title/western-construction-religion (Accessed: 28 February 2019).

[297] Wikipedia Contributors. "Definition of Religion." Wikipedia, The Free Encyclopedia Last Modified 10th October 2018. Accessed November 16, 2018. https://en.wikipedia.org/w/index.php?title=Definition_of_religion&oldid=863386436.

[298] Smith, W. C. (1991) The meaning and end of religion. Fortress Press. ISBN: 9780800624750

[299] Pramod K. Nayar. 2005. "The Western Construction of Religion." Journal for Cultural and Religious Theory 6 (3): 105–8. http://www.jcrt.org/archives/06.3/nayar.pdf.

[300] Dubuisson, D. (2007) *The western construction of religion : myths, knowledge, and ideology.* (Accessed: 28 February 2019).

[301] Nongbri, B. (2013) *Before religion : a history of a modern concept.* Yale University Press. Available at: https://www.jstor.org/stable/j.ctt32bqx9 (Accessed: 20 February 2019). ISBN: 9780300154177

[302] Wikipedia Contributors. "Religion." Wikipedia, The Free Encyclopedia Last Modified 9th December 2018. Accessed December 12, 2018. https://en.wikipedia.org/w/index.php?title=Religion&oldid=872878645.

[303] John Morreall, and Tamara Sonn. 2013. 50 Great Myths about Religions. Wiley-Blackwell.

[304] Nongbri, B. (2013) *Before religion : a history of a modern concept.* Yale University Press. Available at: https://www.jstor.org/stable/j.ctt32bqx9 (Accessed: 20 February 2019). ISBN: 9780300154177

[305] Nongbri, B. (2013) *Before religion : a history of a modern concept.*Pg. 152 (Accessed: 20 February 2019).

[306] E. E. (Edward Evan) Evans-Pritchard. 1965. Theories of Primitive Religion,. Clarendon Press.

[307] Allen, G. (1897) *The Evolution of the Idea of God - An Inquiry into the Origins of Religions.* Available at: http://www.gutenberg.org/files/57581/57581-0.txt (Accessed: 10 November 2018)

[308] Schmidt, W. and Rose, H. J. (2014) The origin and growth of religion : facts and theories. Wythe-North.

[309] Wikipedia Contributors. "Anthropology of Religion." Wikipedia, The Free Encyclopedia Last Modified 7th August 2018. Accessed December 12, 2018. https://en.wikipedia.org/w/index.php?title=Anthropology_of_religion&oldid=853865829.

[310] Religion - The belief in and worship of a superhuman controlling power, especially a personal God or gods.(section 5) Oxford English Dictionary. Oxford England. Oxford University Press 2020.

[311] Weitzman, S. (2017) *The Origin of the Jews: The Quest for Roots in a Rootless Age* Princeton University Press; Available at: http://assets.press.princeton.edu/chapters/i10969.pdf (Accessed: 20 February 2019). ISBN: 9781400884933.

[312] Pals, D. L. (1996). *Seven theories of religion.* Oxford University Press. Retrieved from https://books.google.com.pk/books/about/Seven_Theories_of_Religion.html?id=62g76d_cPDgC&redir_esc=y

[313] Schmidt, W. and Rose, H. J. (2014) Pg. 8. The origin and growth of religion : facts and theories. Wythe-North.

[314] Wikipedia Contributors. "Evolutionary Origin of Religions." Wikipedia, The Free Encyclopedia Last Modified 16th January 2019. Accessed February 1, 2019. https://en.wikipedia.org/w/index.php?title=Evolutionary_origin_of_religions&oldid=878658646. 318

[315] Peoples, H. C., Duda, P., & Marlowe, F. W. (2016). Hunter-Gatherers and the Origins of Religion. *Human Nature, 27*(3), 261–282. https://doi.org/10.1007/s12110-016-9260-0

[316] Violatti, Cristian. "The Meaning of European Upper Paleolithic Rock Art."*Ancient History Encyclopedia.* Ancient History Encyclopedia, 10 Feb 2015. Web. 13 Feb 2019.

[317] Peoples, H. C., Duda, P., & Marlowe, F. W. (2016). Hunter-Gatherers and the Origins of Religion. *Human Nature, 27*(3), 261–282. https://doi.org/10.1007/s12110-016-9260-0

[318] Wikipedia Contributors. "Phylogenetics." Wikipedia, The Free Encyclopedia Last Modified 26th February 2019. Accessed March 21, 2019. https://en.wikipedia.org/w/index.php?title=Phylogenetics&oldid=885150589.

[319] Donald E. (Donald Edward) Brown. 1991. Human Universals. New York City: McGraw-Hill Education.

[320] Fitzgerald, T. (2007) *Discourse on civility and barbarity : a critical history of religion and related categories*. Oxford University Press.

[321] Henning, C. L. (1898) *On the Origin of Religion, Source: American Anthropologist*. Available at: https://www.jstor.org/stable/pdf/658576.pdf?refreqid=excelsior%3Ac614d83bdf9eddc6d54775400d2ab1 3f(Accessed: 6 February 2019).

[322] Wikipedia Contributors. "Evolutionary Origin of Religions." Wikipedia, The Free Encyclopedia Last Modified 16th January 2019. Accessed February 1, 2019. https://en.wikipedia.org/w/index.php?title=Evolutionary_origin_of_religions&oldid=878658646.

[323] Hamer, D. H. (2005) The God gene : how faith is hardwired into our genes. Anchor Books. ISBN: 0385720319

[324] Dawkins, R. (1989) *The selfish gene*. Oxford University Press.ISBN 0192860925

[325] Dennett, D. C. (1998) 'The Evolution Of Religious Memes: Who—or What—Benefits ? on JSTOR', *Method & Theory in the Study of Religion*,10(1), pp.115–128. Available at: https://www.jstor.org/stable/23555104?read-now=1&seq=1#metadata_info_tab_contents (Access: 24 February 2019).

[326] Henning, C. L. (1898). *On the Origin of Religion. Source: American Anthropologist* (Vol. 11). Retrieved from https://www.jstor.org/stable/pdf/658576.pdf?refreqid=excelsior%3Ac614d83bdf9eddc6d54775400d2ab13f.

[327] Howerth, I. W. "Brintons's Theory of the Origen of Religion." *The Monist*, vol. 10, no. 2, 1900, pp. 293–298. *JSTOR*, www.jstor.org/stable/27899127.

[328] Fitzgerald, T. (2007). *Discourse on civility and barbarity : a critical history of religion and related categories*. Oxford University Press.

[329] Wikipedia Contributors. "Theories about Religions." Wikipedia, The Free Encyclopedia Last Modified 8th November 2018. Accessed February 6, 2019. https://en.wikipedia.org/w/index.php?title=Theories_about_religions&oldid=867844443.

[330] Kunin, S. D. (2003). *Religion The modern Theories*. Edinburgh: Edinburgh University Press. Retrieved from https://www.bookdepository.com/search/Religion-Seth-Daniel-Kunin/9780748615223

[331] Pals, D. L. (1996). *Seven theories of religion*. Oxford University Press. Retrieved from https://books.google.com.pk/books/about/Seven_Theories_of_Religion.html?id=62g76d_cPDgC&redir_esc=y.

332 Wikipedia Contributors. "E. E. Evans-Pritchard." Wikipedia, The Free Encyclopedia Last Modified 126th September 2018. Accessed March 22, 2019. https://en.wikipedia.org/w/index.php?title=E._E._Evans-Pritchard&oldid=861366863.

333 Schmidt, W. and Rose, H. J. (2014) Pg.6 The origin and growth of religion : facts and theories. Wythe-North.

334 **Yuval Noah Harari. 2017. Homo Deus - A Brief History of Tomorrow. London: London: Vintage. https://readandlaugh.files.wordpress.com/2018/04/**

335 The 1922 edition of *The Golden Bough* as downloadable and searchable.pdfs. (https://wordandsilence.com/2017/01/21/classic-jam-hits/)

336 Herbert Spencer. 1898. The Principles of Sociology. In Three Volumes. Vol. 1 Pg. 422. New York: D. Appleton and Company,.https://oll.libertyfund.org/titles/2634

337 Herbert Spencer. 1898. The Principles of Sociology. In Three Volumes. Vol. 3 Pg. 7. New York: D. Appleton and Company. https://oll.libertyfund.org/titles/2634. https://oll.libertyfund.org/titles/2634

338 Henning, C. L. (1898). *On the Origin of Religion. Source: American Anthropologist* (Vol. 11). Retrieved from https://www.jstor.org/stable/pdf/658576.pdf?refreqid=excelsior%3Ac614d83bdf9eddc6d54775400d2ab13f.

339 Schmidt, W. and Rose, H. J. (2014) The origin and growth of religion : facts and theories. Wythe-North.

340 Andrew M Mckinnon. 2005. "Opium as Dialectics of Religion: Metaphor, Expression and Protest." Critical Sociology 31 (1–2): 15–38.

341 Stark, R. and Sims Bainbridge, W. (1987) *A Theory of Religion*. New York: Peter Lang. Available at: https://www.questia.com/library/63947769/a-theory-of-religion (Accessed: 5 April 2019).

342 Wikipedia Contributors. "Sociology of Religion/Religious Typologies'." Wikipedia, The Free Encyclopedia Last Modified 2nd January 2018. Accessed April 5, 2019. https://en.wikibooks.org/w/index.php?title=Sociology_of_Religion/Religious_Typologies&oldid=3356535.

343 Bruce, S. (1993) Religion and Rational Choice: A Critique of Economic Explanations of Religious Behavior*, Sociology of Religion. http://socrel.oxfordjournals.org/ (Accessed: 22 March 2019).

344 Singh, D. and Chatterjee, G. (2016) 'The evolution of religious belief in humans: a brief review with a focus on cognition', *Journal of Genetics*, 96(3), pp. 517–524. doi: 10.1007/s12041-017-0794-7.

345 Stephen Jay Gould; Richard Lewontin, J. Maynard Smith, and Robin Holliday. 1979. "The Spandrels of San Marco and the Panglossian Paradigm: A Critique of the Adaptationist Programme." Proceedings of the Royal Society of London. Series B. Biological Sciences 205 (1161): 581–198. https://doi.org/10.1098/rspb.1979.0086.

[346] Singh, D. and Chatterjee, G. (2016) 'The evolution of religious belief in humans: a brief review with a focus on cognition', *Journal of Genetics*, 96(3), pp. 517–524. doi: 10.1007/s12041-017-0794-7.

[347] Rue, L., 2004. Religion is not about God: how spiritual traditions nurture our biological nature and what to expect when they fail. Rutgers University Press, New Brunswick, NJ and London.ISBN 10: 0813535115 ISBN 13: 9780813535111

[348] Seiwert Hubert (2009): Theory of religion as myth - On Loyal Rue (2005), Religion is not about God. Pg. (224-241) in : Stausberg, M. (ed.) (2009) 'Contemporary Theories of Religion A critical companion', in. New York: Routledge. Available at: https://www.gko.uni-leipzig.de/fileadmin/user_upload/religionswissenschaft/Pdf/Publikationen_Seiwert/Seiwert__-__ (Accessed: 27 February 2019).

[349] R. Dawkins. 2006. The God Delusion. Boston: Houghton Mifflin. ISBN: 0-618-68000-4.

[350] Brinton, D. (1898). Religions of Primitive Peoples. Philosophical Review 7:108. 10.2307/2175561

[351] Howerth, I. W. "Brintons's Theory of the Origen of Religion." *The Monist*, vol. 10, no. 2, 1900, pp. 293–298. *JSTOR*, www.jstor.org/stable/27899127.

[352] Wikipedia Contributors. "Cognitive Science of Religion." Wikipedia, The Free Encyclopedia Last Modified 7th June 2018. Accessed April 9, 2019. https://en.wikipedia.org/w/index.php?title=Cognitive_science_of_relCognitive science of religionigion&oldid=844865028.

[353] Lari Launonen. 2018. "The Naturalness of Religion: What It Means and Why It Matters." Neue Zeitschrift Für Systematische Theologie Und Religionsphilosophie. 60 (1): 84–102. https://doi.org/0.1515/nzsth-2018-0005..

[354] Kurt; Gray, and Daniel Wegner. 2010. "Blaming God for Our Pain: Human Suffering and the Divine Mind." Personality and Social Psychology Review. 14 (1): 7–16. https://doi.org/10.1177/1088868309350299

[355] James H Leuba. 1909. "The Psychological Origin of Religion." The Monist 19 (1): 27–35. https://www.jstor.org/stable/pdf/27900158.pdf.

[356] S. Guthrie. 2002. "Animal Animism: Evolutionary Roots of Religious Cognition." In Current Approaches in the Cognitive Science of Religion, edited by Ilkka Pyysiainen and Veikko Anttonen, 1st ed., 38–67. https://scholar.google.com/citations?user=CHQ3wdgAAAAJ&hl=en#d=gs_md_cita-d&u=%2Fcitations%3Fview_op%3Dview_citation%26hl%3Den%26user%3DCHQ3wdgAAAAJ%26citation_for_view%3DCHQ3wdgAAAAJ%3ARHpTSmoSYBkC%26tzom%3D0.

[357] Pals, D. L. (1996). *Seven theories of religion*. Oxford University Press. Retrieved from https://books.google.com.pk/books/about/Seven_Theories_of_Religion.html?id=62g76d_cPDgC&redir_esc=y

358 Wikipedia Contributors."List of Ethnic Religions." Wikipedia, The Free Encyclopedia Last Modified 4th June 2018. Accessed February 23, 2019. https://en.wikipedia.org/w/index.php?title=List_of_ethnic_religions&oldid=844328818.

359 Wikipedia contributors, 'Mythologies of the indigenous peoples of the Americas', *Wikipedia, The Free Encyclopedia,* 19 February 2019, 14:53 UTC, <https://en.wikipedia.org/w/index.php?title=Mythologies_of_the_indigenous_peoples_of_the_Americas&oldid=884099986> [Accessed 23 February 2019]

360 Wikipedia Contributors. "Native American Religion." Wikipedia, The Free Encyclopedia Last Modified 19th February 2019. Accessed February 23, 2019. https://en.wikipedia.org/w/index.php?title=Native_American_religion&oldid=884088537.

361 T. W. (Thomas William) Doane. 1882. "Bible Myths and Their Parallels in Other Religions : Being a Comparison of the Old and New Testament Myths and Miracles with Those of Heathen Nations of Antiquity, Considering Also Their Origin and Meaning : New York : Commonwealth Co Trinity College - University of Toronto. 1882. https://archive.org/details/biblemythsandthe00doanuoft.

362 Kersey. Graves, 1999 The Worlds Sixteen Crucified Saviours (4th Revise and End edition) San Diego.Book Tree. ISBN:0: 1585090182

363 Michael Dowd, 2009. Thank God for Evolution: How the Marriage of Science and Religion Will Transform Your Life and Our World Penguin Random House. (New York) ISBN: 9780452295346

364 Christopher Hitchens, God Is Not Great: How Religion Poisons Everything (London: Atlantic Books, 2007)

365 De Robigne Mortimer Bennett, The Champions of the Church: Their Crimes and Persecutions. Published by D. M. Bennett. Liberal and Scientific Publishing House (New York Public Library) <https://books.google.co.uk/books/about/The_Champions_of_the_Church.html?id=YsEPAAAAIAAJ&redir_esc=y>

366 Huggins, R. V (2013 October) *Krishna as Virgin Born Crucified Savior: A Product of Western Ignorance of Basic Hindu Beliefs* Available at: http://ronaldvhuggins.blogspot.com/2013/10/krishna-as-virgin-born-crucified-savior.html?view=magazine (Accessed: 25 April 2019).

367 Huggins, R. V (2015 September) *The Buddha: Virgin Born? Dying and Rising God?* Available at: http://ronaldvhuggins.blogspot.com/2015/09/the-buddha-virgin-born-dying-and-rising.html?view=magazine (Accessed: 25 April 2019)

368 Huggins, R. V (2015 September) *Was Prometheus a Crucified Savior on the same pattern as many other Gods including Jesus ?* Available at: http://ronaldvhuggins.blogspot.com/2015/09/was-prometheus-crucified-savior-on-same.html?view=magazine (Accessed: 17th January 2020).

369 Butt, Kyle: Thompson, (2001) *Jesus Christ—Unique Savior or Average Fraud?,* *[Part I], Truth According to Scripture.* Available at: https://www.truthaccordingtoscripture.com/documents/apologetics/jesus-unique-savior-pt1.php#.XMGPwy-ZNsM (Accessed: 25 April 2019).

[370] Butt, Kyle: Thompson, B. (2001) *Jesus Christ—Unique Savior or Average Fraud? [Part II]*, *Truth According to Scripture*. Available at: https://www.truthaccordingtoscripture.com/documents/apologetics/jesus-unique-savior-pt2.php#.XMGJXy-ZNsM (Accessed: 25 April 2019).

[371] Ned, (n.d.)'Response to the Book by Doane' <https://www.answering-islam.org/Pagan/doane.html> [Accessed 17 January 2019]

[372] Huggins, R. V (2015 September) *Was Prometheus a Crucified Savior on the same pattern as many other Gods including Jesus ?* Available at: http://ronaldvhuggins.blogspot.com/2015/09/was-prometheus-crucified-savior-on-same.html?view=magazine. (Accessed: 17th January 2020).

[373] Huggins, R. (2015) *Indra as a Virgin Born, Crucified Savior? You're Kidding Right?* Available at:http://ronaldvhuggins.blogspot.com/2015/09/indra-as-virgin-born-crucified-savior.html?view=magazine (Accessed: 27 April 2019).

[374] Wikipedia Contributors."Gilgamesh Flood Myth." Wikipedia, The Free Encyclopedia Last Modified 25th April 2019. Accessed April 26, 2019. https://en.wikipedia.org/w/index.php?title=Gilgamesh_flood_myth&oldid=894134762.

[375] Wikipedia Contributors. "Moses." Wikipedia, The Free Encyclopedia Last Modified 106th April 2019. Accessed April 16, 2019. https://en.wikipedia.org/w/index.php?title=Moses&oldid=892693916.

[376] http://www.godchecker.com/pantheon/chinese-mythology.php?deity=FO-HI

[377] Wikipedia Contributors."Miraculous Births." Wikipedia, The Free Encyclopedia Last Modified 17th February 2019. Accessed April 12, 2019. https://en.wikipedia.org/w/index.php?title=Miraculous_births&oldid=883791876.

[378] Jeter, D. G. (2009) *Mind Over Matter: The Heresy of Gnosticism both Then and Now*, *https://www.insight.org*. Available at: https://www.insight.org/resources/article-library/individual/mind-over-matter-the-heresy-of-gnosticism-both-then-and-now (Accessed: 27 April 2019).

[379] Wikipedia Contributors. "Salvation in Christianity." Wikipedia, The Free Encyclopedia Last Modified 17th April 2019. Accessed April 19, 2019. https://en.wikipedia.org/w/index.php?title=Salvation_in_Christianity&oldid=892811100.

[380] Wikipedia Contributors. "Church Fathers." Wikipedia, The Free Encyclopedia Last Modified 18th April 2019. Accessed April 19, 2019. https://en.wikipedia.org/w/index.php?title=Church_Fathers&oldid=893048723.

[381] Michael J. Murray, and Michael Rea, '"Philosophy and Christian Theology" (Atonement)', Stanford Encyclopedia of Philosophy, Edward N. Zalta (Ed.) Metaphysics Research Lab, Stanford University, 1997. https://plato.stanford.edu/entries/christiantheology-philosophy/#Ato (Accessed: 18th January 2020).

[382] K F Dougherty, 'Atonement and Scripture' (New York: St. Pius X Seminary Garrison, 2012) Accessed 18th January 2020

[383] Lundy, J. P. (1876) Monumental Christianity; or, The art and symbolism of the primitive church as witnesses and teachers of the one Catholic faith and practice,. New York,. Available at: http://hdl.handle.net/2027/hvd.hn5kjf (Accessed: 23 April 2019).

[384] Huggins, R. (2015) *Indra as a Virgin Born, Crucified Savior? You're Kidding Right?* Available at:http://ronaldvhuggins.blogspot.com/2015/09/indra-as-virgin-born-crucified-savior.html?view=magazine (Accessed: 27 April 2019).

[385] D. Yoder, 'Toward a Definition of Folk Religion', Western Folklore, 33 (1974), 2–15. https://www.jstor.org/stable/1498248 387

[386] Wikipedia Contributors, 'History of Hinduism', Wikipedia, The Free Encyclopedia Last Modified 13th April 2019 <https://en.wikipedia.org/w/index.php?title=History_of_Hinduism&oldid=892236618> [Accessed 26 April 2019]

[387] Wikipedia Contributors, 'Bengali Renaissance', Wikipedia, The Free Encyclopedia Last Modified 22nd April 2019 <https://en.wikipedia.org/w/index.php?title=Bengali_Renaissance&oldid=893685372> [Accessed 30 April 2019]

[388] Wikipedia Contributors, 'Vedas', Wikipedia, The Free Encyclopedia Last Modified 29th March 2019 <https://en.wikipedia.org/w/index.php?title=Vedas&oldid=890019870> [Accessed 29 April 2019]

[389] Flood, G. (2009) *BBC - Religions - History of Hinduism.* Available at: https://www.bbc.co.uk/religion/religions/hinduism/history/history_1.shtml (Accessed: 28 April 2019).

[390] *Main Beliefs and Practices of Hinduism* (no date). https://www.hinduwebsite.com/hinduism/h_beliefs.asp (Accessed: 18th January 2020).

[391] Weitzman, S. (2017) *The Origin of the Jews: The Quest for Roots in a Rootless Age - introduction.* Princeton University Press; Available at: http://assets.press.princeton.edu/chapters/i10969.pdf (Accessed: 20 February 2019).

[392] Wikipedia contributors, 'Judaism', *Wikipedia, The Free Encyclopedia,* 28 April 2019, 22:28 UTC, <https://en.wikipedia.org/w/index.php?title=Judaism&oldid=894613716> [Accessed 2 May 2019]

[393] Wikipedia Contributors, 'Judaism', Wikipedia, The Free Encyclopedia Last Modified 128th April 2019 <https://en.wikipedia.org/w/index.php?title=Judaism&oldid=894613716> [Accessed 2 May 2019].

[394] Wikipedia Contributors, 'Israelites', Wikipedia, The Free Encyclopedia Last Modified 3rd March 2019 <https://en.wikipedia.org/w/index.php?title=Israelites&oldid=886027599> [Accessed 2 May 2019]

[395] Melvin Ember, Carol R. Ember, and Ian Skoggard, Encyclopedia of Diasporas: Immigrant and Refugee Cultures Around the World. Volume I: Overviews and Topics; Volume II: Diaspora Communities. (Springer US, 2004)

[396] Jacob Neusner, '"Pharisaic-Rabbinic" Judaism: A Clarification', History of Religions, 12 (1973), 250–70. www.jstor.org/stable/1062026.

[397] Hector M. Patmore, Adam, Satan, and the King of Tyre: The Interpretation of Ezekiel 28:11-19 in Late Antiquity (Jewish and Christian Perspectives), Chapter Six The Hebrew Text (Leiden, Boston, Paderborn and Singapore: Brill, 2012)

[398] Wikipedia Contributors, 'Jewish Principles of Faith', Wikipedia, The Free Encyclopedia Last Modified 10th May 2019 <https://en.wikipedia.org/w/index.php?title=Jewish_principles_of_faith&oldid=896501109> [Accessed 11 May 2019]

[399] Chadwick, H. (1967) *The Early Church.* Penguin Books. ISBN:014-01-3753-X

[400] 'Tertullian: The Apology, translated by Wm' Reeve (1889). Griffith, Farran, Okeden & Welsh Newberry House, London and Sydney

[401] Chadwick, H. (1967) *The Early Church.* Penguin Books. ISBN:014-01-3753-X

[402] Wikipedia Contributors, 'Religion in Africa', Encyclopedia Britannica 2003 ISBN: 9780852299562, p. 306 <https://en.wikipedia.org/w/index.php?title=Religion_in_Africa&oldid=935725189> [Accessed 18 January 2020]

[403] Frankopan, P. (2015) *The silk roads : a new history of the world.* Bloomsbury Publishing. Available at: https://books.google.co.in/books/about/The_Silk_Roads.html?id=M1FFCQAAQBAJ&printsec=frontcover&source=kp_read_button&redir_esc=y#v=onepage&q&f=false (Accessed: 29 September 2018).

[404] John Roxborogh (1995) 'Contextualisation and re-contextualisation: Regional patterns in the history of Southeast Asian Christianity', *Published in Asia Journal of Theology 9(1)*, 9(1), (Pg. 30–46.)

[405] Chadwick, H. (1967) *The Early Church.* Penguin Books.

[406] The Nag Hammadi library (also known as the "Chenoboskion Manuscripts" and the "Gnostic Gospels") is a collection of early Christian and Gnostic texts discovered near the Upper Egyptian town of Nag Hammadi in 1945. Thirteen leather-bound papyrus codices buried in a sealed jar were found by a local farmer named Muhammed al-Samman. The writings in these codices comprise 52 mostly Gnostic treatises, but they also include three works belonging to the *Corpus Hermeticum* and a partial translation/alteration of Plato's *Republic*. The discovery of these texts significantly influenced modern scholarship's pursuit and knowledge of early Christianity and Gnosticism.The contents of the codices were written in the Coptic language. The best-known of these works is probably the Gospel of Thomas, of which the Nag Hammadi codices contain the only complete text. The buried manuscripts date from the 3rd and 4th centuries

[407] Wikipedia Contributors, 'Nag Hammadi Library', Wikipedia, The Free Encyclopedia Last Modified 131st May 2019 <https://en.wikipedia.org/w/index.php?title=Nag_Hammadi_library&oldid=899582104> [Accessed 12 July 2019]

[408] Michael J. Kruger (2014) *Did the Earliest Christians Really Think Jesus Was God? One Important Example – Canon Fodder, Canon fodder.* Available at: https://www.michaeljkruger.com/did-the-earliest-christians-really-think-jesus-was-god-one-important-example/ (Accessed: 22 May 2019).

409 Joseph T Lienhard, (1987) 'The "Arian" Controversy : Some Categories Reconsidered', Theological Studies, 48 (3), 415 <http://cdn.theologicalstudies.net/48/48.3/48.3.1.pdf> [Accessed 22 May 2019]

410 Wikipedia Contributors, 'Arian Controversy', Wikipedia, The Free Encyclopedia Last Modified 15th May 2019 <https://en.wikipedia.org/w/index.php?title=Arian_controversy&oldid=897201265> [Accessed 22 May 2019]

411 Edward Gibbon, The Decline and Fall of the Roman Empire, Volumes 1 to 6, (First Published 1776-1789) (Penguin Random House, 2010) ISBN:9780307700766

412 Wikipedia Contributors, 'Christology', Wikipedia, The Free Encyclopedia Last Modified 9th May 019 <https://en.wikipedia.org/w/index.php?title=Christology&oldid=896277361> [Accessed 18 January 2020]

413 Wikipedia Contributors, 'Christology', Wikipedia, The Free Encyclopedia Last Modified 9th May 019 <https://en.wikipedia.org/w/index.php?title=Christology&oldid=896277361> [Accessed 18 January 2020]

414 John Binns, An Introduction to the Christian Orthodox Churches, Cambridge University Press, UK, 2002, p. 144.

415 Fitzgerald, T. (2007) *Discourse on civility and barbarity : a critical history of religion and related categories*. Oxford University Press.

416 Wikipedia Contributors, 'Dead Sea Scrolls', Wikipedia, The Free Encyclopedia Last Modified 2nd July 2019 <s://en.wikipedia.org/w/index.php?title=Dead_Sea_Scrolls&oldid=904475265> [Accessed 13 July 2019]

417 Geza Vermes, The Dead Sea Scrolls, 3rd edn (London: Penguin, 1990)

418 Baignent, and R. Leigh, The Dead Sea Scrolls Deception, 2nd edn (Corgi, 1992)

419 Armstrong, K. (2004) *Islam A Short History, Phoenix Paperback*. Phoenix Press,The Orion Publishing Group Ltd.

420 Gordon, M. S. (2002) *Understanding Islam*. Duncan Baird Publishers.

[421] Sunnah is the body of literature which discusses and prescribes the traditional customs and practices of the Islamic community, both social and legal - In pre-Islamic Arabia, the term *sunnah* referred to precedents established by tribal ancestors, accepted as normative and practiced by the entire community. The early Muslims did not immediately agree on what constituted their Sunnah. Some looked to the people of Medina for an example, and others followed the behaviour of the companions of the Prophet Muhammad, whereas the provincial legal schools, current in Iraq, Syria, and the Hejaz (in Arabia) in the eighth century CE, attempted to equate Sunnah with an ideal system—based partly on what was traditional in their respective areas and partly on precedents that they themselves had developed. These varying sources, which created differing community practices, were finally reconciled late in the eighth century by the legal scholar Abū ʿAbd Allāh al-Shāfiʿī (767–820), who accorded the Sunnah of the Prophet Muhammad—as preserved in eyewitness records of his words, actions, and approbations (the Hadith)—normative and legal status second only to that of the Qurʾān. - Afsaruddin, Asma. "Sunnah". *Encyclopedia Britannica*. https://www.britannica.com/topic/Sunnah Retrieved 21 April 2018.

[422] Prof. Shahul Hameed (no date) *Islam Online Archive*. https://archive.islamonline.net/?p=5322 (Accessed: 18th January 2020).

[423] Gordon, M. S. (2002) *Understanding Islam*. Duncan Baird Publishers.

[424] Armstrong, K. (2004) *Islam A Short History, Phoenix Paperback*. Phoenix Press,The Orion Publishing Group Ltd.

[425] Wikipaedia Contributors, 'Faith', Wikipedia, The Free Encyclopedia Last Modified 3rd June 2019 <https://en.wikipedia.org/w/index.php?title=Faith&oldid=900159350> [Accessed 12 June 2019]

[426] Paul Tillich, Dynamics of Faith, First Published 1957 (New York: Harperone, 2009)

[427] Lennox, John (2011). Gunning for God: Why the New Atheists Are Missing the Target. United kingdom: Lion. p. 55. ISBN 0-7459-5322-0.

[428] Alister E. McGrath. Inventing the Universe: Why We Can't Stop Talking about Science, Faith and God. London: Hodder & Stoughton, 2016. ISBN:10: 1444798480

[429] Colin Wells. "How Did God Get Started?" BU College of Arts & Sciences, Arion, A Journal of Humanities and the Classics. Accessed January 18, 2020. http://www.bu.edu/arion/files/2009/04/Colin-Wells-how-did-God-get-started.pdf.

[430] Colin Wells. "How Did God Get Started?" BU College of Arts & Sciences, Arion, A Journal of Humanities and the Classics. Accessed January 18, 2020. http://www.bu.edu/arion/files/2009/04/Colin-Wells-how-did-God-get-started.pdf.

[431] Wikipedia Contributors. "Foundationalism." Wikipedia, The Free Encyclopedia Last modified 27th March 2019. Accessed June 18, 2019. https://en.wikipedia.org/w/index.php?title=Foundationalism&oldid=889753509.

[432] Plantinga, Alvin (1993). *Warrant: The Current Debate*. 1. Oxford: Oxford University Press. ISBN 9780195078619.

433 Zwemer, S. M. (1945) *Origin of Religion - Evolution or Revelation.* Third and Revised. New York: Loizeaux Brothers. Available at: http://www.zwemercenter.com/wp-content/uploads/2017/10/Zwemer-Origin-of-Religion.pdf (Accessed: 28 February 2019).

434 The Error of God: Error Management Theory, Religion, and the Evolution of Cooperatio Dominic P. Johnson (Pg. 169) in Levin, S. A. (2009) *Games, Groups, and the Global Good.* 1st edn. Edited by S. A. Levin. Heidelberg: Springer-Verlag Berlin Heidelberg.

435 Wikipedia contributors, 'Pascal's Wager', *Wikipedia, The Free Encyclopedia,* 10 February 2019, 06:50 UTC, <https://en.wikipedia.org/w/index.php?title=Pascal%27s_Wager&oldid=882604417> [Accessed 28 February 2019]

436 Paul Tillich, Dynamics of Faith, First Published 1957 (New York: Harperone, 2009)

437 Hebrews, Chapter 11 Verse 1, King James Version of The Bible.

438 William Wainwright. "Concepts of God." The Stanford Encyclopedia of Philosophy, Edward N. Zalta (Ed.) Metaphysics Research Lab, Stanford University. Spring 2017. hthttps://plato.stanford.edu/archives/spr2017/entries/concepts-god/ Accessed 18th July 2020

439 Albert Schweitzer. 2005. The Quest of the Historical Jesus. A Republication of the. First Published Edition in 1911 by Dover Publications Inc. Mineola, New York:

440 Pals, D. L. (1996) *Seven theories of religion.* Oxford University Press. Available at: https://books.google.com.pk/books/about/Seven_Theories_of_Religion.html?id=62g76d_cPDgC&redir_esc=y (Accessed: 16 December 2018).

441 Herbert Spencer. 1898. The Principles of Sociology. In Three Volumes. Vol. 1 Pg. 422. New York: D. Appleton and Company,.https://oll.libertyfund.org/titles/2634

442 Allen, G. (1897) *The Evolution of the Idea of God - An Inquiry into the Origins of Religions.* Available at: http://www.gutenberg.org/files/57581/57581-0.txt (Accessed: 10 November 2018).

443 Wake, C. Staniland. 'Reviewed Work: The Evolution of the Idea of God: An Inquiry Into the Origins of Religion by Grant Allen.'*The Monist* 8, no. 4 (1898): 627-30. Accessed January 18, 2020. www.jstor.org/stable/27897533.

444 G A Coe. 1898. "'Review: The Evolution of Tf/e Idea of God: An Inquiry into the Origins of Religion. By Grant Allen',." The Philosophical Review 7 (2): 210–13. https://doi.org/10.2307/2175769.

445 Wikipedia Contributors. "Euhemerus." Wikipedia, The Free Encyclopedia Last Modified 10th April 2019. Accessed June 25, 2019. https://en.wikipedia.org/w/index.php?title=Euhemerus&oldid=891897919.

446 Henning, C. L. (1898) *On the Origin of Religion, Source: American Anthropologist.* Available at: https://www.jstor.org/stable/pdf/658576.pdf?

[447] Steven Taylor. n.d. "The Origins of God." Accessed January 18, 2020. https://www.stevenmtaylor.com/essays/

[448] James DeMeo. 2011. Saharasia. The 4000 BCE Origins of Child Abuse, Sex-Repression, Warfare and Social Violence in the Deserts of the Old World. Oregon: Natural Energy Works. ISBN:13: 978-0980231649

[449] Thomas Römer. 2015. The Invention of God. Boston: Harvard University Press. https://doi.org/9780674504974.

[450] Dominic D.P. Johnson. 2009. "The Error of God: Error Management Theory, Religion, and the Evolution of Cooperation." In Games, Groups, and the Global Good - Springer Series in Game Theory, S.A. Levin,(Ed) Pg. 169–80. Heidelberg London New York: Springer Dordrecht. https://doi.org/10.1007/978-3-540-85436-4_10.

[451] Jill G. de Villiers, and Peter A. de Villiers. 2014. "The Role of Language in Theory of Mind Development." Top Lang Disorders 34 (4): 313–28. https://alliedhealth.ceconnection.com/files/TheRoleofLanguageinTheoryofMindDevelopment-1415277302473.pdf.

[452] McGiffert, Arthur Cushman. "Modern Ideas of God." *The Harvard Theological Review*, vol. 1, no. 1, 1908, pp. 10–27. *JSTOR*, www.jstor.org/stable/1507529

[453] Petrescu, A. (2014) 'The Idea of God in Kantian Philosophy', *Procedia - Social and Behavioral Sciences*, 163, pp. 199–203. doi: 10.1016/j.sbspro.2014.12.307.

[454] Cockerill, R. C. (1905) 'Definition of God', *The Monist*. Hegeler Institute, 15(4), pp. 637–638. doi: 10.2307/27899628

[455] Paul Carus. "Paul Carus." Wikipedia, The Free Encyclopedia Last Modified 2nd November 2019. Accessed January 18, 2020. https://en.wikipedia.org/w/index.php?title=Paul_Carus&oldid=924273489.

[456] Carus, P. (2007) *God: An Enquiry Into the Nature of Man's Highest Ideal and a Solution of the Problem from the Standpoint of Science*. Originally. New York: Cosimo Classics. books.google.co.uk/books?id=CFY4A6dMTWIC&pg=PA84&lpg=PA84&dq=tled+God:+An+Inquiry+into+the+Nature+of+Man%27s+Highest+Ideal+and+a+Solution+of+the+Problem+from+the+Standpoint+of+Science&source=bl&ots=3gJhlwk4NK&sig=Y1_wOii6Q7WxjDTkypmi0zw93Dc&hl=en&. ISBN: 9781602063907

[457] Carus, P. The God of Science: In Reply to Rev. Henry Collin Minton. The Monist, Vol. 14, No. 3 (April, 1904), pp. 458-469 Published by: Oxford University Press https://www.jstor.org/stable/27899495 Accessed: 18-01-2020

[458] Wikipedia Contributors. "Apologetics." Wikipedia, The Free Encyclopedia Last Modified 27th August 2019. https://en.wikipedia.org/w/index.php?title=Apologetics&oldid=912681746. Accessed 29th August 2019

[459] Douglas Groothuis (2011) *Christian Apologetics - A Comprehensive Case for Biblical Faith. Inter Varsity Press, Illinois. ISBN* 978-0-8308-6901-5 (digital) ISBN 978-0-8308-3935-3 (print)

460 Pittenger, W.N., 1952. Christian Apologetics. Theology 55, 282–286. doi: 10.1177/0040571x5205538602

461 http://www.existence-of-god.com/index.html

462 Wikipedia Contributors. "God of the Gaps." Wikipedia, The Free Encyclopedia Last Modified 27th April 2019. https://en.wikipedia.org/w/index.php?title=God_of_the_gaps&oldid=894424127. 23rd June 2019

463 https://strangenotions.com/god-exists/#1

464 This statement may seem a big assumption, but all the conventionally recognised belief systems reference some obligations of care and respect towards others.

465 Wikipedia Contributors. n.d. "Null Hypothesis." Wikipedia, The Free Encyclopedia Last Modified 16th August 2018. https://en.wikipedia.org/w/index.php?title=Null_hypothesis&oldid=855248015.Accessed 23rd August 2018

466 Longo, G. S. and Kim-Spoon, J. (2014) 'What Drives Apostates and Converters? The Social and Familial Antecedents of Religious Change among Adolescents.', *Psychology of religion and spirituality*. NIH Public Access, 6(4), pp. 284–291. doi: 10.1037/a0037651.

467 Socialization, Religious (2003) *New Catholic Encyclopaedia - Encyclopedia.com*. Available at: www.encyclopedia.com/religion/encyclopedias-almanacs-transcripts-and-maps/socialization-religious. Updated January 8th 2020 (Accessed: 18th January 2020).

468 Francis, L. J. and Katz, Y. J. (1992) 'The Relationship between Personality and Religiosity in an Israeli Sample', *Journal for the Scientific Study of Religion*. WileySociety for the Scientific Study of Religion, 31(2), p. 153. doi: 10.2307/1387005.

469 Francis, L. *et al.* (1981) 'Are introverts more religious?', *British Journal of Social Psychology*. Wiley/Blackwell (10.1111), 20(2), pp. 101–104. doi: 10.1111/j.2044-8309.1981.tb00481.x.

470 Benjamin. Beit-Hallahmi, and Michael. Argyle. 2014. The Psychology of Religious Behaviour, Belief, and Experience. Routledge. https://books.google.co.uk/books?id=vSciAwAAQBAJ&pg=PA164&lpg=PA164&dq=introversion,+extroversion+and+religious+faith&source=bl&ots=MAI_xm46zH&sig=CFR41OCXQwXcJ8HGMlMw9sZB1Js&hl=en&sa=X&ved=2ahUKEwjt2tyH2YXdAhUHKewKHQbPDZk4UBDoATAEegQIBRAB#v=onepage&q=introversion%2C extroversion and religious faith&f=false.ISBN 1317799046. Accessed 18th January 2020

471 Loewenthal, K. (1986) 'Factors Affecting Religious Commitment', *The Journal of Social Psychology*. Taylor & Francis Group , 126(1), pp. 121–123. doi: 10.1080/00224545.1986.9713579.

472 Cornwall, M. (1987) 'The Social Bases of Religion: A Study of Factors Influencing Religious Belief and Commitment', *Review of Religious Research*. Religious Research Association, Inc., 29(1), p. 44. doi: 10.2307/3511951.

[473] Wikipedia Contributors. "Eysenck Personality Questionnaire." Wikipedia, The Free Encyclopedia Last Modified 3rd February 2018. https://en.wikipedia.org/w/index.php?title=Eysenck_Personality_Questionnaire&oldid=823833223.

[474] Marie Cornwall. 1988. "The Influence of Three Agents of Religious Socialization: Family, Church, and Peers." In The Religion and Family Connection: Social Science Perspectives, edited by Darwin L. Thomas, 207–31. Religious Studies Center, Brigham Young University. https://rsc-legacy.byu.edu/archived/religion-and-family-connection-social-science-perspectives/chapter-11-influence-three (Accessed: 19th January 2020)

[475] Longo, G. S. and Kim-Spoon, J. (2014) 'What Drives Apostates and Converters? The Social and Familial Antecedents of Religious Change among Adolescents.', *Psychology of religion and spirituality*. NIH Public Access, 6(4), pp. 284–291. doi: 10.1037/a0037651.

[476] Wikipedia Contributors. n.d. "Immanence." Wikipedia, The Free Encyclopedia Last Modified 6th June 2019. Accessed July 10, 2019. https://en.wikipedia.org/w/index.php?title=Immanence&oldid=900527042.

[477] Ramdas Lamb. 2002. Rapt in the Name: The Ramnamis, Ramnam, and Untouchable Religion in Central India. New York: State University of New York Press.https://www.sunypress.edu/p-3587-rapt-in-the-name.aspx. ISBN:978-0-7914-5386-5.

[478] Aparna Chawla. 2014. "Why Does Hinduism Have so Many Gods?" USA Today Network's Europe Union Experience August 20th. 2014.

[479] "Hindu Concepts About God." Hindu American Foundation. Accessed January 19, 2020. https://www.hafsite.org/sites/default/files/HinduConceptsAboutGod2.0_2.pdf.

[480] Jay Lakhani. n.d. "An Insight into Hinduism." London Inter Faith Centre. Accessed January 19, 2020. http://londoninterfaith.org.uk/resources/talks-conferences-and-articles/an-insight-into-hinduism/.

[481] Tracey R Rich. n.d. "What Do Jews Believe?" Judaism 101 Last Modified August 2019. Accessed January 19, 2020. http://www.jewfaq.org/beliefs.htm

[482] Talal Itani.(Translator) n.d. "Quran English Translation." ClearQuran Dallas, Beirut. Accessed January 19, 2020. https://m.clearquran.com/downloads/quran-english-translation-clearquran-edition-allah.pdf.

[483] "The 99 Names of Allah." Accessed January 19, 2020. https://99namesofallah.name.

[484] Wikipedia Contributors. n.d. "Names of God in Islam." Wikipedia, The Free Encyclopedia Last Modified 1st August 2019. Accessed August 12, 2019. https://simple.wikipedia.org/w/index.php?title=Names_of_God_in_Islam&oldid=6624664.

[485] Huda. 2019. "'Allah (God) in Islam.'" Learn Religions Apr. 17, 2019. 2019. https://www.learnreligions.com/allah-god-in-islam-2004296.

[486] Yasir-al-Wakeel. n.d. "The Nature of Allah." Al-Islam.Org. Accessed January 19, 2020. https://www.al-islam.org/articles/nature-of-allah-yasir-al-wakeel.

487 Omam Khalid. n.d. "10 Facts about Allah - It's Time for You to Know Your Creator a Bit More!" Islamic Finder. Accessed January 19, 2020. https://www.islamicfinder.org/news/10-facts-about-allah/.

488 Wikipedia contributors, "Image of God," *Wikipedia, The Free Encyclopedia,* https://en.wikipedia.org/w/index.php?title=Image_of_God&oldid=943855502 (accessed March 14, 2020).

489 By prayer, I mean just talking to God as if he was sitting next to you, like you would a friend. It may also include listening and being quiet. It can occur anywhere and anytime and in any situation, desperate or evolving. Taking an opportunity to be quiet and undisturbed is helpful at times when possible, but not essential, especially in the middle of a crisis or acute situation where immediate action is required. Listening allows thoughts to come into your head that may represent God's communication– see later. There are other specific types and methods of prayer for particular occasions, or situations, but when we have a problem, just speak its name and be open to what happens. It isn't rocket science – though explanations of how it all operates might be ! Anyone can do it, you don't need training, though talking about prayer with experienced people can be useful, but can also be unhelpful, if the person has particular fixed ideas about prayer. But if what they say make sense and follows "the general drift"(see later) – then embrace it.

490 Muzammil Siddiqi. n.d. "Why Does Allah Allow Suffering and Evil in the World?I." Islam on Line. Accessed January 18, 2020. https://archive.islamonline.net/?p=885.

491 https://islamqa.info/en/answers/2850/explaining-human-suffering-and-why-Allah-does-not-prevent-it Accessed 10/08/2019

492 "Hinduism on Suffering." Hinduwebsite.Com. https://www.hinduwebsite.com/hinduism/h_suffering.asp. Accessed January 18, 2020

493 Library, Jewish Virtual. n.d. "Suffering." A Project of AICE (American-Israel Cooperative Enterprise). Accessed January 18, 2020. https://www.jewishvirtuallibrary.org/suffering.

494 Paul Tillich. 2009. Dynamics of Faith. First Published 1957. New York: Harperone.

495 "does it follow the general drift ?" - being compatible with the overall / general view and direction of Christ's teaching and his way of love - I have heard this saying ascribed to the Rev. George Mcleod, founder of the Iona Community, though have not been able to find a specific reference. Such a yardstick of the 'general drift' of Christ's teaching should also be used to assess our interpretation of scripture. Does our understanding and interpretation of a scriptural passage lead to a loving and caring conclusion and action, or not ?

496 From a Christian perspective, the Holy Spirit is the "power" that Jesus said would be the "helper" to be present and take his place after his resurrection and ascension into heaven (however that is understood). In Christianity the Holy Spirit is the same as God and forms part of the Trinity - God, Jesus Christ the Son and the Holy Spirit. (see chapter 7). Also see : Wikipedia contributors, 'Holy Spirit in Christianity', *Wikipedia, The Free Encyclopedia,* Last modified 24 July 2018, https://en.wikipedia.org/w/index.php?title=Holy_Spirit_in_Christianity&oldid=851827533 [Accessed 11 August 2018]

[497] Wikipedia Contributors. n.d. "Common Good." Wikipedia, The Free Encyclopedia Last Modified 4th July 2019. Accessed July 26, 2019. https://en.wikipedia.org/w/index.php?title=Common_good&oldid=904820088.

[498] https://www.licc.org.uk/product/life-on-the-frontline-2/

[499] Wikipedia Contributors. n.d. "Free Will." Wikipedia, The Free Encyclopedia Last Modified 20th June 2019. Accessed July 9, 2019. https://en.wikipedia.org/w/index.php?title=Free_will&oldid=902653004

[500] Timothy O'Connor, and Franklin Christopher. n.d. "Free Will." The Stanford Encyclopedia of Philosophy Edward N. Zalta (Ed.) Metaphysics Research Lab, Stanford University. Metaphysics Research Lab, Stanford University. https://plato.stanford.edu/archives/sum2019/entries/freewill/.

[501] Wikipedia Contributors. n.d. "Jesus Bids Us Shine - Written by Susan Bogert Warner (1819-1885)." Wikipedia, The Free Encyclopedia Last Modified 15th May 2019. Accessed July 16, 2019. https://en.wikipedia.org/w/index.php?title=Jesus_Bids_Us_Shine&oldid=895544979.

[502] Wikipedia Contributors. n.d. "John Shelby Spong." Wikipedia, The Free Encyclopedia Last Modified 27th June 2019. Accessed July 17, 2019. https://en.wikipedia.org/w/index.php?title=John_Shelby_Spong&oldid=903668458.

[503] "Profile: The One True Bishop of Durham: Dr David Jenkins, Retiring Scourge of Sacred Cows." Voices - The Independent., July 15, 215AD. http://www.independent.co.uk/voices/profile-the-one-true-bishop-of-durham-dr-david-jenkins-retiring-scourge-of-sacred-cows-1392030.html.

[504] Robinson, J. A. T. (1963). Honest to God (2003rd ed.). John Knox Press.

[505] "ReceptiveEcumenism." Churches Together in England and Wales. Accessed January 19, 2020. https://www.cte.org.uk/Groups/91312/Home/Resources/Theology/Receptive_Ecumenism/What_is_Receptive/What_is_Receptive.aspx.

[506] "Interfaith Dialogue: Sharing Our Faith and Learning From Each Other." PATHEOS. Accessed January 19, 2020. http://www.patheos.com/Topics/Interfaith-Dialogue.

[507] World Council of Churches. 1992. "Issues in Christian-Muslim Relations: Ecumenical Considerations." World Council of Churches. 1992. https://www.oikoumene.org/en/resources/documents/wcc-programmes/interreligious-dialogue-and-cooperation/interreligious-trust-and-respect/issues-in-christian-Muslim-relations-ecumenical-considerations. Accessed 18th January 2020

[508] "New Steps on an Ancient Pilgrimage :Together from Canterbury to Rome." International Anglican- Roman Catholic Commission for Unity and Mission (IARCUM). 2016.

[509] J. A. Simpson, and E. S. C. Weiner. 1989. Ecumenism - The principle or aim of promoting unity among the world's Christian Churches. Oxford English Dictionary. Oxford England: Oxford: Clarendon Press. www.oxforddictionaries.com/definition/English/ecumenism

510 World Council of Churches. (2013). The Church: Towards a Common Vision. In Faith and Order Paper No. 214. Geneva: WCC Publications. http://doi.org/ISBN: 978-2-8254-1587-0

511 Grow, B. (2015). A Lament on the Distance Between the Pew and the Desk Chairs of Seminarian's: A Christian's Grievance. The Evangelical Calvanist. Accessed 18th January 2020 from growrag.wordpress.com/2015/11/11/a-lament-on-the-distance-between-the-pew-and-the-desk-chairs-of-seminarians-a-christians-grievance/?iframe=true&preview=true.

512 Gayla Postma. 216. "A Growing Divide - Proceedings of Synod Church of England 2016." The Banner. Accessed January 19, 2016. http://www.thebanner.org/news/2016/06/synod-2016-a-growing-divide.

513 Power, M. (2007). From Ecumenism to Community Relations: Inter-Church Relationships in Northern Ireland, 1980-2005. Dublin: Irish Academic Press.

514 Matthew Chapter 25: verses 31–46 - God separates the sheep who have responded to the needs of the world from the goats who did not. The Bible

515 Zia H. Shah. 2015. "Forty Hadiths or Sayings of the Prophet Muhammad about Compassionate Living." The Muslim Times 9th February. 2015. https://themuslimtimes.info/2015/02/09/forty- hadith-about-compassionate-living/.

516 Forst Rainer. 2017. "Toleration." Stanford Encyclopedia of Philosophy. Edward N. Zalta (Ed.). Metaphysics Research Lab, Stanford University. https://plato.stanford.edu/archives/fall2017/entries/toleration/.Accessed 19th January 2020

517 Andrew Fiala. n.d. "Toleration." Internet Encyclopedia of Philosophy (IEP). Accessed January 19, 2020. https://www.iep.utm.edu/tolerati/.

518 "How A Christian Community Reacts When Muslims Moves next Door." You Tube. How A Christian Community Reacts When Muslims Moves next door.

519 Carol Kuruvilla. n.d. "Muslim Activist Collects Stories Of Interfaith Solidarity After New Zealand Massacre." Religion 22/03/2019 HUFFPOST. Accessed January 19, 2020. https://www.huffingtonpost.co.uk/entry/new-zealand-interfaith-solidarity_n_5c93d44de4b0e9efc8b5b1f3?ri18n=true.

520 Carol Kuruvilla. n.d. "Jews, Christians, Sikhs And Others Mourn With Muslims After New Zealand Attack." Religion 15/03/2019 HUFFPOST. Accessed January 19, 2020. https://www.huffingtonpost.co.uk/entry/new-zealand-mosque-attack-mourning_n_5c8bd8a7e4b0d7f6b0f314a9?ri18n=true

521 John Vincent. 2019. Jesus The Radical - Saving the World. Sheffield England: Ashram Press. ISBN:978-0-9559073-7-1

522 John Vincent. 2015. Radical Jesus. The Way of Jesus Then and Now. 3rd ed. Sheffield England: Ashram Press. ISBN:978-0-9559073-4-0

523 C.T.R. Hewer: (2018) The Importance of a Paradigm shift in understanding Christianity and Islam. www.chrishewer.org Copyright © 2018 C.T.R. Hewer

[524] Ted Grimsrud (2007) (06) How Does God Communicate? (3.18.07) | Peace Theology, Peace Theology. Available at: https://peacetheology.net/doctrine/06-how-does-god-communicate-3-18-07/ (Accessed: 31 August 2017).

[525] Cathleen Falsani. n.d. "The Worst Ideas of the Decade: The Prosperity Gospel." Washington Post. Accessed January 18, 2020. http://www.washingtonpost.com/wp-srv/special/opinions/outlook/worst-ideas/prosperity-gospel.html

[526] Wikipedia Contributors. n.d. "Prosperity Theology." Wikipedia, The Free Encyclopedia Last Modified 18th July 2018. Accessed January 18, 2020. https://en.wikipedia.org/w/index.php?title=Prosperity_theology&oldid=850874618.

Made in the USA
Monee, IL
12 September 2025

25513823R00292